Real-time Computer Control

Prentice Hall International
Series in Systems and Control Engineering

M. J. Grimble, Series Editor

BAGCHI, A., *Optimal Control of Stochastic Systems*
BENNETT, S., *Real-time Computer Control: An introduction* – 2nd edition
BITMEAD, R. R., GEVERS, M. and WERTZ, V., *Adaptive Optimal Control*
BUTLER, H., *Model Reference Adaptive Control*
ISERMANN, R., LACHMANN, K. H. and MATKO, D., *Adaptive Control Systems*
KUCERA, V., *Analysis and Design of Discrete Linear Control Systems*
LUNZE, J., *Feedback Control of Large-Scale Systems*
McLEAN, D., *Automatic Flight Control Systems*
OLSSON, G. and PIANI, G., *Computer Systems for Automation and Control*
PARKS, P. C. and HAHN, V., *Stability Theory*
PATTON, R., CLARK, R. N. and FRANK, P. M. (editors), *Fault Diagnosis in Dynamic Systems*
PETKOV, P. H., CHRISTOV, N. D. and KONSTANTINOV, M. M., *Computational Methods for Linear Control Systems*
ROGERS, E. and LI, Y., *Parallel Processing in a Control Systems Environment*
SÖDERSTROM, T. and STOICA, P., *System Identification*
SOETERBOEK, A. R. M., *Predictive Control: A unified approach*
STOORVOGEL, A., *The H_∞ Control Problem*
WATANABE, K., *Adaptive Estimation and Control*
WILLIAMSON, D., *Digital Control and Instrumentation*

Real-time Computer Control

An Introduction

Second edition

Stuart Bennett

Department of Automatic Control and Systems Engineering
University of Sheffield

PRENTICE HALL

NEW YORK · LONDON · TORONTO · SYDNEY · TOKYO · SINGAPORE

First published 1994 by
Prentice Hall International (UK) Limited
Campus 400, Maylands Avenue
Hemel Hempstead
Hertfordshire, HP2 7EZ
A division of
Simon & Schuster International Group

Typeset in 10/12 pt Times by
Mathematical Composition Setters Ltd, Salisbury, Wiltshire

Printed and bound in Great Britain by
Redwood Books, Trowbridge, Wilts

Library of Congress Cataloging-in-Publication Data

Bennett, S. (Stuart)
    Real-time computer control : an introduction / Stuart Bennett.
        p.   cm. — (Prentice Hall international series in systems and
control engineering)
    Includes bibliographical references and index.
    ISBN 0-13-764176-1
    1. Real-time control systems. I. Title. II. Series.
TJ217.7.B46 1994
629.8'9—dc20   93-28154
                                                                     CIP

British Library Cataloguing in Publication Data

A catalogue record for this book is available from
the British Library

ISBN 0-13-764176-1   (pbk)

---

1 2 3 4 5   98 97 96 95 94

Contents

Preface

Over the past 30 years digital control of industrial processes has changed from being the exception to the commonplace. Each succeeding year sees an increase in the range of applications and each advance in hardware design widens the potential application areas. Microprocessors are now a normal component employed in a wide range of electronic systems.

Computers in one form or another now form an integral part of most real-time control systems; such computers are generally referred to as *embedded real-time computers* and an understanding of how to design and build systems containing embedded computers is an essential requirement for a systems engineer. The knowledge required covers both hardware and software design and construction, and of the two the software engineering is the most difficult and least understood. The difficulties of specifying, designing and building real-time software have led in recent years to intensive work on development methodologies and one of the major objectives of this book is to introduce the reader to the fundamental ideas underlying these methodologies.

Traditional computing courses for engineers have typically emphasised the hardware and programming language aspects of using computers. An understanding of the basic principles of operation of computer hardware and of programming languages is of course important; however, these are details with respect to the overall system design problem. Also in the past it has frequently been assumed that real-time control software will be written in an assembly language. It is of course true that for many engineering applications use of assembly languages, and languages such as FORTH and real-time BASIC, are legitimate, economical means of producing a viable solution. However, for large-scale, complex and safety-critical systems such approaches are no longer adequate and in this book I attempt to cover some of the important and fundamental ideas underlying a software engineering approach to specifying, designing and building real-time software for computer control applications.

The book is intended for final-year undergraduate students and practising engineers, and is also suitable as a text for computer science students requiring an introduction to real-time software from an application viewpoint as opposed to an operating system viewpoint. It assumes that the reader has some familiarity with at least one high-level programming language, with basic ideas about computer hardware and an understanding of simple feedback control concepts. It does not require familiarity with control design techniques. Examples in the text are given using Modula-2 and should be easily followed by anyone familiar with Pascal.

In Chapters 1 and 2 I introduce basic concepts relating to real-time systems and their characteristics and provide an overview of computer control applications. Chapter 3 provides a brief summary of the important hardware building blocks for computers used for control. Chapter 4 introduces some of the practical problems of implementing control algorithms. I use the simple and widely used PID (Proportional + Integral + Derivative) controller as an example and I explain its operation in detail. The section of this chapter which uses z-transform notation and deals with the implementation of controllers designed using discrete control techniques can be omitted by readers who have no background in control systems. In the final part of Chapter 4, I introduce in general terms how to deal with systems which involve more than just the control algorithm.

Chapter 5 on real-time languages and Chapter 6 on operating systems and concurrency provide basic information about two essential tools that will be required by the real-time system builder. In Chapters 7, 8 and 9 I deal with the methodologies that have been developed in recent years to help in the specification, design and construction of real-time software and real-time systems. I concentrate mainly on two methodologies, MASCOT and the Hatley & Pirbhai variant of the Yourdon method. The standard methodologies provide very little guidance for the actual implementation of the software design and in Chapter 10 I describe some of the implementation problems and possible approaches to solving them. The final chapter deals with the vital topic of building dependable software.

Many people assisted me in producing the first edition of this book: colleagues in the Department of Automatic Control and Systems Engineering at the University of Sheffield, in particular Steve White. In preparing this second edition I have received much useful comment from Les Woolliscroft and from two reviewers.

Stuart Bennett
Sheffield, UK
April 1993

1

Introduction to Real-time Systems

The purpose of this chapter is to:

- Provide a general introduction to computer control and to embedded computer systems.
- Define what is meant by real-time systems.
- Show how real-time systems can be classified.
- Illustrate the difficulties of writing real-time software.

You may not be familiar with some of the terms used in this chapter; you do not need to be concerned at this stage as they will be explained in detail in the later chapters.

1.1 HISTORICAL BACKGROUND

The earliest proposal to use a computer operating in *real time* as part of a control system was made in a paper by Brown and Campbell in 1950. The paper contains a diagram (see Figure 1.1) which shows a computer in both the feedback and feedforward loops. Brown and Campbell assumed that analog computing elements would be used but they did not exclude the use of digital computing elements. The first digital computers developed specifically for real-time control were for airborne operation, and in 1954 a Digitrac digital computer was successfully used to provide an automatic flight and weapons control system.

The application of digital computers to industrial control began in the late 1950s. The initiative came, not from the process and manufacturing industries, but from the computer and electronic systems manufacturers who were looking to extend their markets and to find outlets for equipment which had failed to be adopted by the military (Williams, 1977). The first industrial installation of a computer system was in September 1958 when the Louisiana Power and Light Company installed a Daystrom computer system for plant monitoring at their power station in Sterling, Louisiana. However, this was not a control system: the first industrial computer control installation was made by the Texaco Company who installed an RW-300 (Ramo-Wooldridge Company) system at their Port Arthur refinery in Texas. The refinery was run under *closed-loop control* on 15 March 1959 (Anon, 1959).

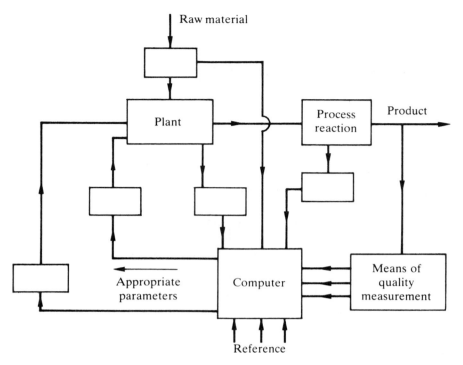

Raw material

Plant

Process
reaction

Product

Appropriate
parameters

Computer

Means of
quality
measurement

Reference

Figure 1.1   Computer used in control of plant (redrawn from Brown and Campbell,
Mechanical Engineering, 72: 124 (1950)).

During 1957–8 the Monsanto Chemical Company, in co-operation with the Ramo-Wooldridge Company, studied the possibility of using computer control and in October 1958 decided to implement a scheme on the ammonia plant at Luling, Louisiana. Commissioning of this plant began on 20 January 1960 and closed-loop control was achieved on 4 April 1960 after an almost complete rewrite of the control algorithm part of the program. They also experienced considerable problems with *noise* on the measurement signals. This scheme, like the system installed by the B. F. Goodrich Company on their acrylanite plant at Calvert City, Kentucky, in 1959–60, and some 40 other systems based on the RW-300 were *supervisory control* systems used for *steady-state optimisation* calculations to determine the *set points* for standard analog controllers; that is, the computer did not control directly the movement of the valves or other *plant actuators*.

The first *direct digital control* (DDC) computer system was the Ferranti Argus 200 system installed in November 1962 at the ICI ammonia-soda plant at Fleetwood, Lancashire, UK, the planning for which had begun in 1959 (Burkitt, 1965). It was a large system with provision for 120 control loops and 256 measurements, of which 98 and 224 respectively were actually used on the Fleetwood system. In 1961 the

Monsanto Company also began a DDC project for a plant in Texas City and a *hierarchical* control scheme for the petrochemical complex at Chocolate Bayou.

The Ferranti Argus 200 used a ferrite core store to hold the control program rather than a rotating drum store as used by the RW-300 computer. The program was held in a programmable read-only memory (PROM). It was loaded by physically inserting pegs into a plug board, each peg representing one bit in the memory word. Although laborious to set up initially, the system proved to be very reliable in that destruction of the memory contents could only be brought about by the physical dislodgement of the pegs. In addition, security was enhanced by using special power supplies and switch-over mechanisms to protect information held in the main core store. This information was classified as follows:

Set points	Loss most undesirable
Valve demand	Presence after controlled stoppage allows computer to gain control of plant immediately and without disturbing the plant (referred to as *bumpless transfer*)
Memory calculations	Loss is tolerable, soon will be updated and only slight disturbance to plant
Future development	Extension to allow for optimisation may require information to be maintained for long periods of time

In addition to improved reliability the Argus system provided more rapid memory access than the drum stores of the RW-300 and similar machines and as such represented the beginning of the second phase of application of computers to real-time control.

The computers used in the early 1960s combined magnetic core memories and drum stores, the drums eventually giving way to hard disk drives. They included the General Electric 4000 series, IBM 1800, CDC 1700, Foxboro FOX 1 and 1A, the SDS and Xerox SIGMA series, Ferranti Argus series and Elliot Automation 900 series. The attempt to resolve some of the problems of the early machines led to an increase in the cost of systems: the increase was such that frequently their use could be justified only if both DDC and supervisory control were performed by the one computer. A consequence of this was the generation of further problems particularly in the development of the software. The programs for the early computers had been written by specialist programmers using machine code; this was manageable because the tasks were clearly defined and the quantity of code relatively small. In combining DDC and supervisory control, not only did the quantity of code for a given application increase, but the complexity of the programming also increased. The two tasks have very different time-scales and the DDC control programs have to be able to interrupt the supervisory control programs. The increase in the size of the programs meant that not all the code could be stored in core memory: provision had to be made for the swapping of code between the drum memory and core.

The solution appeared to lie in the development of general purpose *real-time operating systems* and high-level languages. In the late 1960s real-time operating systems were developed and various PROCESS FORTRAN compilers made their appearance. The problems and the costs involved in attempting to do everything in one computer led users to retreat to smaller systems for which the newly developing minicomputer (DEC PDP-8, PDP-11, Data General Nova, Honeywell 316, etc.) was to prove ideally suited. The cost of the minicomputer was small enough to avoid the need to load a large number of tasks onto one machine: indeed by 1970 it was becoming possible to consider having two computers on the system, one simply acting as a stand-by in the event of failure. Throughout the 1970s the technological developments in integrated circuits and construction techniques for circuit boards led to an increase in reliability of computer systems, a large reduction in cost and major increases in processor power and in the amount of fast memory that could be provided. These changes began to force attention more and more on the problems of writing correct and dependable software. The advent of the microprocessor in 1974 led to a further reappraisal of approaches and made economically possible the use of *distributed computer control systems*. These developments are considered in more detail in Chapter 2.

1.2 ELEMENTS OF A COMPUTER CONTROL SYSTEM

As a simple example which illustrates the various operations of a computer control system, let us consider the hot-air blower shown in Figure 1.2. A centrifugal fan blows air over a heating element and into a tube. A thermistor bead is placed at the outlet end of the tube and forms one arm of a bridge circuit. The amplified output of the bridge circuit is available at B and provides a voltage, in the range 0 to 10 volts, proportional to temperature. The current supplied to the heating element can be varied by supplying a dc voltage in the range 0 to 10 volts to point A.

The position of the air-inlet cover to the fan is adjusted by means of a reversible motor. The motor operates at constant speed and is turned on or off by a logic signal applied to its controller; a second logic signal determines the direction of rotation. A potentiometer wiper is attached to the air-inlet cover and the voltage output is proportional to the position of the cover. Microswitches are used to detect when the cover is fully open and fully closed.

The operator is provided with a panel from which the control system can be switched from automatic to manual control. In manual mode the heat output and fan cover position can be adjusted using potentiometers. Switches are provided to operate the fan and heater. Panel lights indicate *fan on, heater on, cover fully open, cover fully closed* and *auto/manual* status. The desired output temperature (this is

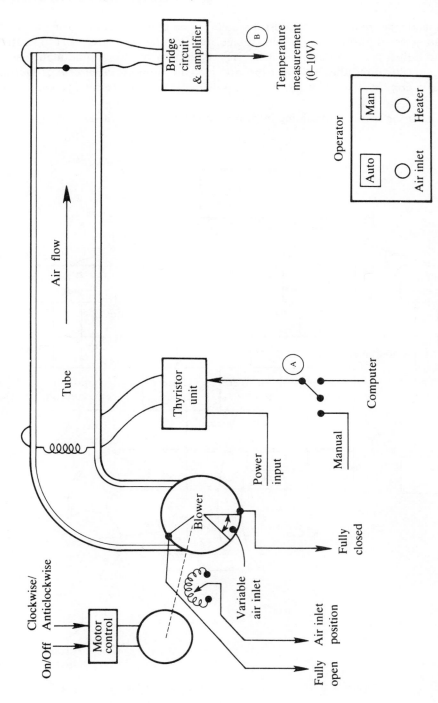

Figure 1.2 A simple plant – a hot-air blower.

known as the *set point* for the control system) is set by the operator using a slider potentiometer, the setting of which can be read by the computer. The operator can also adjust the fan cover position. The operation of this simple plant using a computer requires that software be provided to support *monitoring*, *control* and *actuation* of the plant. A general schematic of the system is shown in Figure 1.3.

Monitoring involves obtaining information about the current state of the plant.

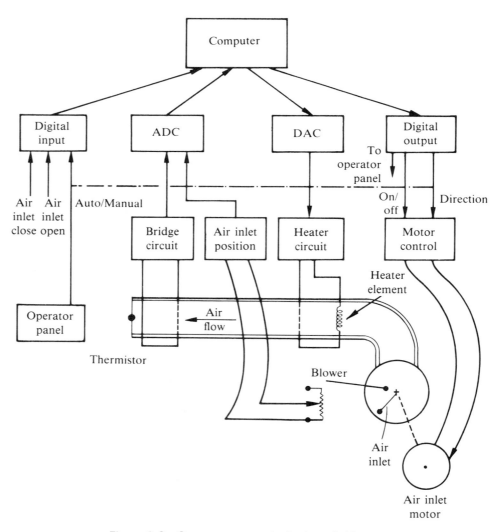

Figure 1.3   Computer control of a hot-air blower.

In the above example the information is available from the plant instruments in the following two forms:

Analog signals	Air temperature Fan-inlet cover position
Digital (logic) signals	Fan-inlet cover position (fully open, fully closed); status signals: auto/manual, fan motor on, heater on

Control involves the digital equivalent of continuous feedback control for the control of temperature (direct digital control, DDC) and for position control of the fan-inlet cover. Sequence and interlock control operations are also required since, for example, the heater should not be on if the fan is not running. The computer has also to handle automatic change-over from simply tracking (monitoring) the manual control operations to controlling the system when the operator requests a change from manual to automatic control. This change-over should be carried out

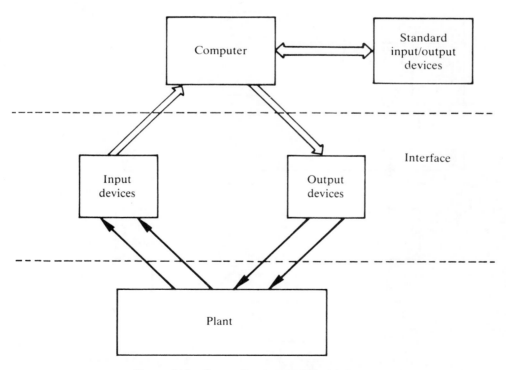

Figure 1.4 Generalised computer system.

without disturbing the temperature of the air at the output of the tube (a change-over which does not cause a plant disturbance is referred to as a *bumpless transfer*). These actions may involve parallel logic operations, time-sequential control and timing of operations.

Actuation requires the provision of a voltage proportional to the demanded heat output to drive the heater control; logic signals indicating on/off and the direction in which the fan-inlet cover is to be moved; and logic signals for the operator display.

The monitoring and actuation tasks thus involve a range of interface devices including analog-to-digital converters (ADCs), digital-to-analog converters (DACs), digital input and digital output lines and pulse generators. Details of the different types of interfaces are given in Chapter 3. For the present we will represent them simply as input devices and output devices as shown in Figure 1.4. Note that it is normal practice to use the terms 'input' and 'output' with reference to the computer;

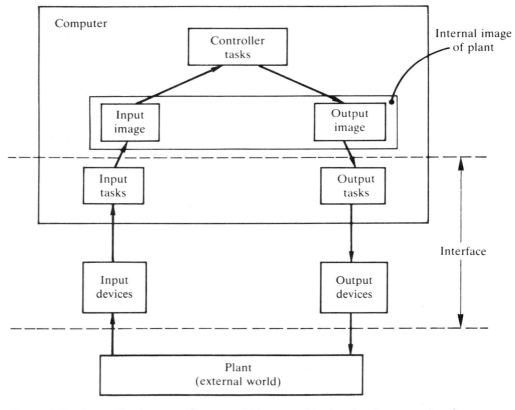

Figure 1.5 Generalised computer control system showing hardware and software interface.

thus input devices are devices which transmit information to the computer. Each of the various types of device will require software to operate it and we will represent this software as input tasks and output tasks. This generalised picture of a computer control system is shown in Figure 1.5. You should note that the interface boundary in Figure 1.5 is shown inside the computer block; this is to indicate that software (program code) is required to operate the interface devices.

The input devices plus the input software gather the information needed to create an *input image* of the plant. The input image is a snapshot of the status of the plant and this snapshot is renewed at specified intervals. (All of it may be renewed at the same time or some parts may be renewed less frequently than others.) Where the external information is in analog form the process of obtaining the snapshot will involve *digitisation* (conversion of a continuous measure into a discrete measure) as well as *sampling*.

The *output image* represents the current set of outputs generated by the control calculations. The output image will be updated periodically by the control tasks. It is the job of the output tasks to convey data contained in the output image to the plant. The control software within the computer can thus be considered as operating on an input image (or model) of the plant to produce the output image.

Figure 1.6 shows a simplified block diagram of the feedback control part of the system. In this diagram some of the symbols and terms that we shall use are defined. The diagram represents a *continuous* control system. The equivalent *sampled* (computer-controlled) version is shown in Figure 1.7.

In the simple computer control model we have described above we divided the software tasks to be performed into three major areas:

- plant input tasks;
- plant output tasks; and
- control tasks.

In doing so we have assumed that communication with the operator is treated as part of the plant input and plant output tasks. However, in many applications communication will extend beyond simple indicators and switches. Plant engineers, plant managers, pilots, air traffic controllers, drivers, and machine operators, for example, will require detailed information on all aspects of the operation of the

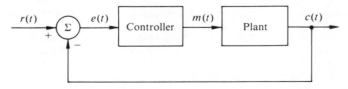

Figure 1.6 Simplified block diagram of continuous feedback control system: $r(t) =$ set point, $c(t) =$ controlled variable, $e(t) = r(t) - c(t) =$ error, and $m(t) =$ manipulated variable.

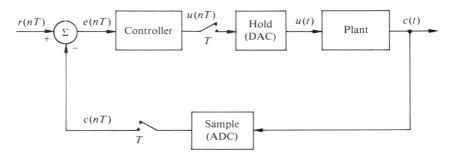

Figure 1.7 Simplified block diagram of a sampled feedback control system: $c(nT)$, $r(nT)$, $e(nT)$, $u(nT)$ are sampled values of $c(t)$, $r(t)$, $e(t)$, $u(t)$ at sample times nT where n is an integer and T is the sampling interval.

plant, aircraft, car, radar system, etc. Control of the system may be shared between several computers, not necessarily all on the same site, and hence information may have to be transmitted between computers. We must therefore extend the model to include communication tasks as shown in Figure 1.8. The communication tasks are assumed to cover input and output from keyboards/VDUs, remote transmission links, backing store (disks, etc.), output to printers, chart recorders, graph plotters, local-area networks (LANs) and wide-area networks (WANs).

The computer system may operate in a purely sequential way with the tasks – plant input, control, plant output and communication – being carried out in turn and with the sequence then being repeated indefinitely; or it may operate in a parallel manner, with the various tasks being carried out *concurrently*. In the case of a computer system with a single processing unit full parallel operation cannot be achieved and the system is said to operate *pseudo-concurrently*. The use of concurrency (or pseudo-concurrency) gives rise to problems relating to synchronising the various tasks and the sharing of resources between the tasks. Concurrency and its associated problems are covered in more detail in Chapters 5 and 6.

1.3 REAL-TIME SYSTEMS – DEFINITION

The title of this book includes the words *Real Time* and throughout the book we will use phrases such as real-time systems, real-time control: what do we mean by real time? The *Oxford Dictionary of Computing* offers the definition:

> Any system in which the time at which the output is produced is significant. This is usually because the input corresponds to some movement in the physical world, and the output has to relate to that same movement. The lag from input time to output time must be sufficiently small for acceptable timeliness.

This definition covers a very wide range of systems; for example, from workstations running under the UNIX operating system from which the user expects to receive

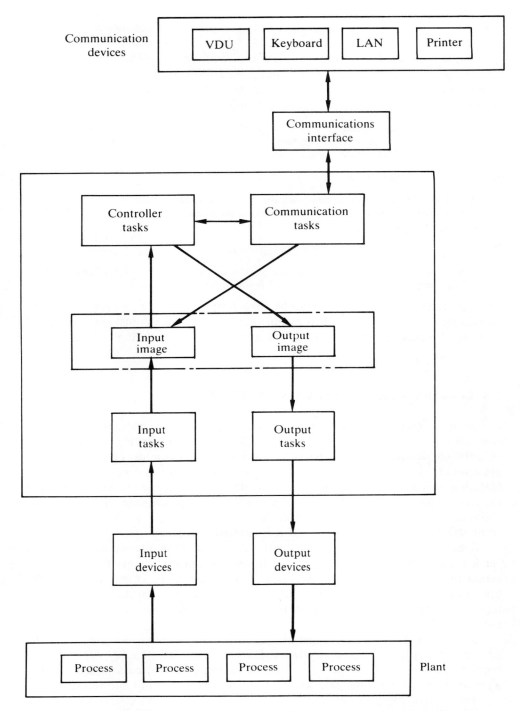

Figure 1.8 Computer control system showing communication tasks.

a response within a few seconds through to aircraft engine control systems which must respond within a specified time and failure to do so could cause the loss of control and possibly the loss of many lives. We are concerned with the latter type of system and Cooling (1991) offers the definition:

> Real-time systems are those which must produce correct responses within a definite time limit. Should computer responses exceed these time bounds then performance degradation and/or malfunction results.

An alternative definition is:

> A real-time system reads inputs from the plant and sends control signals to the plant at times determined by plant operational considerations − not at times limited by the capabilities of the computer system.

We can therefore define a real-time program as:

> A program for which the correctness of operation depends both on the logical results of the computation and the time at which the results are produced.

The computer systems that we shall be considering fall into a category known as *embedded* computers by which is meant that the computer is just one functional element of a real-time system and is not a stand-alone computing machine.

1.4 CLASSIFICATION OF REAL-TIME SYSTEMS

A common feature of real-time systems and embedded computers is that the computer is connected to the environment within which it is working by a wide range of interface devices and receives and sends a variety of stimuli. For example, the plant input, plant output, and communication tasks shown in Figure 1.8 have one feature in common − they are connected by physical devices to processes which are external to the computer. These external processes all operate in their own time-scales and the computer is said to operate in real time if actions carried out in the computer relate to the time-scales of the external processes.

Synchronisation between the external processes and the internal actions (tasks) carried out by the computer may be defined in terms of the passage of time, or the actual time of day, in which case the system is said to be *clock based* and the operations are carried out according to a time schedule. Or it may be defined in terms of events, for example the closure of a switch, in which case it is said to be *event based*.

There is a third category, *interactive*, in which the relationship between the actions in the computer and the system is much more loosely defined. Typically, the requirement is that a set of operations in the computer should be completed within a predetermined time. The majority of communication tasks fall into this category.

The control tasks, although not obviously and directly connected to the external environment, also need to operate in real time, since time is usually involved in determining the parameters of the algorithms used (see Chapter 4). It is useful to divide tasks to be carried out by embedded computers into the above categories and you should learn to recognise the characteristics of each class.

1.4.1 Clock-based Tasks (cyclic, periodic)

A process plant operates in real time and thus we talk about the plant *time constants* (a time constant is a measure of the time taken by a plant to respond to a change in input or load and is used as a *characteristic* of the plant). Time constants may be measured in hours for some chemical processes or in milliseconds for an aircraft system. For feedback control the required *sampling rate* will be dependent on the time constant of the process to be controlled. The shorter the time constant of the process, the faster the required sampling rate. The computer which is used to control the plant must therefore be synchronised to real time and must be able to carry out all the required operations – measurement, control and actuation – within each sampling interval.

The completion of the operations within the specified time is dependent on the number of operations to be performed and the speed of the computer. Synchronisation is usually obtained by adding a clock to the computer system – normally referred to as a *real-time* clock – and using a signal from this clock to *interrupt* the operations of the computer at some predetermined fixed time interval.

The computer may carry out the plant input, plant output and control tasks in response to the clock interrupt or, if the clock interrupt has been set at a faster rate than the sampling rate, it may count each interrupt until it is time to run the tasks. In larger systems the tasks may be subdivided into groups for controlling different parts of the plant and these may need to run at different sampling rates. The clock interrupt is frequently used to keep a clock and calendar so that the computer system is aware of both the time and the date. Clock-based tasks are typically referred to as *cyclic* or *periodic* tasks where the terms can imply either that the task is to run once per time period T (or cycle time T), or is to run at exactly T unit intervals.

1.4.2 Event-based Tasks (aperiodic)

There are many systems where actions have to be performed not at particular times or time intervals but in response to some event. Typical examples are: turning off a pump or closing a valve when the level of a liquid in a tank reaches a predetermined value (consider what happens when the level of the fuel in a car petrol tank reaches the pump nozzle); or switching a motor off in response to the closure of a microswitch indicating that some desired position has been reached. Event-based systems are also used extensively to indicate alarm conditions and initiate alarm actions, for example as an indication of too high a temperature or too great a pressure. The specification of event-based systems usually includes a requirement that the system must respond within a given maximum time to a particular event.

Event-based systems normally employ interrupts to inform the computer system that action is required. Some small, simple systems may use *polling*; that is, the computer periodically asks (polls) the various sensors to see if action is required.

Events occur at non-deterministic intervals and event-based tasks are frequently referred to as *aperiodic* tasks. Such tasks may have deadlines expressed in terms of having start times or finish times (or even both). For example, a task may be required to start within 0.5 seconds of an event occurring, or alternatively it may have to produce an output (end time) within 0.5 seconds of the event.

1.4.3 Interactive Systems

Interactive systems probably represent the largest class of real-time systems and cover such systems as automatic bank tellers; reservation systems for hotels, airlines and car rental companies; computerised tills, etc. The real-time requirement is usually expressed in terms such as 'the average response time must not exceed. . .'. For example, an automatic bank teller system might require an average response time not exceeding 20 seconds. Superficially this type of system seems similar to the event-based system in that it apparently responds to a signal from the plant (in this case usually a person), but it is different because it responds at a time determined by the internal state of the computer and without any reference to the environment. An automatic bank teller does not know that you are about to miss a train, or that it is raining hard and you are getting wet: its response depends on how busy the communication lines and central computers are (and of course the state of your account).

Many interactive systems give the impression that they are clock based in that they are capable of displaying the date and time; they do indeed have a real-time clock which enables them to keep track of time. The test as to whether or not they are clock based as described above is to ask, 'Can the system be tightly synchronised to an external process?' If the answer is 'yes' they are clock based; if it is 'no' then they are interactive.

1.5 TIME CONSTRAINTS

It is now common practice to divide real-time systems (and real-time tasks) into two categories:

- *Hard real-time*: these are systems that must satisfy the deadlines on each and every occasion.
- *Soft real-time*: these are systems for which an occasional failure to meet a deadline does not comprise the correctness of the system.

A typical example of a hard real-time control system is the temperature control loop of the hot-air blower system described above. In control terms, the temperature loop is a *sampled data* system which can be represented in block diagram form as shown in Figure 1.9. Design of a suitable control algorithm for this system involves the choice of the sampling interval T_s. If we assume that a suitable sampling interval is 10 ms, then at 10 ms intervals the input value must be read, the control calculation carried out and the output value calculated, and the output value sent to the heater drive.

As an example of hard time constraints associated with event-based tasks let us assume that the hot-air blower is being used to dry a component which will be damaged if exposed to temperatures greater than $50°C$ for more than 10 seconds. Allowing for the time taken for the air to travel from the heater to the outlet and the cooling time of the heater element — and for a margin of safety — the alarm response requirement may be, say, that overtemperature should be detected and the heater switched off within seven seconds of the overtemperature occurring. The general form of this type of constraint is that the computer must respond to the event within some specified maximum time.

An automatic bank teller provides an example of a system with a soft time constraint. A typical system is event initiated in that it is started by the customer placing their card in the machine. The time constraint on the machine responding will be specified in terms of an average response time of, say, 10 seconds, with the average being measured over a 24 hour period. (Note that if the system has been carefully specified there will also be a maximum time, say 30 seconds, within which the system should respond.) The actual response time will vary: if you have used such a system you will have learned that the response time obtained between 12 and 2 p.m. on a Friday is very different from that at 10 a.m. on a Sunday.

In practice the above categories are only guides: for example, in the hot-air blower an occasional missed sample would not seriously affect the performance; neither would a variation in sampling interval such that the inequality $9.95 \leqslant T_s \leqslant 10.05$ ms with a mean of $T_s = 10$ ms was satisfied. The system would not be satisfactory, however, if the sampling interval T_s was in the range $1 < T_s < 1000$ ms with a mean of 10 ms over a 24 hour period. Similarly, the automatic bank teller would not be satisfactory if at busy times customers had to wait 10 minutes, even if it achieved a mean response measured over a 24 hour period of 20 seconds. A typical specification might be a mean response measured over a 24 hour period of 15 seconds, with 95% of requests being satisfied within 30 seconds and no response time greater than 60 seconds.

A hard time constraint obviously represents a much more severe constraint on the performance of the system than a soft time constraint and such systems present a difficult challenge both to hardware and to software designers. Most real-time systems contain a mixture of activities that can be classified as clock based, event based, and interactive with both hard and soft time constraints (they will also contain activities which are not real time). A system designer will attempt to reduce the number of activities (tasks) that are subject to a hard time constraint.

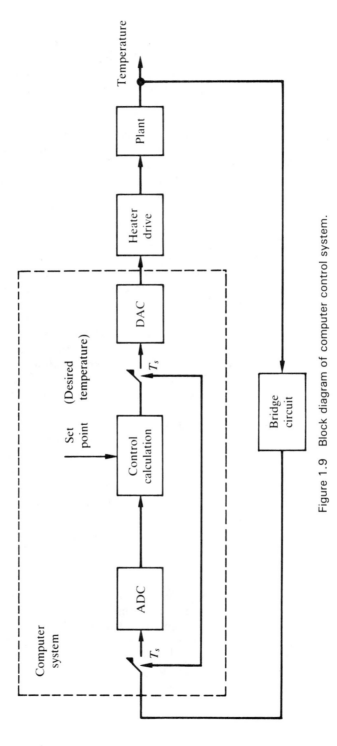

Figure 1.9   Block diagram of computer control system.

We can formally define the constraints as follows:

Hard		Soft	
Periodic (cyclic)	*Aperiodic (event)*	*Periodic (cyclic)*	*Aperiodic (event)*
$t_c(i) = t_s \pm a$	$t_e(i) \leqslant T_e$	$\dfrac{1}{n}\displaystyle\sum_{i=1}^{n} t_c(i) = t_s \pm a$	$\dfrac{1}{n}\displaystyle\sum_{i=1}^{n} t_e(i) \leqslant T_a$
		$n = T/t_s$	$n = T/t_s$

$t_c(i)$ the interval between the i and $i-1$ cycles,
$t_e(i)$ the response time to the ith occurrence of event e,
t_s the desired periodic (cyclic) interval,
T_e the maximum permitted response time to event e,
T_a the average permitted response time to event e measured over some time interval T,
n the number of occurrences of event e within the time interval T, or the number of cyclic repetitions during the time interval T,
a a small timing tolerance.

For some systems and tasks the timing constraints may be combined in some form or other, or relaxed in some way. For example, soft aperiodic (event) time constraints are often combined with a hard time constraint such that $t_c(i) \leqslant T_m$ where $T_m \gg T_a$. In some circumstances an occasional missed periodic hard time constraint might not be serious; for example, in a feedback control system missing the occasional sample will not result in a major disturbance of the controlled system (note: this is the case only if the manipulated variable is held constant at the previous value).

1.6 CLASSIFICATION OF PROGRAMS

The importance of separating the various activities carried out by computer control systems into real-time and non-real-time tasks, and in subdividing real-time tasks into the two different types, arises from the different levels of difficulty of designing and implementing the different types of computer program. Experimental studies have shown clearly that certain types of program, particularly those involving real-time and interface operations, are substantially more difficult to construct than, for instance, standard data processing programs (Shooman, 1983; Pressman, 1992). As we shall discuss later, the division of software into small, coherent *modules* is an important design technique and one of the guidelines for module division that we shall introduce is to put activities with different types of time constraints into separate modules.

Theoretical work on mathematical techniques for proving the correctness of a program, and the development of formal specification languages, such as 'Z' and VDM, has clarified the understanding of differences between different types of program. Pyle (1979), drawing on the work of Wirth (1977), presented definitions identifying three types of programming:

- sequential;
- multi-tasking; and
- real-time.

The definitions are based on the kind of arguments which would have to be made in order to verify, that is to develop a formal proof of correctness for programs of each type.

1.6.1 Sequential

In classical sequential programming *actions* are strictly ordered as a time sequence: the behaviour of the program depends only on the effects of the individual *actions* and their *order*; the time taken to perform the action is not of consequence. Verification, therefore, requires two kinds of argument:

1. that a particular statement defines a stated action; and
2. that the various program structures produce a stated sequence of events.

1.6.2 Multi-tasking

A multi-task program differs from the classical sequential program in that the actions it is required to perform are not necessarily disjoint in time; it may be necessary for several actions to be performed in parallel. Note that the *sequential relationships* between the actions may still be important. Such a program may be built from a number of parts (processes or tasks are the names used for the parts) which are themselves partly sequential, but which are executed concurrently and which communicate through shared variables and synchronisation signals.

Verification requires the application of arguments for sequential programs with some additions. The task (processes) can be verified separately only if the constituent variables of each task (process) are distinct. If the variables are shared, the potential concurrency makes the effect of the program unpredictable (and hence not capable of verification) unless there is some further rule that governs the sequencing of the several actions of the tasks (processes). Note that the use of a synchronising procedure means the time taken for each individual action is not relevant to the verification procedure. The task can proceed at any speed: the correctness depends on the actions of the synchronising procedure. (Synchronisation techniques are discussed extensively in Chapters 5 and 6.)

1.6.3 Real-time

A real-time program differs from the previous types in that, in addition to its actions not necessarily being disjoint in time, the sequence of some of its actions is not determined by the designer but by the environment – that is, by events occurring in the outside world which occur in real time and without reference to the internal operations of the computer. Such events cannot be made to conform to the intertask synchronisation rules. A real-time program can still be divided into a number of tasks but communication between the tasks cannot necessarily wait for a synchronisation signal: the environment task cannot be delayed. (Note that in process control applications the main environment task is usually that of keeping real time, that is a real-time clock task. It is this task which provides the timing for the scanning tasks which gather information from the outside world about the process.) In real-time programs, in contrast to the two previous types of program, the *actual time taken* by an action is an essential factor in the process of verification.

In the rest of the book we shall assume that we are concerned with real-time software and references to sequential and multi-tasking programs should be taken to imply that the program is real time. Non-real-time programs will be referred to as *standard* programs.

Consideration of the types of reasoning necessary for the verification of programs is important, not because we, as engineers, are seeking a method of formal proof, but because we are seeking to understand the factors which need to be considered when designing real-time software. Experience shows that the design of real-time software is significantly more difficult than the design of sequential software. The problems of real-time software design have not been helped by the fact that the early high-level languages were sequential in nature and they did not allow direct access to many of the detailed features of the computer hardware. As a consequence, real-time features had to be built into the operating system which was written in the assembly language of the machine by teams of specialist programmers. The cost of producing such operating systems was high and they had therefore to be general purpose so that they could be used in a wide range of applications in order to reduce the unit cost of producing them. These operating systems could be *tailored*, that is they could be reassembled to exclude or include certain features, for example to change the number of tasks which could be handled, or to change the number of input/output devices and types of device. Such changes could usually only be made by the supplier.

1.7 SUMMARY

In this chapter a brief history of computer control has been given; information on further development and the reasons for particular computer structures is given in Chapter 2. The simple example of a hot-air blower has been used to illustrate the breadth of knowledge required to design and implement a computer control system. The engineer is required to have knowledge of:

> the plant;
> transducers;
> actuators;
> computer hardware;
> interface techniques;
> communication systems;
> software design;
> programming languages;
> control algorithm design; and
> signal processing.

In this book we do not attempt to cover all these areas in detail, but instead concentrate on software design and the implementation of software.

As should have already been apparent, there will be an emphasis on dividing the operations to be performed into separate tasks. The detail of operations within the tasks will be hidden within the software for that task. This is a process known as *abstraction*. The organisation of the tasks to form an overall system can then be carried out without the distraction the detail of their operations would otherwise provide.

So far we have used the terms process, plant, program and task loosely and without defining them. In subsequent chapters they are used as follows:

> *Process, plant*: The physical system that is to be controlled.
> *Program*: The complete software package used to provide the control. Sometimes also referred to as the *application* program.
> *Module*: A subdivision of the program.
> *Task*: A subdivision of the program (or of a module of the program) whose execution can be separately scheduled from other parts of the program. A program or module may be divided into several tasks and because the tasks can be separately scheduled they have the potential to run concurrently.

The key ideas described in this chapter are:

- Computer control involves many different activities that have to be carried out concurrently.
- Digital computers are sequential devices and an individual computer can give, because of its speed of operation, the appearance of carrying out activities concurrently.

- Real-time systems have to carry out both periodic and aperiodic activities.
- Real-time systems have to satisfy time constraints that can be either a hard constraint or a soft (average value) constraint.
- Real-time software is more difficult to specify, design and construct than non-real-time software.

EXERCISES

1.1 You have been asked to design a computer-based system to control all the operations of a retail petrol (gasoline) station (control of pumps, cash receipts, sales figures, deliveries, etc.). What type of real-time system would you expect to use?

1.2 An automatic bank teller works by polling each teller in turn. Some tellers are located outside buildings and others inside. How could the polling system be organised to ensure that the waiting time at the outside locations was less than at the inside locations?

1.3 Would you classify any of the following systems as real-time? In each case give reasons for your answer and classify the real-time systems as hard or soft.
(a) A simulation program run by an engineer on a personal computer.
(b) An airline seat-reservation system with on-line terminals.
(c) A microprocessor-based automobile ignition and fuel injection system.
(d) A computer system used to obtain and record measurements of force and strain from a tensile strength testing machine.
(e) An aircraft autopilot.

1.4 For (a) the petrol (gas) station system in Exercise 1.1 and (b) the automatic bank teller in Exercise 1.2, list the activities which have to be performed and estimate the time requirements of each activity.

1.5 For the hot-air blower described in section 1.2, estimate the precision required for the analog-to-digital and digital-to-analog converters.

1.6 In the passage given below identify the different types of time constraints.

> The boiler pressure alarm lamp must be lit within 0.6 seconds of the boiler pressure high signal being set true. The temperature of the steam leaving the boiler must be read every 0.5 seconds. The operator display should be updated on average at 2-second intervals and on no occasion should the update interval exceed 10 seconds. The response to a request for a printout of the current plant status should typically be completed within 2 minutes.

1.7 Explain the difference between a real-time program and a non-real-time program. Why are real-time programs more difficult to verify than non-real-time programs?

2

Concepts of Computer Control

In this chapter we first use some typical process control applications to illustrate the range of operations that a computer-controlled system has to carry out. We then consider some of the differences between the process control applications and other embedded system applications.

In the final sections of the chapter we examine some of the computer configurations that are used for implementing computer control systems.

The aims of this chapter are to:

- Explain the terms used in process control applications and to provide examples of particular types of application.
- Describe in general terms the various types of control strategies that are used.
- Explain and compare basic computer configurations used for control.
- Explain the importance of the human–computer interface.

2.1 INTRODUCTION

Computers are now used in so many different ways that we could take up the whole book by simply describing various applications. However, when we examine the applications closely we find that there are many common features. The basic features of computer control systems are illustrated in the following sections using examples drawn from industrial process control. In this field applications are typically classified under the following headings:

- batch;
- continuous; and
- laboratory (or test).

The categories are not mutually exclusive: a particular process may involve activities which fall into more than one of the above categories; they are, however, useful for describing the general character of a particular process.

2.1.1 Batch

The term *batch* is used to describe processes in which a sequence of operations are carried out to produce a quantity of a product − the batch − and in which the sequence is then repeated to produce further batches. The specification of the product or the exact composition may be changed between the different runs. A typical example of batch production is rolling of sheet steel. An ingot is passed through the rolling mill to produce a particular gauge of steel; the next ingot may be either of a different composition or rolled to a different thickness and hence will require different settings of the rolling mill.

An important measure in batch production is *set-up* time (or *change-over* time), that is, the time taken to prepare the equipment for the next production batch. This is *wasted* time in that no output is being produced; the ratio between *operation* time (the time during which the product is being produced) and set-up time is important in determining a suitable batch size. In mechanical production the advent of the NC (Numerically Controlled) machine tool which can be set up in a much shorter time than the earlier automatic machine tool has led to a reduction in the size of batch considered to be economic.

2.1.2 Continuous

The term *continuous* is used for systems in which production is maintained for long periods of time without interruption, typically over several months or even years. An example of a continuous system is the catalytic cracking of oil in which the crude oil enters at one end and the various products − fractionates − are removed as the process continues. The ratio of the different fractions can be changed but this is done without halting the process. Continuous systems may produce batches, in that the product composition may be changed from time to time, but they are still classified as continuous since the change in composition is made without halting the production process. A problem which occurs in continuous processes is that during change-over from one specification to the next, the output of the plant is often not within the product tolerance and must be scrapped. Hence it is financially important that the change be made as quickly and smoothly as possible.

There is a trend to convert processes to continuous operation − or, if the whole process cannot be converted, part of the process. For example, in the baking industry bread dough is produced in batches but continuous ovens are frequently used to bake it whereby the loaves are placed on a conveyor which moves slowly through the oven. An important problem in mixed mode systems, that is systems in which batches are produced on a continuous basis, is the tracking of material through the process; it is obviously necessary to be able to identify a particular batch at all times.

2.1.3 Laboratory Systems

Laboratory-based systems are frequently of the operator-initiated type in that the computer is used to control some complex experimental test or some complex equipment used for routine testing. A typical example is the control and analysis of data from a vapour phase chromatograph. Another example is the testing of an audiometer, a device used to test hearing. The audiometer has to produce sound levels at different frequencies; it is complex in that the actual level produced is a function of frequency since the sensitivity of the human ear varies with frequency. Each audiometer has to be tested against a sound-level meter and a test certificate produced. This is done by using a sound-level meter connected to a computer and using the output from the computer to drive the audiometer through its frequency range. The results printed out from the test computer provide the test certificate.

As with attempts to classify systems as batch or continuous so it can be difficult at times to classify systems solely as laboratory. The production of steel using the electric arc furnace involves complex calculations to determine the appropriate mix of scrap, raw materials and alloying additives. As the melt progresses samples of the steel are taken and analysed using a spectrometer. Typically this instrument is connected to a computer which analyses the results and calculates the necessary adjustment to the additives. The computer used may well be the computer which is controlling the arc furnace itself.

In whatever way the application is classified the activities being carried out will include:

- data acquisition;
- sequence control;
- loop control (DDC);
- supervisory control;
- data analysis;
- data storage; and
- human–computer interfacing (HCI).

The objectives of using a computer to control the process will include:

- efficiency of operation;
- ease of operation;
- safety;
- improved products;
- reduction in waste;
- reduced environmental impact; and
- a reduction in direct labour.

2.1.4 General Embedded Systems

If we examine the general range of systems which use embedded computers – from domestic appliances, through hi-fi systems, automobile management systems, intelligent instruments, active control of structures, to large flexible manufacturing systems and aircraft control systems – we will find that the activities that are carried out in the computer and the objectives of using a computer are similar to those listed above. The major differences will lie in the balance between the different activities, the time-scales involved, and the emphasis given to the various objectives.

2.2 SEQUENCE CONTROL

Although sequence control occurs in some part of most systems it often predominates in batch systems and hence a batch system is used to illustrate it. Batch systems are widely used in the food processing and chemical industries where the operations carried out frequently involve mixing raw materials, carrying out some process, and then discharging the product. A typical reactor vessel for this purpose is shown in Figure 2.1. A chemical is produced by the reaction of two other chemicals at a specified temperature. The chemicals are mixed together in a sealed vessel (the reactor) and the temperature of the reaction is controlled by feeding hot or cold water through the water jacket which surrounds the vessel. The water flow

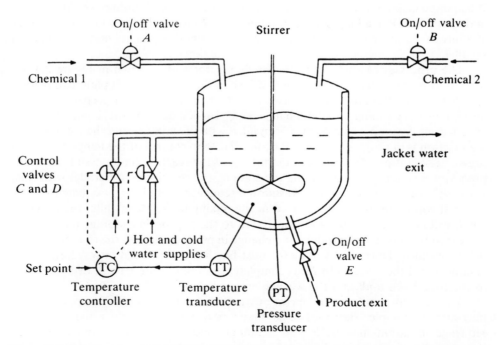

Figure 2.1 A simple chemical reactor vessel (redrawn from Bennett and Linkens, *Real-time Computer Control*, Peter Peregrinus (1984)).

is controlled by adjusting valves C and D. The flow of material into and out of the vessel is regulated by the valves A, B and E. The temperature of the contents of the vessel and the pressure in the vessel are monitored.

The procedure for the operation of the system may be as follows:

1. Open valve A to charge the vessel with chemical 1.
2. Check the level of the chemical in the vessel (by monitoring the pressure in the vessel); when the correct amount of chemical has been admitted, close valve A.
3. Start the stirrer to mix the chemicals together.
4. Repeat stages 1 and 2 with valve B in order to admit the second chemical.
5. Switch on the three-term controller and supply a set point so that the chemical mix is heated up to the required reaction temperature.
6. Monitor the reaction temperature; when it reaches the set point, start a timer to time the duration of the reaction.
7. When the timer indicates that the reaction is complete, switch off the controller and open valve C to cool down the reactor contents. Switch off the stirrer.
8. Monitor the temperature; when the contents have cooled, open valve E to remove the product from the reactor.

When implemented by computer all of the above actions and timings would be based upon software. For a large chemical plant such sequences can become very lengthy and intricate and, to ensure efficient operating, several sequences may take place in parallel.

The processes carried out in the single reactor vessel shown in Figure 2.1 are often only part of a larger process as is shown in Figure 2.2. In this plant two reactor vessels ($R1$ and $R2$) are used alternately, so that the processes of preparing for the next batch and cleaning up after a batch can be carried out in parallel with the actual production. Assuming that $R1$ has been filled with the mixture and the catalyst, and the reaction is in progress, there will be for $R1$: loop control of the temperature and pressure; operation of the stirrer; and timing of the reaction (and possibly some in-process measurement to determine the state of the reaction). In parallel with this, vessel $R2$ will be cleaned – the washdown sequence – and the next batch of raw material will be measured and mixed in the mixing tank. Meanwhile, the previous batch will be thinned down and transferred to the appropriate storage tank and, if there is to be a change of product or a change in product quality, the thin-down tank will be cleaned. Once this is done the next batch can be loaded into $R2$ and then, assuming that the reaction in $R1$ is complete, the contents of $R1$ will be transferred to the thin-down tank and the washdown procedure for $R1$ initiated.

The various sequences of operations required can become complex and there may also be complex decisions to be made as to when to begin a sequence. The sequence initiation may be left to a human operator or the computer may be programmed to supervise the operations (*supervisory control* – see below). The decision to use human or computer supervision is often very difficult to make. The

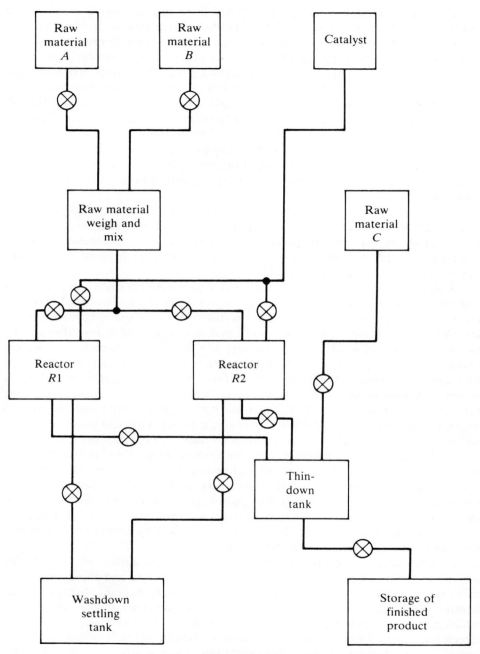

Figure 2.2   Typical chemical batch process.

aim is usually to minimise the time during which the reaction vessels are idle since this is unproductive time. The calculations needed and the evaluation of options can be complex, particularly if, for example, reaction times vary with product mix, and therefore it would be expected that decisions made using computer supervisory control would give the best results. It is certainly true that completely automatic control of the process would avoid the quite natural tendency for operators to avoid starting the next batch sequence close to the end of their shift; however, it is difficult using computer control to obtain the same flexibility that can be achieved using a human operator (and to match the ingenuity of good operators). As a consequence many supervisory systems are mixed; the computer is programmed to carry out the necessary supervisory calculations and to present its decisions for confirmation or rejection by the operator, or alternatively it presents a range of options to the operator.

In most batch systems there is also, in addition to the sequence control, some continuous feedback control: for example, control of temperatures, pressures, flows, speeds or currents. In process control terminology continuous feedback control is referred to as *loop control* or *modulating control* and in modern systems this would be carried out using DDC.

A similar mixture of sequence, loop and supervisory control can be found in continuous systems. Consider the float glass process shown in Figure 2.3. The raw material – sand, powdered glass and fluxes (the frit) – is mixed in batches and fed into the furnace. It melts rapidly to form a molten mixture which flows through the furnace. As the molten glass moves through the furnace it is refined. The process requires accurate control of temperature in order to maintain quality and to keep fuel costs to a minimum – heating the furnace to a higher temperature than is necessary wastes energy and increases costs. The molten glass flows out of the furnace and forms a ribbon on the float bath; again, temperature control is important as the glass ribbon has to cool sufficiently so that it can pass over rollers without damaging its surface. The continuous ribbon passes into the lehr where it is annealed and where temperature control is again required. From the lehr the glass ribbon moves down the line towards the cut-up stations at a speed which is too great for manual inspection so automatic inspection is used, faults being marked by spraying paint onto the ribbon. It then passes under the cutters which cut it into sheets of the required size; automatic stackers then lift the sheets from the production line. The whole of this process is controlled by several computers and involves loop, sequence and supervisory control.

Sequence control systems can vary from the large – the start-up of a large boiler turbine unit in a power station when some 20 000 operations and checks may have to be made – to the small – starting a domestic washing machine. Most sequence control systems are simple and frequently have no loop control. They are systems which in the past would have been controlled by relays, discrete logic, or integrated circuit logic units. Examples are simple presses where the sequence might be: locate blank, spray lubricant, lower press, raise press, remove article, spray lubricant. Special computer systems known as *programmable logic controllers* (PLCs)

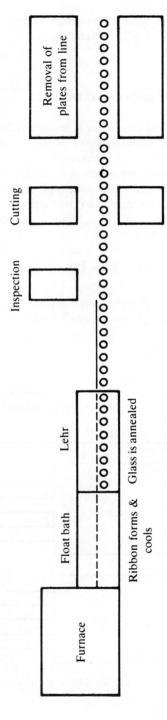

Figure 2.3 Schematic of float glass process.

have been developed for these simple sequence systems together with special programming languages (Henry, 1987; Kissell, 1986).

2.3 LOOP CONTROL (DIRECT DIGITAL CONTROL)

In direct digital control (DDC) the computer is in the feedback loop as is shown in Figure 2.4. This is a generalised representation of the system shown in Figures 1.4 and 1.5; the system shown in Figure 2.4 is assumed to involve several control loops all of which are handled within one computer. A consequence of the computer being in the feedback loop is that it forms a *critical* component in terms of the reliability of the system and hence great care is needed to ensure that, in the event of the failure or malfunctioning of the computer, the plant remains in a safe condition. The usual means of ensuring safety are to limit the DDC unit to making *incremental* changes to the actuators on the plant; and to limit the rate of change of the actuator settings (the actuators are labelled *A* in Figure 2.4). The advantages claimed for DDC over analog control are:

1. Cost – a single digital computer can control a large number of loops. In the early days the break-even point was between 50 and 100 loops, but now with the introduction of microprocessors a single-loop DDC unit can be cheaper than an analog unit.

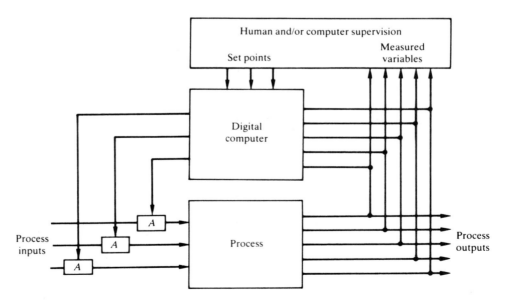

Figure 2.4   Direct digital control.

2. Performance – digital control offers simpler implementation of a wide range of control algorithms, improved controller accuracy and reduced drift.
3. Safety – modern digital hardware is highly reliable with long mean-time-between-failures and hence can improve the safety of systems. However, the software used in programmable digital systems may be much less reliable than the hardware.

The development of integrated circuits and the microprocessor have ensured that in terms of cost the digital solution is now cheaper than the analog. Single-loop controllers used as stand-alone controllers are now based on the use of digital techniques and contain one or more microprocessor chips which are used to implement DDC algorithms.

The adoption of improved control algorithms has, however, been slow. Many computer control implementations have simply taken over the well-established analog PID (Proportional + Integral + Derivative) algorithm.

2.3.1 PID Control

The PID control algorithm has the general form

$$m(t) = K_p[e(t) + 1/T_i \int_0^t e(t)dt + T_d de(t)/dt] \tag{2.1}$$

where $e(t) = r(t) - c(t)$ and $c(t)$ is the measured variable, $r(t)$ is reference value or set point, and $e(t)$ is error; K_p is the overall controller gain; T_i is the integral action time; and T_d is the derivative action time. This algorithm can be expressed in many other forms and we will deal with the details of it and methods of implementation in Chapter 4.

For a wide range of industrial processes it is difficult to improve on the control performance that can be obtained by using either PI or PID control (except at considerable expense) and it is for this reason that the algorithms are widely used. For the majority of systems PI control is all that is necessary. Using a control signal that is made proportional to the error between the desired value of an output and the actual value of the output is an obvious and (hopefully) a reasonable strategy. The ratio between the control signal and the error signal can be adjusted using the proportionality constant (gain) K_p. Choosing the value of K_p involves a compromise: a high value of K_p gives a small steady-state error and a fast response, but the response will be oscillatory and may be unacceptable in many applications; a low value gives a slow response and a large steady-state error. By adding the integral action term the steady-state error can be reduced to zero since the integral term, as its name implies, integrates the error signal with respect to time. For a given error value the rate at which the integral term increases is determined by the integral action time T_i. The major advantage of incorporating an integral term arises from the fact that it compensates for changes that occur in the process being controlled.

A purely proportional controller operates correctly only under one particular set of process conditions: changes in the load on the process or some environmental condition will result in a steady-state error; the integral term compensates for these changes and reduces the error to zero. For a few processes which are subjected to sudden disturbances the addition of the derivative term can give improved performance. Because derivative action produces a control signal that is related to the rate of change of the error signal, it *anticipates* the error and hence acts to reduce the error that would otherwise arise from the disturbance.

There is much more to computer control than simple DDC and the use of DDC is not limited to the PID control algorithm; algorithms developed using various digital control design techniques can be equally effective and a lot more flexible than the three-term controller. However, the art of tuning PID controllers is well established and the technique gives a well-behaved controller so that the introduction of new techniques has been slow. In fact, because the PID controller copes perfectly adequately with 90% of all control problems, it provides a strong deterrent to the adoption of new control system design techniques.

2.3.2 DDC Applications

DDC may be applied either to a single-loop system implemented on a small microprocessor or to a large system involving several hundred loops. The loops may be cascaded, that is with the output or actuation signal of one loop acting as the set point for another loop, signals may be added together (ratio loops) and conditional switches may be used to alter signal connections. A typical industrial system is shown in Figure 2.5. This is a steam boiler control system. The steam pressure is controlled by regulating the supply of fuel oil to the burner, but in order to comply with the pollution regulations a particular mix of air and fuel is required. We are not concerned with how this is achieved but with the elements which are required to implement the chosen control system.

The steam pressure control system generates an actuation signal which is fed to an auto/manual bias station. If the station is switched to auto then the actuation signal is transmitted; if it is in manual mode a signal which has been entered manually (say, from a keyboard) is transmitted. The signal from the bias station is connected to two units, a high signal selector and a low signal selector each of which has two inputs and one output. The high selector transmits the higher of the two input signals, the low selector transmits the lower of the two inputs. The signal from the low selector provides the set point for the DDC loop controlling the oil flow, the signal from the high selector provides the set point for the air flow controller (two cascade loops). A ratio unit is installed in the air flow measurement line. A signal from the controller which monitors the combustion flames directly (using an optical pyrometer) is added to the air flow signal to provide the input to the air flow controller.

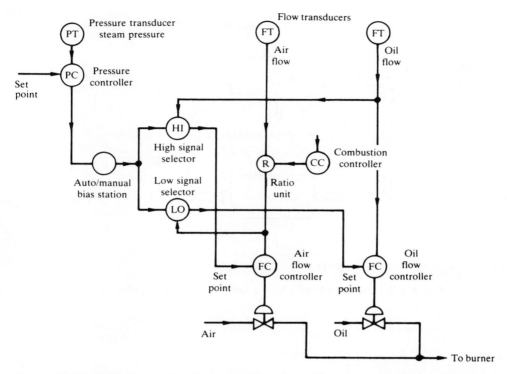

Figure 2.5 A boiler control scheme (redrawn from Bennett and Linkens, *Real-time Computer Control*, Peter Peregrinus (1984)).

DDC is not necessarily limited to simple feedback control as shown in Figure 2.6. It is possible to use techniques such as inferential, feedforward and adaptive or self-tuning control. Inferential control, illustrated in Figure 2.7, is the term applied to control where the variables on which the feedback control is to be based cannot be measured directly, but have to be 'inferred' from measurements of some other quantity. In Figure 2.7, some of the outputs can be measured and used directly in the feedback control; other outputs required by the controller cannot be measured directly and hence some other process measurement is made and from this the value of the controlled variable is inferred.

Inferential measurements are frequently used in distillation column control. A schematic of a binary distillation column is shown in Figure 2.8. The four independent variables usually controlled are the liquid levels H_a and H_b in the accumulator and the reboiler, and the compositions X_a and X_b of the top and bottom products. The compositions can be measured directly by spectrographic techniques but it is more usual to measure the temperatures at points Y_a and Y_b near the top and bottom of the column and the pressure P in the column. The temperatures represent the boiling points of the mixture at the position in the

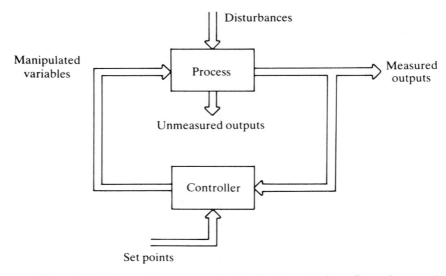

Figure 2.6   General structure of a feedback control configuration.

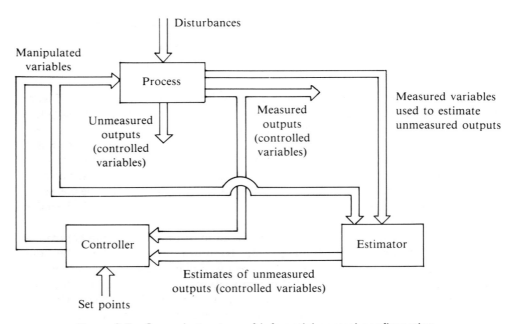

Figure 2.7   General structure of inferential control configuration.

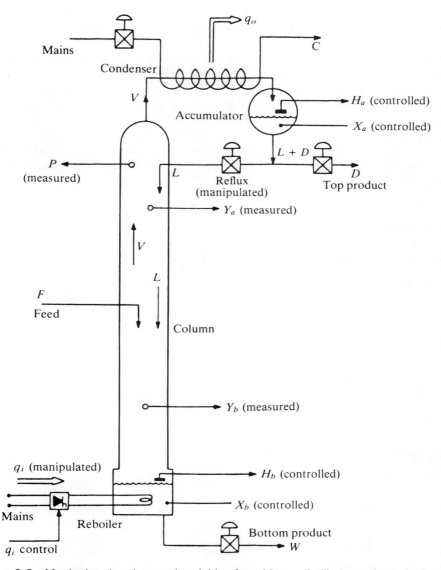

Figure 2.8 Manipulated and control variables for a binary distillation column (redrawn from Bennett and Linkens, *Computer Control of Industrial Processes*, Peter Peregrinus (1982)).

column and from measurements of pressure and temperature the compositions can be inferred (Edwards, 1982).

Feedforward control is frequently used in the process industries; it involves measuring the disturbances on the system rather than measuring the outputs and is illustrated in Figure 2.9. For example, in the hot rolling of sheet steel, if the temperature of the billet is known as it approaches the first-stage mill, the initial setting of the roll gap can be calculated accurately and estimates of the reduction at each stage of the mill can be made; hence the initial gaps for the subsequent stages can also be calculated. If this is done the time taken to get the gauge of the steel within tolerance can be much reduced and hence the quantity of scrap (out of tolerance) steel reduced. The effect of introducing feedforward control is to speed up the response of the system to disturbances; it can, however, only be used for disturbances which can be measured, and in plants where the effects of the disturbances can be predicted accurately.

2.3.3 Adaptive Control

Adaptive control can take several forms. Three of the most common are:

- preprogrammed adaptive control (gain scheduled control);
- self-tuning; and
- model-reference adaptive control.

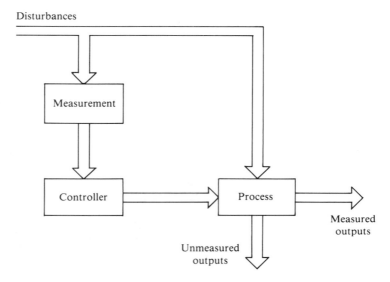

Figure 2.9   General structure of a feedforward control configuration.

Programmed adaptive control is illustrated in Figure 2.10a. The adaptive, or adjustment, mechanism makes preset changes on the basis of changes in auxiliary process measurements. For example, in a reaction vessel a measurement of the level of liquid in the vessel (an indicator of the volume of liquid in the vessel) might be used to change the gain of the temperature controller; in many aircraft controls the measured air speed is used to select controller parameters according to a preset schedule. An alternative form is shown in Figure 2.10b in which measurements of changes in the external environment are used to select the gain or other controller parameters. For example, in an aircraft autostabiliser, control parameters may be changed according to the external air pressure. Another example is the use of

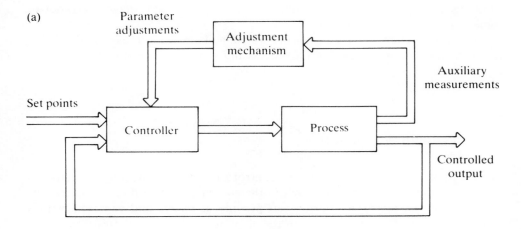

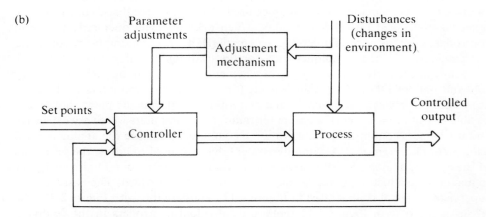

Figure 2.10 Programmed adaptive control (gain scheduled): (a) auxiliary process measurements; (b) external environment (open loop).

measurements of external temperature and wind velocities to adjust control parameters for a building environment control system.

Adaptive control using self-tuning is illustrated in Figure 2.11 and uses identification techniques to achieve continual determination of the parameters of the process being controlled; changes in the process parameters are then used to adjust the actual controller.

An alternative form of self-tuning is frequently found in commercial PID controllers (usually called autotuning). At operator-initiated intervals or periodically, the controller injects a small disturbance to the process and measures the response, the response is compared to some desired response and controller parameters are adjusted to bring the response closer to the desired response. The comparison may be based on a simple measure such as percentage overshoot or some more complex comparators.

The model reference technique is illustrated in Figure 2.12; it relies on the ability to construct an accurate model of the process and to measure the disturbances which affect the process.

2.4 SUPERVISORY CONTROL

The adoption of computers for process control has increased the range of activities that can be performed, for not only can the computer system directly control the operation of the plant, but also it can provide managers and engineers with a comprehensive picture of the status of the plant operations. It is in this *supervisory* role and in the presentation of information to the plant operator − large rooms full of dials and switches have been replaced by VDUs and keyboards − that the major changes have been made: the techniques used in the basic feedback control of the plant have changed little from the days when pneumatically operated three-term controllers were the norm. Direct digital control (DDC) is often simply the computer implementation of the techniques used for the traditional analog controllers.

Many of the early computer control schemes used the computer in a supervisory role and not for DDC. The main reasons for this were (a) computers in the early days were not always very reliable and caution dictated that the plant should still be able to run in the event of a computer failure; (b) computers were very expensive and it was not economically viable to use a computer to replace the analog control equipment in current use. A computer system that was used to adjust the set points of the existing analog control system in an optimum manner (to minimise energy or to maximise production) could perhaps be economically justified. The basic idea of supervisory control is illustrated in Figure 2.13 (compare this with Figure 2.4). The circles labelled *C* in Figure 2.13 represent individual controllers in the feedback loop; these can be themselves digital computers (or some other form of controller), but their operation is supervised by a digital computer.

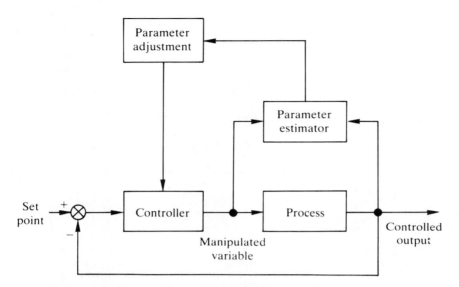

Figure 2.11   Self-tuning adaptive control.

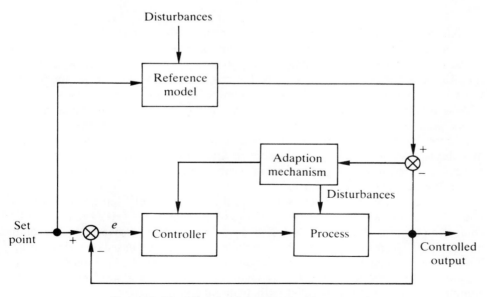

Figure 2.12   Model-reference adaptive control.

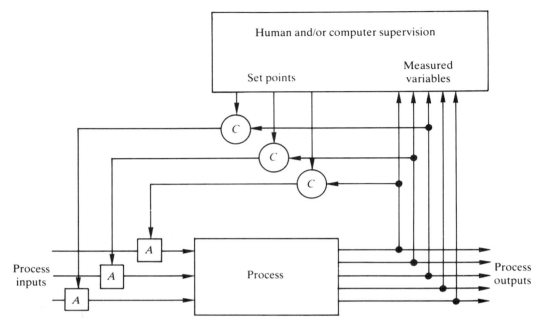

Figure 2.13 Supervisory control.

An example of supervisory control is shown in Figure 2.14. Two evaporators are connected in parallel and material in solution is fed to each unit. The purpose of the plant is to evaporate as much water as possible from the solution. Steam is supplied to a heat exchanger linked to the first evaporator and the steam for the second evaporator is supplied from the vapours boiled off from the first stage. To achieve maximum evaporation the pressures in the chambers must be as high as safety permits. However, it is necessary to achieve a balance between the two evaporators; if the first is driven at its maximum rate it may generate so much steam that the safety thresholds for the second evaporator are exceeded. A supervisory control scheme can be designed to balance the operation of the two evaporators to obtain the best overall evaporation rate.

Most applications of supervisory control are very simple and are based upon knowledge of the steady-state characteristics of the plant. In a few systems complex control algorithms have been used and have been shown to give increased plant profitability. The techniques used have included optimisation based on hill climbing, linear programming and simulations involving complex non-linear models of plant dynamics and economics. In these applications the complex algorithms have to be computed in real time in parallel with plant operation. There is now a growing interest in the use of *expert* systems to provide supervisory control (see Efstathiou, 1989; Bennett, 1992).

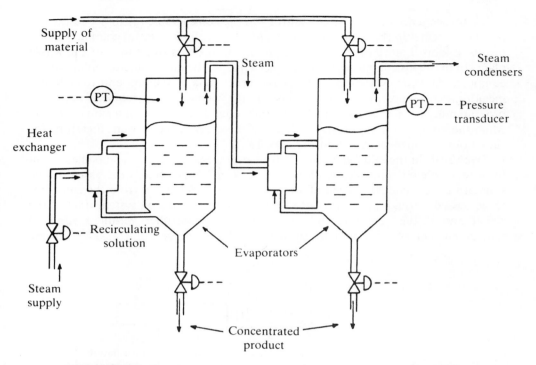

Figure 2.14 An evaporation plant (redrawn from Bennett and Linkens, *Real-time Computer Control*, Peter Peregrinus (1984)).

2.5 CENTRALISED COMPUTER CONTROL

Throughout most of the 1960s computer control implied the use of one central computer for the control of the whole plant. The reason for this was largely financial: computers were expensive. From the previous sections it should now be obvious that a typical computer-operated process involves the computer in performing many different types of operations and tasks. Although a general purpose computer can be programmed to perform all of the required tasks the differing time-scales and security requirements for the various categories of task make the programming job difficult, particularly with regard to the testing of software. For example, the feedback loops in a process may require calculations at intervals measured in seconds while some of the alarm and switching systems may require a response in less than 1 second; the supervisory control calculations may have to be repeated at intervals of several minutes or even hours; production management will want summaries at shift or daily intervals; and works management will require weekly or monthly analyses. Interrelating all the different time-scales can cause serious difficulties.

A consequence of centralised control was the considerable resistance to the use of DDC schemes in the form shown in Figure 2.4; with one central computer in the feedback loop, failure of the computer results in the loss of control of the whole plant. In the 1960s computers were not very reliable: the mean-time-to-failure of the computer hardware was frequently of the order of a few hours and to obtain a mean-time-to-failure of 3 to 6 months for the whole system required *defensive* programming to ensure that the system could continue running in a safe condition while the computer was repaired. Many of the early schemes were therefore for supervisory control as shown in Figure 2.13.

However, in the mid 1960s the traditional process instrument companies began to produce digital controllers with analog back-up. These units were based on the standard analog controllers but allowed a digital control signal from the computer to be passed through the controller to the actuator: the analog system tracked the signal and if the computer did not update the controller within a specified (adjustable) interval the unit dropped on to local analog control. This scheme

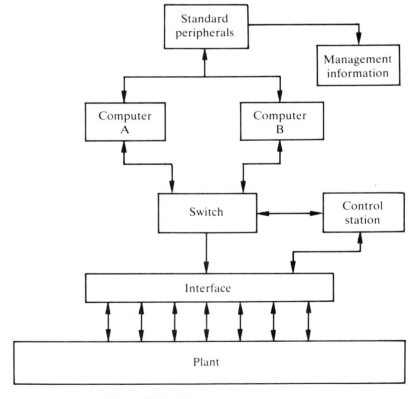

Figure 2.15 Dual computer scheme.

enabled DDC to be used with the confidence that if the computer failed, the plant could still be operated. The cost, however, was high in that two complete control systems had to be installed.

By 1970 the cost of computer hardware had reduced to such an extent that it became feasible to consider the use of dual computer systems (Figure 2.15). Here, in the event of failure of one of the computers, the other takes over. In some schemes the change-over is manual, in others automatic failure detection and change-over is incorporated. Many of these schemes are still in use. They do, however, have a number of weaknesses: cabling and interface equipment is not usually duplicated, neither is the software – in the sense of having independently designed and constructed programs – so that the lack of duplication becomes crucial. Automatic failure and change-over equipment when used becomes in itself a critical component. Furthermore, the problems of designing, programming, testing and maintaining the software are not reduced: if anything they are further complicated in that provision for monitoring ready for change-over has to be provided.

The continued reduction of the cost of hardware and the development of the microprocessor has made multi-computer systems feasible. These fall into two types:

1. Hierarchical – Tasks are divided according to function, for example with one computer performing DDC calculations and being subservient to another which performs supervisory control.
2. Distributed – Many computers perform essentially similar tasks in parallel.

The above can of course be combined to give a hierarchical, distributed system.

2.6 HIERARCHICAL SYSTEMS

This is the most natural development in that it follows the typical company decision-making structure shown in the pyramid in Figure 2.16: each decision element receives commands from the level above and sends information back to that level and, on the basis of information received from the element or elements below and from constraints imposed by elements at the same level, sends commands to the element(s) below and information to element(s) at the same level. This structure also follows a natural division of the production process in terms of the time-response requirements of the different levels. At the bottom of the pyramid, or hierarchy, a fast response (measured in milliseconds or seconds) to simple problems is required: as one progresses up the hierarchy the complexity of the calculations increases as does the time allowed for the response.

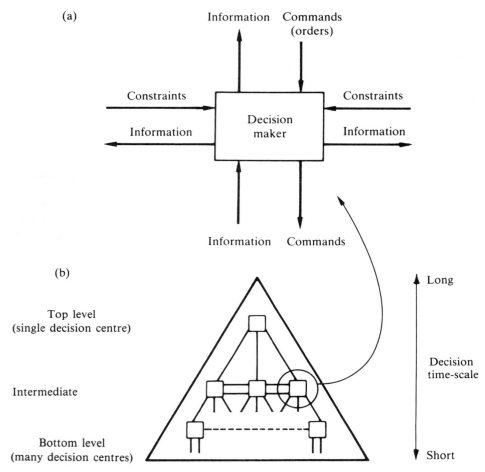

Figure 2.16 Hierarchical decision-making: (a) decision-making function; (b) decision-making structure.

A typical example of a hierarchical system is the batch system shown in Figures 2.17 and 2.18. This system has three levels which we have called manager, supervisor and unit control. It is assumed that single computers are used for the manager and supervisor functions and that for each processing unit a single unit control computer is used. At the manager level the functions such as resource allocation, production scheduling and production accounting are carried out. The input information may be, for example, sales orders (actual and forecast), stock levels, selling cost and production costs (or profit margins) on each product, operating costs for each process unit and scheduled maintenance plans for each operating unit, and the current state of production units. On the basis of this information the production schedule, that is the list of products to be produced and the quantities and process

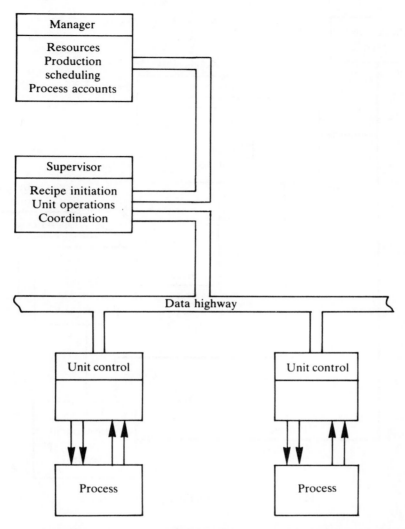

Figure 2.17   Batch control using a hierarchical system.

unit to be used, will be calculated. This may be done daily or at some other interval depending on production times, etc.

The information regarding the production schedule is transferred to the supervisor. It is assumed that the supervisor has a store containing the product recipe (that is, how to make a particular product) and a store of the operation sequences for making the product. When the appropriate unit, as selected by the production plan, is ready the information on the product – set points, alarm conditions, tolerances, etc. – is loaded down into the unit controller as are the

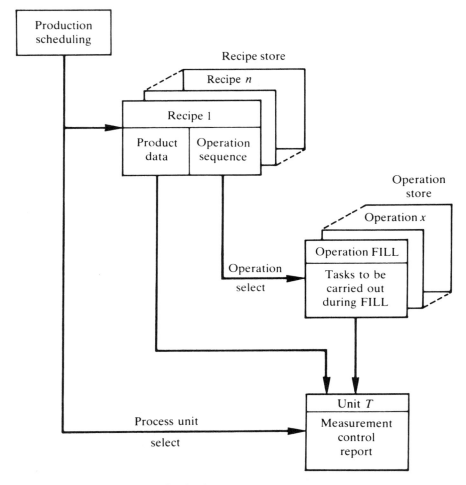

Figure 2.18 Batch control (software).

particular operations to be carried out during each stage of production. In addition the supervisor will receive regular reports of the progress of each process unit and will resolve any conflicts in the demand for resources, such as those that can arise where the units share a weigher (the supervisor will decide which is to use it at a particular time) or where the process requires energy input and the total available is limited, either because of the boiler capacity, or because penalties are imposed for exceeding a specified kV A rating (it will be the responsibility of the supervisor to decide, either automatically or through a request to a human supervisor, which unit is to reduce consumption or in what proportion consumption is to be reduced).

At the lowest level, the unit controllers are responsible for operating the plant:

opening and closing valves and switches, controlling temperatures, pressures, speeds, flows, monitoring alarms, and reporting plant conditions.

Most hierarchical systems will involve some form of distributed network and hence most systems will be a mixture of hierarchical and distributed control.

2.7 DISTRIBUTED SYSTEMS

The underlying assumptions of the distributed approach are that:

1. each unit is carrying out essentially similar tasks to all the other units; and
2. in the event of failure or overloading of a particular unit all or some of the work can be transferred to other units.

In other words, the work is not divided by function and allocated to a particular computer as in hierarchical systems: instead, the total work is divided up and spread across several computers. This is a conceptually simple and attractive approach — many hands make light work — but it poses difficult hardware and software problems since, in order to gain the advantages of the approach, allocation of the tasks between computers has to be dynamic, that is there has to be some mechanism which can assess the work to be done and the present load on each computer in order to allocate work. Because each computer needs access to all the information in the system, high-bandwidth data highways are necessary. There has been considerable progress in developing such highways and the various types are discussed below: computer scientists and engineers are also carrying out considerable research on multi-processor computer systems and this work could lead to totally distributed systems becoming feasible.

There is also a more practical approach to distributing the computing load whereby no attempt is made to provide for the dynamic allocation of resources but instead a simple *ad hoc* division is adopted with, for example, one computer performing all non-plant input and output, one computer performing all DDC calculations, another performing data acquisition and yet another performing the control of the actuators.

In most modern schemes a mixture of distributed and hierarchical approaches is used as shown in Figure 2.19. The tasks of measurement, DDC, operator communications, etc., are distributed among a number of computers which are linked together via a common serial communications highway and are configured in a hierarchical command structure. Five broad divisions of function are shown:

Level 1   All computations and plant interfacing associated with measurement and actuation. This level provides a measurement and actuation database for the whole system.

Level 2   All DDC calculations.

Level 3   All sequence calculations.
Level 4   Operator communications.
Level 5   Supervisory control.
Level 6   Communications with other computer systems.

It is not necessary to preserve rigid boundaries; for example, a DDC unit may perform some sequencing or may interface directly to plant.

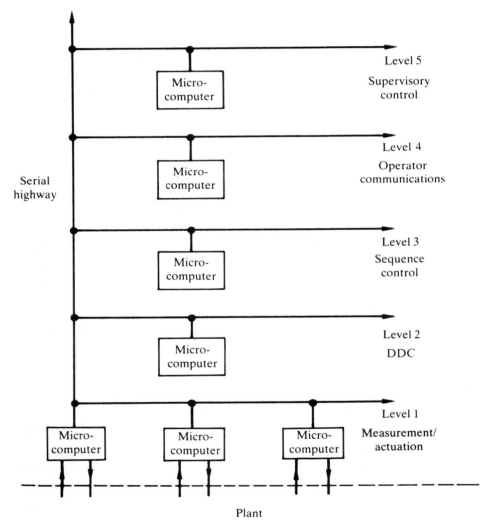

Figure 2.19   A distributed and hierarchical system.

The major advantages of this approach are:

1. The system capabilities are greatly enhanced by the sharing of tasks between processors – the burden of computation for a single processor becomes very great if all of the described control features are included. One of the main computing loads is that of measurement scanning, filtering and scaling, not because any one calculation is onerous but because of the large number of signals involved and the frequency at which the calculations have to be repeated. Separation of this aspect from the DDC, even if only into two processors, greatly enhances the number of control loops that can be handled. The DDC computer will collect measurements, already processed, via the communications link at a much lower frequency than that at which the measurement computer operates.

2. The system is much more flexible than the use of a single processor: if more loops are required or an extra operator station is needed, all that is necessary is to add more boxes to the communication link – of course the other units on the link will need to be updated to be aware of the additional items. It also allows standardisation, since it is much easier to develop standard units for well-defined single tasks than for overall control schemes.

3. Failure of a unit will cause much less disruption in that only a small portion of the overall system will not be working. Provision of automatic or semi-automatic transfer to a back-up system is much easier.

4. It is much easier to make changes to the system, in the form of either hardware replacements or software changes. Changing large programs is hazardous because of the possibility of unforeseen side-effects: with the use of small modules such effects are less likely to occur and are more easily detected and corrected.

5. Linking by serial highway means that the computer units can be widely dispersed: hence it is unnecessary to bring cables carrying transducer signals to a central control room.

2.8 HUMAN–COMPUTER INTERFACE (HCI)

The key to the successful adoption of a computer control scheme is often the facilities provided for the plant operator or user of the system. A simple and clear system for the day-to-day operation of the plant must be provided. All the information relevant to the current state of its operation should be readily available and facilities to enable interaction with the plant – to change set points, to adjust actuators by hand, to acknowledge alarm conditions, etc. – should be provided. A large proportion of the design and programming effort goes into the design and construction of operator facilities and the major process control equipment companies have developed extensive schemes for the presentation of information.

A typical operator station has specially designed keyboards and several display and printer units; extensive use is made of colour displays and mimic diagrams; video units are frequently provided to enable the operator to see parts of the plant (Jovic, 1986).

The standard software packages typically provide a range of display types: an alarm overview presenting information on the alarm status of large areas of the plant; a number of area displays presenting information on the control systems associated with each area; and loop displays giving extensive information on the details of a particular control loop. The exact nature of the displays is usually determined by the engineer responsible for the plant or part of the plant. Additional displays, including trends and summaries of past operations, are frequently available to the engineer, often in the form of hard copy. In addition the plant engineer (or maintenance engineer) will require information on which to base decisions about maintenance schedules and instrument, actuator and plant component replacements.

The plant manager requires access to different information: hard copy printouts – including graphs – that summarise the day-to-day operation of the plant and also provide a permanent plant operating history. Data presented to the manager will frequently have been analysed statistically to provide more concise information and to make decision-making more straightforward. The manager will be interested in assessing the economic performance of the plant and in determining possible improvements in plant operation.

The design of user interfaces is a specialist area. The safe operation of complex systems such as aircraft, nuclear power stations, chemical plants, air traffic control systems and other traffic control systems can be crucially affected by the way in which information is presented to the operator.

2.9 THE CONTROL ENGINEER

Assuming that a decision has been made on the most suitable computer system, the control engineer's responsibility is as follows:

1. To define the appropriate control strategy to meet the system requirements.
2. To define the measurements and actuations and to set up scaling and filter constants, alarm and actuator limits, sampling intervals, etc.
3. To define the DDC controllers, the interlinking or cascading of such controllers and the connections with any other elements in the control scheme.
4. To tune the control scheme, that is to select the appropriate gains so that it performs according to some desired specification.
5. To define and program the sequence control procedures necessary for the automation of plant operation.
6. To determine and implement satisfactory supervisory control schemes.

For a large project all the above requirements are too great for any one person to handle and in such cases a team of engineers would be involved. Additionally, if the programming of the system had to be done from scratch for each individual case, the task would be very burdensome and costly. However, process control applications have many features in common and the major suppliers of process control computer systems offer an extensive range of software packages, so that for each application the main task of the engineer/programmer is to select the appropriate modules and to assemble the required database describing the particular plant. The major omission of much of the standard software is that it fails to provide assistance with the tuning of the plant.

Software for the supervisory and sequencing operations has to be much more flexible in that the range of actions that may be required for any plant is much greater than for the DDC part. Frequently, it is necessary to program this part of the system specifically for each plant.

With the increasing use of microprocessors in a wide variety of applications the standard software approach used by the process controllers is becoming inadequate. It was based largely on the requirements of large plants with long time constants, controlled by a single large computer. Systems now have a computer or computers *embedded* as part of the plant or as part of a piece of equipment and are programmed to carry out only those functions required, that is they do not have within them general purpose software, a large proportion of which may be irrelevant to the particular application. This development has changed the approach to the design of real-time computer control systems and made it necessary for the control engineer to have a greater knowledge of programming languages and operating systems, as well as the techniques of distributed computing and communications.

2.10 ECONOMICS AND BENEFITS OF COMPUTER CONTROL SYSTEMS

Before the widespread availability of microprocessors, computer control was expensive and a very strong case was needed to justify the use of computer control rather than conventional instrumentation. In some cases computers were used because otherwise plant could not have been made to work profitably: this is particularly the case with large industrial processes that require complex sequencing operations. The use of a computer permits the repeatability that is essential, for example, in plants used for the manufacture of drugs. In many applications flexibility is important − it is difficult with conventional systems to modify the sequencing procedure to provide for the manufacture of a different product. Flexibility is particularly important when the product or the product specification may have to be changed frequently: with a computer system it is simple to maintain a database containing the product recipes and thus to change to a new recipe quickly and reliably.

The application of computer control systems to many large plants has frequently been justified on the grounds that even a small increase in productivity (say 1 or 2%) will more than pay for the computer system. After installation it has frequently been difficult to establish that an improvement has been achieved; sometimes production has decreased, but the computer proponents have then argued that but for the introduction of the computer system production would have decreased by a greater amount!

Some of the major benefits to accrue from the introduction of computer systems have been in the increased understanding of the behaviour of the process that has resulted from the studies necessary to design the computer system and from the information gathered during running. This has enabled supervisory systems to keep the plant running at an operating point closer to the desired point to be designed. The other main area of benefit has been in the control of the starting and stopping of batch operations in that computer-based systems have generally significantly reduced the dead time associated with batch operations.

The economics of computer control have been changed drastically by the microprocessor in that the reduction in cost and the improvement in reliability have meant that computer-based systems are the first choice in many applications. Indeed, microprocessor-based instrumentation is frequently cheaper than the equivalent analog unit. The major costs of computer control are now no longer the computer hardware, but the system design and the cost of software: as a consequence attention is shifting towards greater standardisation of design and of software products and the development of improved techniques for design (particularly software design) and for software construction and testing.

The availability of powerful, cheap and highly reliable computer hardware and communications systems makes it possible to conceive and construct large, complex, computer-based control systems. The complexity of such systems raises concern about their dependability and safety. A major concern arises from the difficulty of verifying the correctness of software and of validating a system which contains software. Verification is concerned with answering the question: are we building the product *correctly*? Validation is concerned with the question: are we building the *right* product?

2.11 SUMMARY

In this chapter we have described in some detail typical process control applications in which computer control is used. From these examples we have shown that the control aspects can be subdivided into several main activities which include:

- Data acquisition and processing.
- Sequence control.
- Loop control (modulating control).

- Supervisory control.
- Human—computer interfacing.

These activities are to be found in one form or another in all embedded computer applications and so although in this chapter we have concentrated on process control applications the ideas (and problems) are common to a wide range of other applications.

Also covered were the ways in which several computers can be configured for control applications. These include dual computer systems to increase reliability, and distributed and hierarchical configurations. A brief mention was made of some advanced control strategies.

EXERCISES

2.1   List the characteristics of (a) batch processes and (b) continuous processes.

2.2   You are the manager of a plant which can produce ten different chemical products in batches which can be between 500 and 5000 kg. What factors would you expect to consider in calculating the optimum batch size? What arguments would you put forward to justify the use of an on-line computer to calculate optimum batch size?

2.3   What are the advantages/disadvantages of using a continuous oven? How will the control of the process change from using a standard oven on a batch basis to using an oven in which the batch passes through on a conveyor belt? Which will be the easier to control?

2.4   List the advantages and disadvantages of using DDC.

2.5   List the advantages of using several small computers instead of one large computer in control applications. Are there any disadvantages that arise from using several computers?

2.6   In the section on human—computer interfacing we made the statement 'the design of user interfaces is a specialist area'. Can you think of reasons to support this statement and suggest what sort of background and training a specialist in user interfaces might require?

3

Computer Hardware Requirements for Real-time Applications

This chapter provides a brief overview of some of the basic ideas relating to computer hardware. A brief description of the various types of computers such as microprocessors, microcomputers and special purpose computers is given. A detailed explanation of the standard methods for data transfer including consideration of the use of interrupts is provided. Also given is a brief overview on communication methodologies. The emphasis throughout is on the principles involved and not on the characteristics of a particular microprocessor or microprocessor support chips.

The aims of the chapter are to provide:

- A basic description of the major features of microprocessors.
- A description of the standard interfacing techniques.
- An overview of the standard communication methodologies.

3.1 INTRODUCTION

Although almost any digital computer can be used for real-time computer control and other real-time operations, they are not all equally easily adapted for such work. In the majority of embedded computer-based systems the computer used will be a microprocessor, a microcomputer or a specialised digital processor. Specialised digital processors include fast digital signal processors, parallel computers such as the transputer, and special RISC (Reduced Instruction Set Computers) for use in safety-critical applications (for example, the VIPER (Cullyer and Pygott, 1987)).

3.2 GENERAL PURPOSE COMPUTER

The general purpose microprocessors include the Intel XX86 series, Motorola 680XX series, National 32XXX series and the Zilog Z80 and Z8000 series.

A characteristic of computers used in control systems is that they are modular: they provide the means of adding extra units, in particular specialised input and output devices, to a basic unit. The capabilities of the basic unit in terms of its processing power, storage capacity, input/output bandwidth and interrupt structure determine the overall performance of the system. A simplified block diagram of the basic unit is shown in Figure 3.1; the arithmetic and logic, control, register, memory and input/output units represent a general purpose digital computer.

Of equal importance in a control computer are the input/output channels which provide a means of connecting process instrumentation to the computer, and also the displays and input devices provided for the operator. The instruments are not usually connected directly but by means of interface units. Also of importance is the

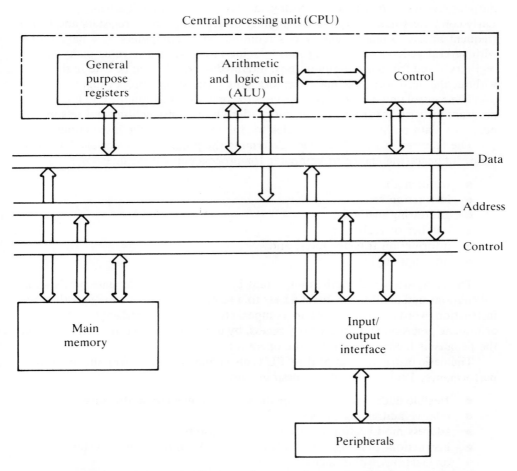

Figure 3.1   Schematic diagram of a general purpose digital computer.

ability to communicate with other computers since, as we discussed in the previous chapter, many modern computer control systems involve the use of several interconnected computers.

3.2.1 Central Processing Unit

The arithmetic and logic unit (ALU) together with the control unit and the general purpose registers make up the central processing unit (CPU). The ALU contains the circuits necessary to carry out arithmetic and logic operations, for example to add numbers, subtract numbers and compare two numbers. Associated with it may be hardware units to provide multiplication and division of fixed point numbers and, in the more powerful computers, a floating point arithmetic unit. The general purpose registers can be used for storing data temporarily while it is being processed. Early computers had a very limited number of general purpose registers and hence frequent access to main memory was required. Most computers now have CPUs with several general purpose registers – some large systems have as many as 256 registers – and for many computations, intermediate results can be held in the CPU without the need to access main memory thus giving faster processing.

The control unit continually supervises the operations within the CPU: it fetches program instructions from main memory, decodes the instructions and sets up the necessary data paths and timing cycles for the execution of the instructions.

The features of the CPU which determine the processing power available and hence influence the choice of computer for process control include:

- wordlength;
- instruction set;
- addressing methods;
- number of registers;
- information transfer rates; and
- interrupt structure.

The computer wordlength is important both in ensuring adequate precision in calculations and in allowing direct access to a large area of main storage within one instruction word. It is possible to compensate for short wordlengths, both for arithmetic precision and for memory access, by using multiple word operations, but the penalty is increased time for the operations.

The basic instruction set of the CPU is also important in determining its overall performance. Features which are desirable are:

- flexible addressing modes for direct and immediate addressing;
- relative addressing modes;
- address modification by use of index registers;
- instructions to transfer variable length blocks of data between storage units or locations within memory; and
- single commands to carry out multiple operations.

These features reduce the number of instructions required to perform 'housekeeping' operations and hence both reduce storage requirements and improve overall speed of operation by reducing the number of accesses to main memory required to carry out the operations. A consequence of an extensive and powerful instruction set is, however, that efficient programming in the assembly language becomes more difficult because the language can become complex; thus it is desirable to be able to program the system using a high-level language which has a compiler designed to make optimum use of the special features of the instruction set. There are many other reasons for *not* using assembly languages and we will discuss some of them in Chapter 5.

Another area which must be considered carefully when selecting a computer for process control is information transfer, both within the CPU and between the backing store and the CPU, and also with the input/output devices. The rate at which such transfers can take place, the ability to carry out operations in parallel with the processing of data, and the ability to communicate with a large range of devices can be crucial to the application to process control. A vital requirement is also a flexible and efficient multi-level interrupt structure.

3.2.2 Storage

The storage used on computer control systems divides into two main categories: fast access storage and auxiliary storage. The fast access memory is that part of the system which contains data, programs and results which are currently being operated on. The major restriction with current computers is commonly the addressing limit of the processor. In addition to RAM (random access memory – read/write) it is now common to have ROM (read-only memory), PROM (programmable read-only memory) or EPROM (electronically programmable read-only memory) for the storage of critical code or predefined functions.

The use of ROM has eased the problem of memory protection to prevent loss of programs through power failure or corruption by the malfunctioning of the software (this can be a particular problem during testing). An alternative to using ROM is the use of memory mapping techniques that trap instructions which attempt to store in a protected area. This technique is usually only used on the larger systems which use a memory management system to map program addresses onto the physical address space. An extension of the system allows particular parts of the physical memory to be set as read only, or even locked out altogether: write access can be gained only by the use of 'privileged' instructions.

The auxiliary storage medium is typically disk or magnetic tape. These devices provide bulk storage for programs or data which are required infrequently at a much lower cost than fast access memory. The penalty is a much longer access time and the need for interface boards and software to connect them to the CPU. Auxiliary or backing store devices operate asynchronously to the CPU and care has to be taken in deciding on the appropriate transfer technique for data between the CPU, fast access memory and the backing store. In a real-time system use of the CPU to carry out the transfer is not desirable as it is slow and no other computation can take place during transfer. For efficiency of transfer it is sensible to transfer large blocks of data rather than a single word or byte and this can result in the CPU not being available for up to several seconds in some cases.

The approach frequently used is direct memory access (DMA). For this the interface controller for the backing memory must be able to take control of the address and data buses of the computer.

3.2.3 Input and Output

The input/output (I/O) interface is one of the most complex areas of a computer system; part of the complication arises because of the wide variety of devices which have to be connected and the wide variation in the rates of data transfer. A printer may operate at 300 baud whereas a disk may require a rate of 500 kbaud. The devices may require parallel or serial data transfers, analog-to-digital or digital-to-analog conversion, or conversion to pulse rates.

The I/O system of most control computers can be divided into three sections:

- process I/O;
- operator I/O; and
- computer I/O.

3.2.4 Bus Structure

Buses are characterised in three ways:

- mechanical (physical) structure;
- electrical; and
- functional.

In mechanical or physical terms a bus is a collection of conductors which carry electrical signals, for example tracks on a printed circuit board or the wires in a ribbon cable. The physical form of the bus represents the *mechanical characteristic* of the bus system. The *electrical characteristics* of the bus are the signal levels, loading (that is, how many loads the line can support), and type of output gates (open-collector, tri-state). The *functional characteristics* describe the type of information which the electrical signals flowing along the bus conductors represent. The bus lines can be divided into three functional groups:

- address lines;
- data lines; and
- control and status lines.

These can be thought of as *where*, *what* and *when*. The address lines provide information on where the information is to be sent (or where it is to be obtained from); the data lines show what the information is; and the control and status lines indicate when it is to be sent.

3.3 SINGLE-CHIP MICROCOMPUTERS AND MICROCONTROLLERS

Many integrated circuit manufacturers produce microcomputers in which all the components necessary for a complete computer are provided on one single chip. A typical single-chip device is shown in Figure 3.2. With only a small amount of EPROM and an even smaller amount of RAM this type of device is obviously intended for small, simple systems. The memory can always be extended by using external memory chips.

The microcontroller is similarly a single-chip device that is specifically intended for embedded computer control applications. The main difference between it and a microcomputer is that it typically will have on board the chip a multi-plexed ADC and some form of process output, for example a pulse width modulator unit. The chip may also contain a real-time clock generator and a watch-dog timer.

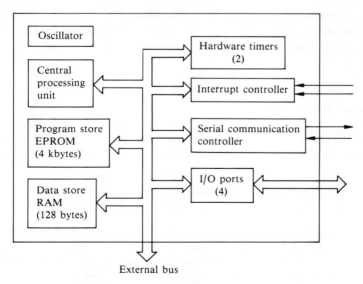

Figure 3.2   A typical single-chip computer.

3.4 SPECIALISED PROCESSORS

Specialised processors have been developed for two main purposes:

- safety-critical applications; and
- increased computation speed.

For safety-critical applications the approach has been to simplify the instruction set – the so-called reduced instruction set computer (RISC). The advantage of simplifying the instruction set is the possibility of formal verification (using mathematical proofs) that the logic of the processor is correct. The second advantage of the RISC machine is that it is easier to write assemblers and compilers for the simple instruction set. An example of such a machine is the VIPER (Cullyer, 1988; Dettmer, 1986), the main features of which are:

- Formal mathematical description of the processor logic.
- Integer arithmetic (32 bit) and no floating point operations (it is argued that floating point operations are inexact and cannot be formally verified).
- No interrupts – all event handling is done using polling (again interrupts make formal verification impossible).
- No dynamic memory allocation.

(There is an unresolved dispute regarding the validity and completeness of the formal verification procedures used for the VIPER processor (MacKenzie, 1993).)

The traditional Von Neumann computer architecture with its one CPU through which all the data and instructions have to pass sequentially results in a bottleneck. Increasing the processor speed can increase the throughput but eventually systems will reach a physical limit because of the fundamental limitation on the speed at which an electronic signal can travel. The search for increased processing speed has led to the abandonment of the Von Neumann architecture for high-speed computing.

3.4.1 Parallel Computers

Many different forms of parallel computer architectures have been devised; however, they can be summarised as belonging to one of three categories:

SIMD   Single instruction stream, multiple data stream.
MISD   Multiple instruction stream, single data stream.
MIMD   Multiple instruction stream, multiple data stream.

These are illustrated in Figure 3.3 where the traditional architecture characterised as SISD (Single instruction stream, single data stream) is also shown.

MIMD systems are obviously the most powerful class of parallel computers in that each processor can potentially be executing a different program on a different data set. The most widely available MIMD system is the INMOS transputer. Each

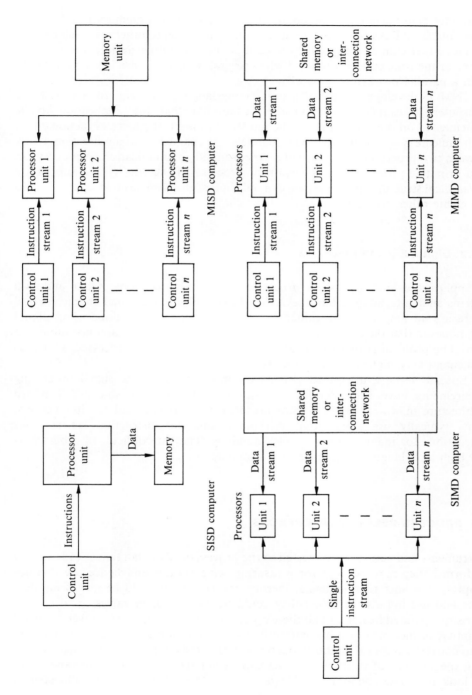

Figure 3.3 Computer system architectures.

transputer chip has a CPU, on-board memory, an external memory interface and communication links for direct point-to-point connection to other transputer chips. An individual chip can be used as a stand-alone computing device; however, the power of the transputer is obtained when several transputers are interconnected to form a parallel processing network.

INMOS developed a special programming language, occam, for use with the transputer. Occam is based on the assumption that the application to be implemented on the transputer can be modelled as a set of processes (actions) that communicate with each other via channels. A channel is a unidirectional link between two processes which provides synchronised communication. A process can be a primitive process, or a collection of processes; hence the system supports a hierarchical structure. Processes are dynamic in that they can be created, can die and can create other processes.

3.4.2 Digital Signal Processors

In applications such as speech processing, telecommunications, radar and hi-fi systems analog techniques have been used for modifying the signal characteristics. There are advantages to be gained if such processing can be done using digital techniques in that the digital devices are inherently more reliable and not subject to drift. The problem is that the bandwidth of the signals to be processed is such as to demand very high processing speeds.

Special purpose integrated circuits optimised to meet the signal processing requirements have been developed. They typically use the so-called Harvard architecture in which separate paths are provided for data and for instructions. DSPs typically use fixed point arithmetic and the instruction set contains instructions for manipulating complex numbers. They are difficult to program as few high-level language compilers are available.

3.5 PROCESS-RELATED INTERFACES

Instruments and actuators connected to the process or plant can take a wide variety of forms: they may be used for measuring temperatures and hence use thermo-couples, resistance thermometers, thermistors, etc.; they could be measuring flow rates and use impulse turbines; they could be used to open valves or to control thyristor-operated heaters. In all these operations there is a need to convert a digital quantity, in the form of a bit pattern in a computer word, to a physical quantity, or to convert a physical quantity to a bit pattern. Designing a different interface for each specific type of instrument or actuator is not sensible or economic and hence we look for some commonality between them. Most devices can be allocated to one of the following four categories:

1. *Digital quantities*: These can be either binary, that is a valve is open or closed, a switch is on or off, a relay should be opened or closed, or a generalised digital quantity, that is the output from a digital voltmeter in BCD (binary coded decimal) or other format.

2. *Analog quantities*: Thermocouples, strain gauges, etc., give outputs which are measured in millivolts; these can be amplified using operational amplifiers to give voltages in the range -10 to $+10$ volts; conventional industrial instruments frequently have a current output in the range 4 to 20 mA (current transmission gives much better immunity to noise than transmission of low-voltage signals). The characteristic of these signals is that they are continuous variables and have to be both sampled and converted to a digital value.

3. *Pulses and pulse rates*: A number of measuring instruments, particularly flow meters, provide output in the form of pulse trains; similarly the increasing use of stepping motors as actuators requires the provision of pulse outputs. Many traditional controllers have also used pulse outputs: for example, valves controlling flows are frequently operated by switching a dc or ac motor on and off, the length of the on pulse being a measure of the change in valve opening required.

4. *Telemetry*: The increasing use of remote outstations, for example electricity substations and gas pressure reduction stations, has increased the use of telemetry. The data may be transmitted by landline, radio or the public telephone network: it is, however, characterised by being sent in serial form, usually encoded in standard ASCII characters. For small quantities of data the transmission is usually asynchronous. Telemetry channels may also be used on a plant with a hierarchy of computer systems instead of connecting the computers by some form of network. An example of this is the CUTLASS system used by the Central Electricity Generating Board, which uses standard RS232 lines to connect a hierarchy of control computers.

The ability to classify the interface requirements into the above categories means that a limited number of interfaces can be provided for a process control computer. The normal arrangement is to provide a variety of interface cards which can be added to the system to make up the appropriate configuration for the process to be controlled; for example, for a process with a large number of temperature measurements several analog input boards may be required.

3.5.1 Digital Signal Interfaces

A simple digital input interface is shown in Figure 3.4. It is assumed that the plant outputs are logic signals which appear on lines connected to the digital input

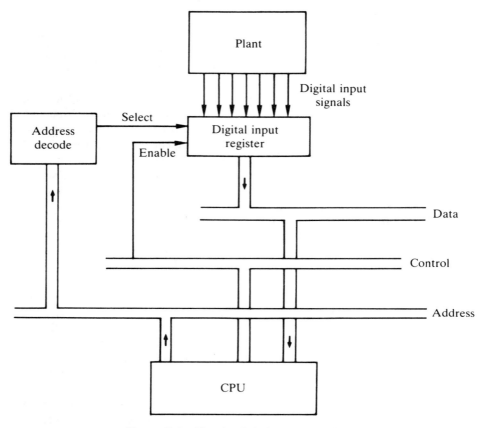

Figure 3.4   Simple digital input interface.

register. It is usual to transfer one word at a time to the computer, so normally the digital input register will have the same number of input lines as the number of bits in the computer word. The logic levels on the input lines will typically be 0 and + 5 V; if the contacts on the plant which provide the logic signals use different levels then conversion of signal levels will be required.

To read the lines connected to the digital input register the computer has to place the address of the register on the address bus and decoding circuitry is required in the interface (address decoder) to select the digital input register. In addition to the 'select' signal an 'enable' signal may also be required; this could be provided by the 'read' signal from the computer control bus. In response to both the 'select' and 'enable' signals the digital input register enables its output gates and puts data onto the computer data bus. Note that for proper operation of the data bus the digital input register must connect its output gates to the data bus only when it is selected and enabled; if it connects at any other time it will corrupt data intended for other devices.

The timing of the transfer of information will be governed by the CPU timing. A typical example is shown in Figure 3.5. For this system it is assumed that the transfer requires three cycles of the system clock, labelled T_1, T_2, and T_3. The address lines begin to change at the beginning of the cycle T_1 and they are guaranteed to be valid by the start of cycle T_2; also at the start of cycle T_2 the READ line becomes active. For the correct read operation the digital input register has to provide stable data at the negative-going edge (or earlier) of the clock during the T_3 cycle and the data must remain on the data lines until the negative-going edge of the following clock cycle. Note that the actual time taken to transfer the data from the data bus to the CPU may be much shorter than the time for which the data is valid. The requirement that it remain valid from the negative-going edge of cycle T_3 until the negative-going edge of the following cycle is to provide for the worst case condition arising from variations in the performance of the various components.

Figure 3.4 shows a system that provides information only on demand from the computer: it cannot indicate to the computer that information is waiting. There are many circumstances in which it is useful to indicate a change of status of input lines to the computer. To do this a status line which the computer can test, or which can be used as an interrupt, is needed.

A simple digital output interface is shown in Figure 3.6. Digital output is the simplest form of output: all that is required is a register or latch which can hold the

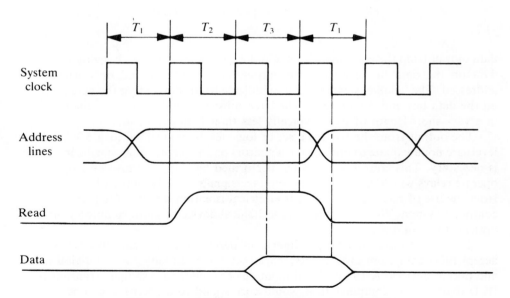

Figure 3.5 Simplified READ (INPUT) timing diagram.

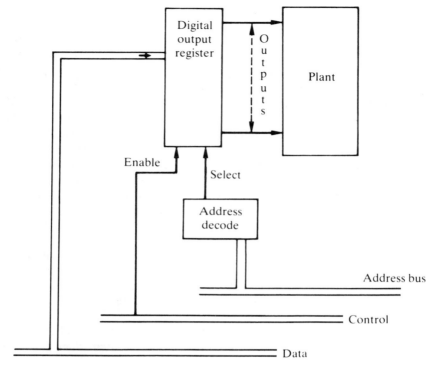

Figure 3.6   Simple digital output interface.

data output from the computer. To avoid the data in the register changing when the data on the data bus changes, the output latch must respond only when it is addressed. The 'enable' signal is used to indicate to the device that the data is stable on the data bus and can be read. The latch must be capable of accepting the data in a very short length of time, typically less than 1 microsecond.

The output from the latch is a set of logic levels, typically 0 to +5 V: if these levels are not adequate to operate the actuators on the plant, some signal conversion is necessary. This conversion is often performed by using the low-level signals to operate relays which carry the higher-voltage signals: an advantage which is gained from the use of relays is that there is electrical isolation between the plant and the computer system. The relay can be a mechanical device or more commonly now an optical isolation device.

The digital input and output interfaces described above can also be used to accept BCD data from instruments, since they are essentially parallel digital input and output devices. A 16 bit digital input device could, for example, transmit four BCD digits to the computer (this would correspond to a precision of one part in 10 000 − 0 to 9999).

Because digital input and output is a frequently required operation, many microprocessor manufacturers produce integrated circuits which provide such an interface.

3.5.2 Pulse Interfaces

In its simplest form a pulse input interface consists of a counter connected to a line from the plant. The counter is reset under program control and after a fixed length of time the contents are read by the computer. A typical arrangement is shown in Figure 3.7, which also shows a simple pulse output interface. The transfer of data from the counter to the computer uses techniques similar to those for the digital input described above.

The measurement of the length of time for which the count proceeds can be

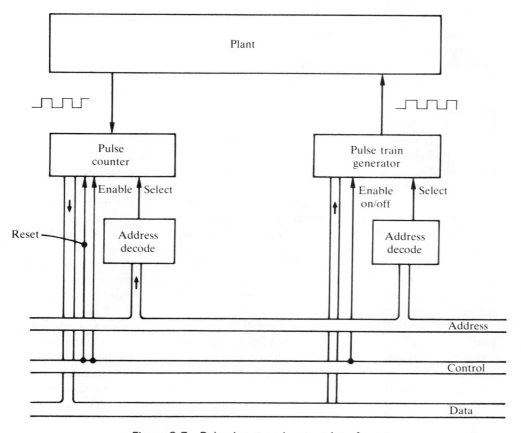

Figure 3.7   Pulse input and output interface.

carried out either by a logic circuit in the counter interface or by the computer. If the timing is done by the computer then the 'enable' signal must inhibit the further counting of pulses. If the computing system is not heavily loaded, the external interface hardware required can be reduced by connecting the pulse input to an interrupt and counting the pulses under program control.

Pulse outputs can take a variety of forms:

1. a series of pulses of fixed duration;
2. a single pulse of variable length (time-proportioned output); and
3. pulse width modulation – a series of pulses of different widths sent at a fixed frequency.

For type 1 the computer turns a pulse generator on or off, or loads a register with the number of pulses to be transmitted. The pulse output is sent to the process and used to decrement the register contents; when the register reaches zero the pulse output is turned off. A system of this type could be used, for example, to control the movement of a stepping motor.

For type 2 the computer raises or lowers a logic line and thus sends a variable length pulse to the plant, or loads a register with a number specifying the length of pulse required and interface logic is used to generate the pulse. The variable length pulse system is used typically to operate process control valves. Using the computer to turn the pulse train on or off directly is usual only on small systems with a few input or output lines, or when the pulse rate is low. For large systems, or for high pulse rates, it is normal to arrange for the interface logic to generate the actual pulse train or to control the duration of the pulse.

For type 3, pulse width modulation (PWM), special purpose interface chips are used to generate the pulses. Normally with this type of output a fast PWM stream will be produced and this is then converted into a linear analog output by using a low-pass filter.

Closely related to pulse counters are *hardware timers*. If the pulse counter is made to count down from a preset value at a fixed rate it can act as a timer. For example, if the clock rate used to decrement the counter is set at one count per millisecond, it can be used as a timer with a precision of 1 ms. The counter is loaded with a binary number corresponding to the desired time interval and the count started; when the counter reaches zero it generates an interrupt to indicate that it has 'timed out', that is that the interval has elapsed. The computer can then take the appropriate action.

The input to a hardware timer is normally a continuously running accurate pulse generator which either may have a fixed frequency or may be programmable to give a range of frequencies, for example thousandths, hundredths, tenths and seconds. The unit is programmed either by setting external switches or by commands sent by the computer. Hardware timers can be used to set the maximum time allowed for the response from an external device: the computer requests a response from a device and at the same time starts a hardware timer; if the device has not responded by the time the hardware timer interrupts then an error signal is generated.

A special form of this is the *watch-dog timer* which is often used on process control computers. The timer is reset at fixed intervals, usually when the operating system kernel is entered: if the watch-dog timer 'times out' it indicates that for some reason the operating system kernel has not been entered at the correct time, either because of some hardware malfunction, or because the normal interrupts have been locked out by a software error. A hardware timer can be used as a real-time clock (see section 3.5.4 below).

3.5.3 Analog Interfaces

The conversion of analog measurements to digital measurements involves two operations: sampling and quantisation. The sampling rate necessary for controlling a process is discussed in the next chapter. As is shown in Figure 3.8 many analog-to-digital converters (ADCs) include a 'sample–hold' circuit on the input to the device. The sample time of this unit is much shorter than the sample time required for the process; this sample–hold unit is used to prevent a change in the quantity being measured while it is being converted to a discrete quantity.

To operate the analog input interface the computer issues a 'start' or 'sample' signal, typically a short pulse (1 microsecond), and in response the ADC switches the 'sample–hold' into SAMPLE for a short period after which the quantisation process commences. Quantisation may take from a few microseconds to several milliseconds. On completion of the conversion the ADC raises a 'ready' or 'complete' line which is either polled by the computer or is used to generate an interrupt.

Use of separate ADCs for each analog input is expensive, despite the reduction in price in recent years, and typically a multiplexer is used to switch the inputs from several input lines to a single ADC (see Figure 3.8). For high-level (0–10 V) signals the multiplexer is usually a solid-state device (typically based on the use of field effect transistor switches); for low-level signals in the millivolt range, for example from thermocouples or strain gauges, mercury-wetted reed-relay switch units are used. For low-level signals, a programmable gain amplifier is usually used between the multiplexer and the sample–hold unit. With a multiplexed system the sequence of operations is more complex than with a single-channel device as the program has to arrange for the selection of the appropriate input channel.

For the simplest systems a single channel-select signal is used which causes the multiplexer to step to the next channel: the channels are thus sampled in sequence. A more elaborate arrangement is to provide random channel selection by connecting channel address inputs to the computer data bus. The sequence of events is then: select the channel address, send the start conversion command and then wait for the conversion complete signal. In some high-speed converters it is possible to send the next channel address during the period in which the present input is being quantised. This technique is also frequently used with reed-relay switching since a delay to allow time for the signal to stabilise is required between selecting a channel and sampling.

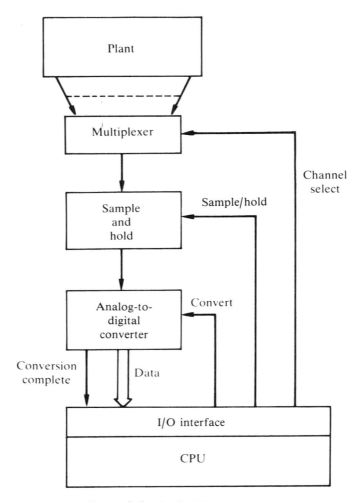

Figure 3.8   Analog input system.

Digital-to-analog conversion is simpler (and hence cheaper) than analog-to-digital conversion and as a consequence it is normal to provide one converter for each output. (It is possible to produce a multiplexer in order to use a single digital-to-analog converter (DAC) for analog output. Why would this solution not be particularly useful?) Figure 3.9 shows a typical arrangement. Each DAC is connected to the data bus and the appropriate channel is selected by putting the channel address on the computer address bus. The DAC acts as a latch and holds the previous value sent to it until the next value is sent. The conversion time is typically from 5 to 20 ms and typical analog outputs are −5 to +5 V, −10 to +10 V, or a current output of 0 to 20 mA.

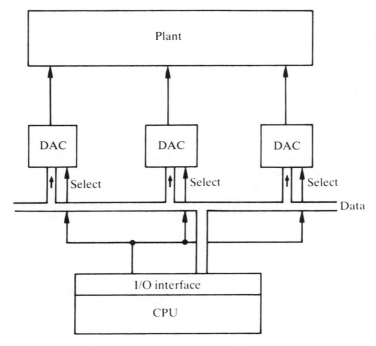

Figure 3.9   Analog output system.

3.5.4 Real-time Clock

A real-time clock is a vital auxiliary device for control computer systems. The hardware unit given the name 'real-time clock' may or may not be a clock; in many systems it is nothing more than a pulse generator with a precisely controlled frequency.

A common form of clock is based on using the ac supply line to generate pulses at 50 (or 60) times per second. By using slightly more complicated circuitry higher pulse rates can be generated, for example 100 (or 120) pulses per second. The pulses are used to generate interrupts and the interrupt handling software counts the interrupts and hence keeps time. If a greater precision in the time measurement than can be provided from the power supply is required then a hardware timer is used. A fixed frequency pulse generator (usually crystal-driven) decrements a counter which, when it reaches zero, generates an interrupt and reloads the count value. The interrupt activates the real-time clock software. The interval at which the timer generates an interrupt, and hence the precision of the clock, is controlled by the count value loaded into the hardware timer.

The choice of the basic clock interval, that is the clock precision, has to be a compromise between the timing accuracy required and the load on the CPU. If too small an interval is chosen, that is high precision, then the CPU will spend a large proportion of its time simply servicing the clock and will not be able to perform any other work.

The real-time clock based on the use of an interval timer and interrupt-driven software suffers from the disadvantage that the clock stops when the power is lost and on restart the current value of real time has to be entered. Real-time clocks are now becoming available in which the clock and date function are carried out as part of the interface unit, that is the unit acts like a digital watch. Real time can be read from the card and the card can be programmed to generate an interrupt at a specified frequency. These units are usually supplied with battery back-up so that even in the absence of mains power the clock function is not lost.

Real-time clocks are also used in batch processing and on-line computer systems. In the former, they are used to provide date and time on printouts and also for accounting purposes so that a user can be charged for the computer time used; the charge may vary depending on the time of day or day of the week. In on-line systems similar facilities to those of the batch computer system are required, but in addition the user expects the terminal to appear as if it is the only terminal connected to the system. The user may expect delays when the program is performing a large amount of calculation but not when it is communicating with the terminal. To avoid any one program causing delays to other programs, no program is allowed to run for more than a fraction of a second; typically timings are 200 ms or less. If further processing for a particular program is required it is only performed after all other programs have been given the opportunity to run. This technique is known as time slicing.

3.6 DATA TRANSFER TECHNIQUES

Although the meaning of the data transmitted by the various processes, the operator and computer peripherals differs, there are many common features which relate to the transfer of the data from the interface to the computer. A characteristic of most interface devices is that they operate synchronously with respect to the computer and that they operate at much lower speeds. Direct control of the interface devices by the computer is known as 'programmed transfer' and involves use of the CPU. Programmed transfer gives maximum flexibility of operation but because of the difference in operating speeds of the CPU and many interface devices it is inefficient. An alternative approach is to use direct memory access (DMA); the transfer requirements are set up using program control but the data transfers take place directly between the device and memory without disturbing the operation of the CPU (except that bus cycles are used).

With the reduction in cost of integrated circuits and microprocessors, detailed control of the input/output operations is being transferred to I/O processors which provide buffered entry. For a long time in on-line computing, buffers have been used to collect information (for example, a line) before invoking the program requesting the input. This approach is now being extended through the provision of I/O processors for real-time systems. For example, an I/O processor can be used to control the scanning of a number of analog input channels, only requesting main computer time when it has collected data from all the channels. This can be extended so that the I/O processor checks the data to test if any values are outside preset limits set by the main system.

A major problem in data transfer is *timing*. It may be thought that under programmed transfer, the computer can read or write at any time to a device, that is, can make an unconditional transfer. For some process output devices, for example switches and indicator lights connected to a digital output interface, or for DACs, unconditional transfer is possible since they are always ready to receive data. For other output devices, for example printers and communications channels, which are not fast enough to keep up with the computer but must accept a sequence of data items without missing any item, unconditional transfer cannot be used. The computer must be sure that the device is ready to accept the next item of data; hence either a timing loop to synchronise the computer to the external device or *conditional transfer* has to be used. Conditional transfer can be used for digital inputs but not usually for pulse inputs or analog inputs. Where unconditional transfer is used to read the digital value or an analog signal, or the value of a digital instrument rather than simply the pattern of logic indicators, then Gray code or some other form of cyclic binary code should be used to avoid the possibility of large transient errors.

3.6.1 Polling

A simple example of conditional transfer is shown in Figure 3.10. Assuming that the data is being transferred to a printer which operates at 40 characters per second, the computer will find that the device is ready once every 25 milliseconds. The three instructions involved in performing the test will take approximately 5 μs (the actual time will depend on the speed of the processor); thus the conditional test will be carried out 5000 times for each character transmitted. The computer will spend 99.98% of its time in checking to see if the device is ready and only 0.02% of the time doing useful work; this is clearly inefficient.

A software timing loop can be used as an alternative to a status line on the interface. For example, a delay can be created by loading a register with a number and repeatedly decrementing the register until it reads zero:

```
        LD B, 25     ;load register B with time delay
   LOOP DEC B        ;decrement B
        JR NZ, LOOP  ;repeat until B is zero
```

To ensure that no transfer is made before the peripheral is ready the time delay must be slightly greater than the maximum delay expected in the peripheral; thus

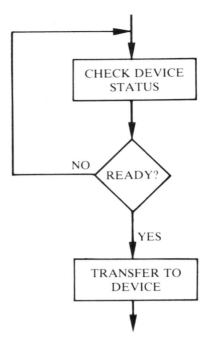

Figure 3.10   Conditional transfer (busy wait).

in terms of use of the CPU this method is even more inefficient than the use of the conditional wait. However, it does slightly simplify and reduce the cost of the interface.

An alternative arrangement for conditional transfer, which allows the computer to continue doing useful work if the device is busy, is shown in Figure 3.11. In this method a check is made to see if the device is ready: if it is ready then the transfer is made; if it is not the computer continues with other work and returns at some later time to check the device again. The technique avoids the inefficiency of waiting in a loop for a device to become ready, but presents the programmer with the difficult task of arranging the software such that all devices are checked at frequent intervals.

Conditional transfer techniques involve *polling*, which is using the computer to check whether a device is ready for a data transfer. The problems of polling using conditional waits can be avoided if the computer can respond to an interrupt signal.

3.6.2 Interrupts

An interrupt is a mechanism by which the flow of the program can be temporarily stopped to allow a special piece of software – an interrupt service routine (also called an interrupt handler) – to run. When this routine has finished, the

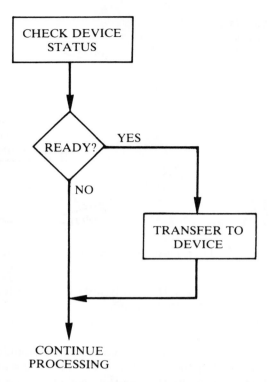

Figure 3.11 Conditional transfer.

program which was temporarily suspended is resumed. The process is illustrated in Figure 3.12.

Interrupts are essential for the correct operation of most real-time computer systems; in addition to providing a solution to the conditional wait problem they are used for:

Real-time clock: The external hardware provides a signal at regularly spaced intervals of time; the interrupt service routine counts the signals and keeps a clock.

Alarm inputs: Various sensors can be used to provide a change in a logic level in the event of an alarm. Since alarms should be infrequent, but may need rapid response times, the use of an interrupt provides an effective and efficient solution.

Manual override: Use of an interrupt can allow external control of a system to allow for maintenance and repair.

Hardware failure indication: Failure of external hardware or of interface units can be signalled to the processor through the use of an interrupt.

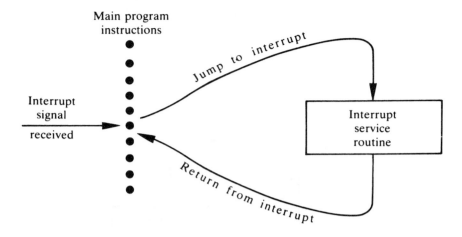

Figure 3.12   Interrupt-driven program control transfer.

Debugging aids: Interrupts are frequently used to insert breakpoints or traces in the program during program testing.

Operating system: Interrupts are used to force entry to the operating system before the end of a time slice.

Power failure warning: It is simple to include in the computer system a circuit that detects very quickly the loss of power in the system and provides a few milliseconds' warning before the loss is such that the system stops working. If this circuit is connected to an interrupt which takes precedence over all other operations in the computer there can be sufficient time to carry out a few instructions which could be sufficient to close the system down in an orderly fashion.

3.6.2.1 Saving and restoring registers

Since an interrupt can occur at any point in a program, precautions have to be taken to prevent information which is being held temporarily in the CPU registers from being overwritten. All CPUs automatically save the contents of the program counter; this is vital. If the contents were not saved then a return to the point in the program at which the interrupt occurred could not be made. Some CPUs, however, do more and save all the registers. The methods commonly used are:

● Store the contents of the registers in a specified area of memory. (Note that this implies that an interrupt cannot be interrupted − see below.)
● Store the registers on the memory stack. This is a simple, widely used method which permits multi-level interrupts; the major disadvantage is the danger of stack overflow.

- Use of an auxiliary set of registers. Some processors provide two sets of the main registers and an interrupt routine can switch to the alternative set. If only two sets are provided multi-level interrupts cannot be handled. An alternative method is to use a designated area of memory as the working registers and then an interrupt only requires a pointer to be changed to change the working register set.

The use of automatic storage of the working registers is an efficient method if all registers are to be used; it is inefficient if only one or two will be used by the interrupt routine. For this reason fully automatic saving is usually restricted to CPUs with only a few working registers in the CPU; systems with many working registers provide an option either to save or not to save. Unless response time is critical it is good engineering practice to save all registers: in this way there is no danger, in a subsequent modification to the interrupt service routine, of using a register which is not being saved, and failing to add it to the list of registers to be saved. The resulting error would be difficult to find since it would cause random malfunctioning of the system.

The machine status must of course be restored on exit from the interrupt routine; this is straightforward for all methods except that which uses the stack to save registers; in this case the registers are restored in the opposite order than that in which they were saved. Systems providing automatic saving also provide automatic restore on exit from the interrupt.

An example of the framework of an interrupt service routine is shown below.

```
INT1:   CALL SAVREG  ;SAVREG is routine which saves
                      ;working registers
;
;code for interrupt handling is inserted here
;
        CALL RESREG  ;RESREG is routine which
                     ;restores working registers
        EI           ;enable interrupts
        RETI         ;return from interrupt routine
```

The above routine is suitable for a system in which interrupts are not allowed to be interrupted; hence the EI instruction which enables interrupts is not executed until immediately prior to the return from interrupt. The return from an interrupt routine has to be handled with care to prevent unwanted effects. For example, the EI instruction does not re-enable interrupts until after the execution of the instruction which immediately follows it. Therefore, by using the EI/RETI combination a pending interrupt cannot take effect until after the return from the previous one has been completed. In this example it is assumed that the microprocessor automatically disables interrupts on acknowledgement of an interrupt and they remain disabled until an EI instruction is executed. Some CPUs operate by disabling interrupts for only one instruction following an interrupt

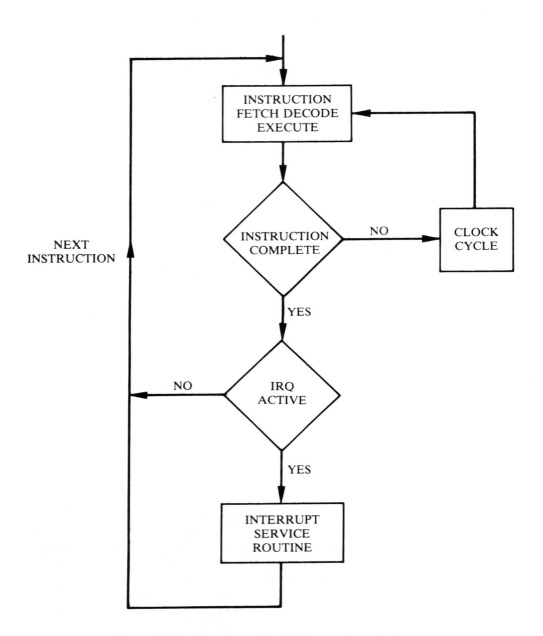

Figure 3.13   Flowchart of basic interrupt mechanism.

acknowledge; it thus becomes the responsibility of the programmer to disable interrupts as the first instruction of the interrupt service routine if only a single level of interrupt is to be supported.

3.6.2.2 Interrupt input mechanisms

A simple form of interrupt input is shown in Figure 3.13. In between each instruction the CPU checks the IRQ line. If it is active, an interrupt is present and the interrupt service routine is entered; if it is not active the next instruction is fetched and the cycle repeats. Note that an instruction involves more than one CPU clock cycle and that the interrupt line is checked only between instructions. Because several clock cycles may elapse between successive checks of the interrupt line, the interrupt signal must be latched and only cleared when the interrupt is acknowledged.

A common arrangement is to have two interrupt lines as shown in Figure 3.14: one of the lines, IRQ, can be enabled and disabled using software and hence the computer can run in a mode in which external events cannot disturb the processing. A second interrupt line is provided; this interrupt cannot be turned off by software and hence it is said to be a non-maskable interrupt (NMI). A typical use would be to provide the power failure detect interrupt.

Although most modern computer CPUs have only one or two interrupt lines, a large number of interrupts can be connected by means of an OR gate. It then

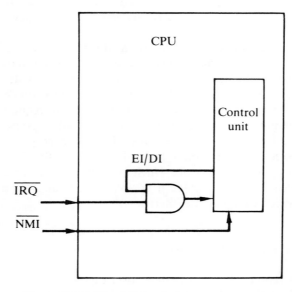

Figure 3.14   Typical basic interrupt system.

becomes a problem to determine which of the many external interrupt lines has generated the CPU interrupt.

3.6.2.3 Interrupt response mechanisms

The CPU may respond to the interrupt in a variety of ways; some of the more popular methods are given below:

1. Transfer control to a specified address – usually in the form of a 'call' instruction.
2. Load the program counter with a new value from a specified register or memory location.
3. Execute a 'call' instruction but to an address supplied from the external system.
4. Use an output signal – an Interrupt Acknowledge – to fetch an instruction from an external device.

Methods 1 and 2 are said to be *software biased*, in that they require little in the way of external hardware and rely instead on software to determine the interrupt source and the appropriate interrupt service routine. Methods 3 and 4 are *hardware biased* in that they require more external hardware but can identify the interrupt source and can transfer program control directly to the appropriate interrupt service routine.

In method 2 the address of the interrupt response routine is stored in specified memory locations; the address stored here is called the *interrupt vector* or *interrupt response vector*. Once the interrupt is detected control is passed to an interrupt response routine and polling must be used to determine which device has caused the interrupt.

The use of polling in interrupt systems has the advantage over normal polling systems that at least one of the inputs is guaranteed to be active. It is clearly, however, not a very satisfactory system if large numbers of devices have to be checked. The load can be reduced by testing the devices which interrupt most frequently first, but this may conflict with response time requirements in that a device which interrupts infrequently may require a rapid response time and hence should be checked first. If an equal response time, on average, for each device is required it will be necessary to rotate the order in which the devices are checked. Doing this can also prevent one device which interrupts frequently from locking out all others. The method does provide a flexible way of allocating priority to the various devices which can generate interrupts.

3.6.2.4 Hardware vectored interrupts

Methods 3 and 4 require the use of some form of vectored interrupt structure to identify which of the external devices has generated an interrupt. They also require a mechanism to arbitrate between the possible sources of the interrupt to prevent more than one interrupt activating the IRQ at any one time. The process of arbitration involves assigning priorities to the various interrupts.

A frequently used arrangement is the daisy chain in which an 'acknowledge' signal is propagated through the devices until it is blocked by the interrupting device. Figure 3.15 shows a typical arrangement. Each unit has an IEI (Interrupt Enable In) pin and an IEO (Interrupt Enable Out) pin; it is assumed that on both pins the active signal is high. The first IEI in the chain is set permanently on 'high'. For any given

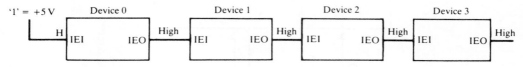

(a) No interrupt condition

(b) Device 2 generates interrupt and is acknowledged

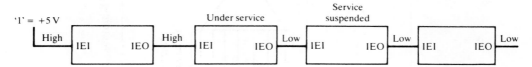

(c) Device 1 generates interrupt, servicing of device 2 is suspended

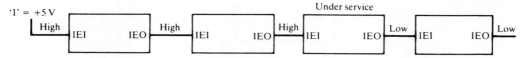

(d) Device 1 servicing completed, 'RET' instruction executed, servicing of device 2 resumed

(e) Servicing of device 2 completed and 'RET' executed

Figure 3.15 Daisy-chain interrupt structure: (a) no interrupt condition; (b) device 2 generates interrupt and is acknowledged; (c) device 1 generates interrupt, servicing of device 2 is suspended; (d) device 1 servicing completed, 'RET' instruction executed, servicing of device 2 resumed; (e) servicing of device 2 completed and 'RET' executed.

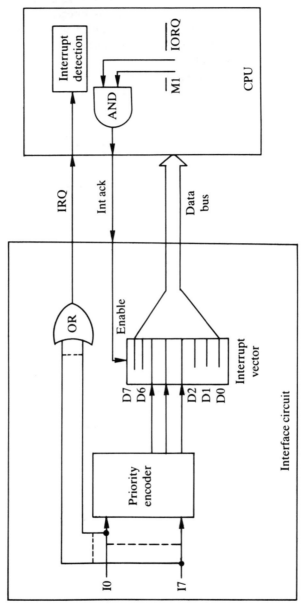

Figure 3.16   Interrupt vectoring using priority encoding circuit.

unit the output pin IEO is high if, and only if, the input IEI is high and the unit is not requesting an interrupt. If a device is requesting an interrupt and IEI is high, that device should set IEO low and in response to an 'interrupt acknowledge' signal send its interrupt vector. If a device is requesting an interrupt but the IEI is low then it should not respond to the interrupt acknowledge signal.

In Figure 3.15a the system is shown with no interrupts active, so all the signals are at high. The effect of Device 2 generating an interrupt and it being acknowledged is shown in Figure 3.15b: Device 2 is serviced, Device 3 is locked out and Devices 0 and 1 can still interrupt. If Device 1 now generates an interrupt the servicing of Device 2 will be suspended in favour of Device 1 (see Figure 3.15c). On completion of the servicing of Device 1, the servicing of Device 2 resumes (Figure 3.15d). Finally, when this is completed, all the IEI/IEO signals return to 'high' and all devices are enabled. The assumption implicit in this arrangement is that one interrupt service routine can itself be interrupted by a higher-priority interrupt. This is known as a multi-level interrupt structure and is dealt with below.

In a daisy-chain arrangement the device priority is determined by the position of the device in the chain and cannot be changed by the software. Great care has to be given to timing considerations; in particular, care has to be taken to allow the IEI signal time to propagate along the chain.

The determination of interrupt priority can be performed using priority encoder circuits (see Figure 3.16). In this system an interrupt occurring on any line causes the interrupt line (IRQ) to become active and also places a three-bit code specifying the number of the interrupt line which is active on the data bus. In the event of more than one line being active the priority encoder supplies the number of the highest-priority interrupt — the usual arrangement is that the line with the lowest number is considered to be of highest priority.

3.6.2.5 Interrupt response vector

The interrupt response vector in the above systems can take a variety of forms: it may be an instruction, the address of the interrupt service routine, the address of a pointer to an interrupt service routine, or part of the address of the interrupt service routine or pointer.

A widely used method is to employ an interrupt mechanism in which the interrupting device supplies the address of the location in which the pointer to the start of the interrupt routine is stored. When a device interrupts it supplies this address and the CPU loads the program counter with the contents of the interrupt vector location and hence control of the CPU passes to the first instruction of the interrupt service routine.

3.6.2.6 Multi-level interrupts

In most real-time systems a single interrupt level is unacceptable; the whole purpose of interrupts is to get a fast response and this would be prevented if a low-priority interrupt could lock out a high-priority one. A typical picture of multi-level interrupts is shown in Figure 3.17. An application program (the main task) is interrupted at regular intervals by the clock interrupt which is the highest-priority interrupt (level 0). When the interrupt occurs control is passed to the clock interrupt service routine (ISR 0) – transfers 1, 2 and 3 in Figure 3.17. During the servicing of the clock interrupt, the printer generates an interrupt request (4), but since the printer is of lower priority than the clock the interrupt is not dealt with until the clock routine ISR 0 has finished. When this occurs, instead of control returning to the main program it passes to the printer service routine ISR 1 (5). The printer service routine does not complete before the next clock interrupt, so it is suspended (6) while the next interrupt from the clock is dealt with. At the termination of the clock routine return is made to the printer (7) and finally, when the printer ISR finishes, a return is made to the main program (8).

It should be obvious that the ability to interrupt an interrupt service routine should be restricted to interrupts which are of higher priority than the routine executing. In order to do this there has to be some facility for masking out (or inhibiting) interrupts of lower priority. Masking is achieved automatically in the daisy-chain system, since a device which wishes to interrupt lowers its IEO line thus preventing all lower-priority devices from responding to the interrupt acknowledge signal. In the daisy-chain system, however, the device must receive a signal from the

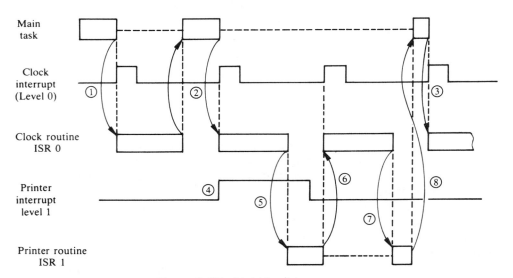

Figure 3.17   Multi-level interrupts.

CPU on return from the interrupt service routine in order that it can set the IEO line high and hence permit access to the system by lower-priority devices.

An alternative scheme used is to have a mask register which can be loaded from software and used to inhibit the lower-priority interrupt lines. Figure 3.18 shows a system which uses a priority encoder and the software sequence is outlined in Figure 3.19. Note that with the mask system it is possible to mask out any interrupt, not just ones with lower priority. This can have advantages if, for example, a high-priority alarm interrupt is continually being generated because of a fault on the plant; once the fault condition has been recognised it is desirable to mask out the interrupt to avoid the computer spending all its time simply servicing the interrupt. The ability to mask out selected levels provides software-controlled priority reallocation.

Figure 3.19 shows typical functions performed by an interrupt service routine. The first requirement is to save the working environment; the current mask register must also be saved and then the new mask register sent out. The interrupts can now be enabled and the actual servicing of the interrupt commenced. When the servicing is completed the interrupts are disabled, the previous mask register restored and the working environment restored. The interrupts can now be enabled and a return from

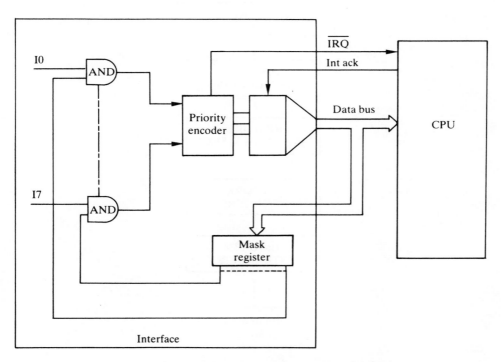

Figure 3.18 Interrupt masking.

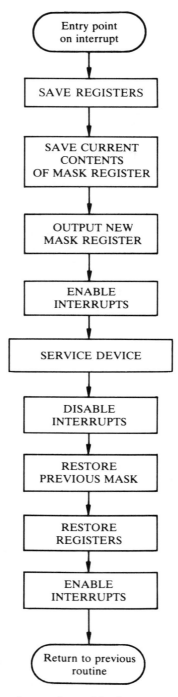

Figure 3.19   Function performed by interrupt servicing routine.

interrupt executed. Note that some computer systems automatically disable all lower-priority interrupts and hence the need to save and restore mask registers is avoided.

3.6.3 Direct Memory Access

Three methods are normally used – burst mode, distributed mode and cycle stealing. In burst mode, the DMA controller takes over the data highways of the computer and locks out the CPU for the period of time necessary to transfer, say, 256 bytes between fast memory and backing memory. The use of burst mode can seriously affect the response time of a real-time system to an external event and because of this may not be acceptable.

In distributed mode the DMA controller takes occasional machine cycles from the CPU's control and uses each cycle to transfer a byte of information between fast memory and backing memory. In a non-real-time system the loss of these machine cycles to the CPU is not noticeable. However, in a real-time system which uses software timing loops the loss of machine cycles will then affect the time taken to complete the timing loop. The program is unaware of the machine cycles used by the DMA controller and hence will still cycle through the same number of instructions; however, the elapsed time may be the equivalent of 200 machine cycles rather than the expected 100 cycles.

The cycle-stealing method only transfers data during cycles when the CPU is not using the data bus. Therefore the program proceeds at the normal rate and is completely unaffected by the DMA data transfers. This is, however, the slowest method of transfer between fast memory and backing store.

3.6.4 Comparison of Data Transfer Techniques

Polling, with either busy wait or periodic checks on device status, provides the simplest method of data transfer, in terms of the programming requirements and in the testing of programs. The use of interrupts results in software which is much less structured than a program with explicit transfers of control; there are potential transfers of control at every point in the program.

Interrupt-driven systems are much more difficult to test since many of the errors may be time dependent. A simple rule is to check the interrupt part of the program if irregular errors are occurring. The generation of appropriate test routines for interrupt systems is difficult; for proper testing it is necessary to generate random interrupt patterns and to carry out detailed analysis of the results.

At high data transfer rates the use of interrupts is inefficient because of the overheads involved in the interrupt service routine – saving and restoring the environment – hence polling is often used. An alternative for high rates of transfer is to substitute hardware for software control and use direct memory access techniques.

3.7 COMMUNICATIONS

The use of distributed computer systems implies the need for communication: between instruments on the plant and the low-level computers (see Figure 3.20); between the Level 1 and Level 2 computers; and between the Level 2 and the higher-level computers. At the plant level communications systems typically involve parallel analog and digital signal transmission techniques since the distances over which communication is required are small and high-speed communication is usually required. At the higher levels it is more usual to use serial communication methods since, as communication distances extend beyond a few hundred yards, the use of parallel cabling rapidly becomes cumbersome and costly.

As the distance between the source and receiver increases it becomes more difficult, when using analog techniques, to obtain a high signal-to-noise ratio; this is particularly so in an industrial environment where there may be numerous sources of interference. Analog systems are therefore generally limited to short distances. The use of parallel digital transmission provides high data transfer rates but is expensive in terms of cabling and interface circuitry and again is normally only used over short distances (or when very high rates of transfer are required).

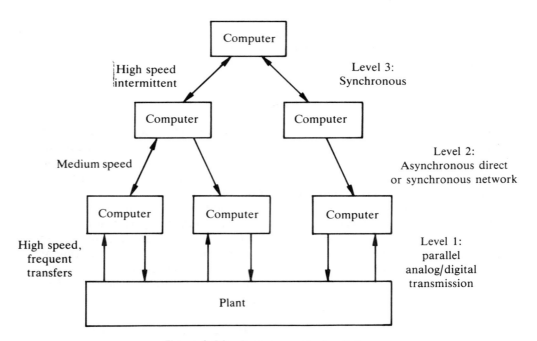

Figure 3.20   Data transmission links.

Serial communication techniques can be characterised in several ways:

1. Mode
    (a) asynchronous
    (b) synchronous
2. Quantity
    (a) character by character
    (b) block
3. Distance
    (a) local
    (b) remote, that is wide area
4. Code
    (a) ASCII
    (b) other

3.7.1 Asynchronous and Synchronous Transmission Techniques

Asynchronous transmission implies that both the transmitter and receiver circuits use their own local clock signals to gate data on and off the data transmission line. So that the data can be interpreted unambiguously there must be some agreement between the transmitter and receiver clock signals. This agreement is forced by the transmitter periodically sending synchronisation information down the transmission line. An alternative approach is to use an additional physical connection – a clock wire – and periodically to send a synchronising signal.

The most common form of asynchronous transmission is the character-by-character system which is frequently used for connecting terminals to computer equipment and was introduced for the transmission of information over telegraph lines. It is sometimes called the stop–start system. In this system each character transmitted is preceded by a 'start' bit and followed by one or two 'stop' bits (see Figure 3.21). The start bit is used by the receiver to synchronise its clock with the incoming data; for correct transfer of data the clock and data signals must remain synchronised for the time taken to receive the following eight data bits and two stop bits. The transmission is thus *bit* synchronous but *character* asynchronous. The advantage of the stop–start system is that, particularly at the lower transmission rates, the frequencies of the clock signal generators do not have to be closely matched. The disadvantage is that for each character transmitted (seven bits) three or four extra bits of information have also to be transmitted and thus the overall information ratio is not very high.

To overcome the problem of transmitting redundant bits, synchronous systems designed to transmit large volumes of data over short periods of time use block-synchronous transmission techniques. The characters are grouped into records, for example blocks of 80 characters, and each record is preceded by a synchronisation signal and terminated with a stop sequence. The synchronisation sequence is used to enable the receiver to synchronise with the transmitter clock.

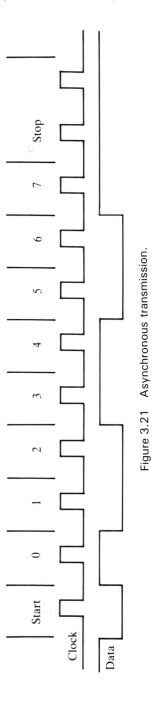

Figure 3.21 Asynchronous transmission.

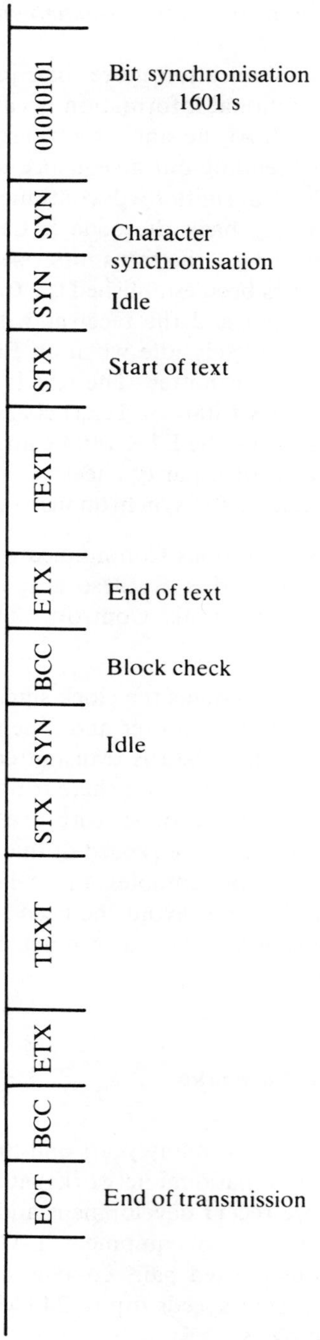

01010101	Bit synchronisation 1601 s
SYN	Character synchronisation
SYN	Idle
STX	Start of text
TEXT	
ETX	End of text
BCC	Block check
SYN	Idle
STX	
TEXT	
ETX	
BCC	
EOT	End of transmission

Figure 3.22 Synchronous transmission.

To establish effective communication more than just a synchronisation signal must be transmitted – the additional information is called the *protocol*. A simple protocol is shown in Figure 3.22. At the start of a transmission, bit synchronisation is achieved by the transmitter sending out a sequence of zeros and ones, followed by the ASCII code 'SYN'. The transmitter will continue to send the SYN code until the receiver responds by sending back the code 'ACK' or a preset time elapses (device time out); if time out occurs, the transmitter sends the bit pattern of zeros and ones again. Once contact has been established the transmitter will send out SYN characters during any idle period and the receiver will respond by sending back ACK; the line will only be completely idle when an EOT (End Of Transmission) character has been sent by the transmitter. The text is broken up into blocks and each block is preceded by an STX (Start of TeXt) character and ended by an ETX (End of TeXt) character. Following the ETX will be an integrity check on the data; typically this will take the form of a parity check.

There are two main standards for synchronous transmission systems:

1. BISYNC (BInary SYNchronous Communication). This is the older system used in IBM equipment and is obsolescent.
2. HDLC (High-level Data Link Control). This is used in most new equipment.

In synchronous transmission systems the clock signal for the timing of the data transfer is provided solely by the transmitter and is sent to the receiver even when no data is being transmitted. When data is transmitted it is superimposed on the clock signal. With synchronous transmission there is no need to transmit extra bits to enable the receiver clock to synchronise with the transmitter and hence the effective data transmission rate for a given speed of line is higher. The disadvantage is that the interface circuitry is more complex and hence more expensive. The use of synchronous transmission does not avoid the need for a transmission protocol. The advantage of block transmission is that a much higher ratio of data bits to control bits can be obtained.

3.7.2 Local- and Wide-area Networks

Wide-area networks have existed for many years and they operate over a very wide geographical area (many are international networks) at moderate speeds. The local-area network (LAN) is a more recent development and it is having a considerable impact on the design of process control equipment. LANs make use of a wide range of transmission media such as twisted pair, co-axial cable, and fibre optics; they operate at a range of transmission speeds (up to 240 Mbit s^{-1}) and use a range of different protocols and topologies.

Typical topologies are shown in Figure 3.23. For computer control applications no one topology represents the best solution: the particular application will

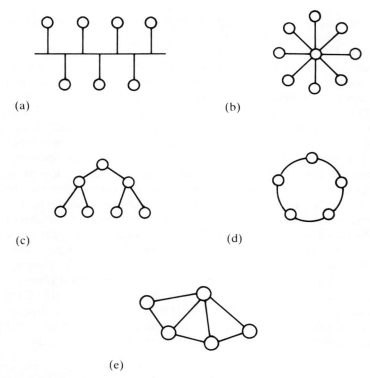

Figure 3.23 LAN topologies: (a) data bus; (b) star; (c) hierarchy (or tree); (d) ring; (e) mesh.

govern the most appropriate one. The characteristics of each are briefly outlined below.

Data bus: This is the simplest of all the LAN topologies. The bus is normally passive and all the devices are simply plugged into the transmitting medium. The bus is inherently reliable because of its passive nature but there may be a limitation on the length of a bus in that any transmitting device connected has to be able to transmit for the full length of the bus. It is a broadcast system and hence a packet of data placed on the bus is available to all devices.

Star: The star network is not very widely used; it depends on a central switching node to which all other nodes are connected by a bidirectional link. Data sent to the central switch can be forwarded either in the broadcast mode, that is to all other nodes, or only to a specified node. The computer-controlled PABX (Private Automatic Branch eXchanges) used in many businesses operate in the star mode – all the telephone lines connect to the central unit – while the forwarding system used is to a specified node. A weakness of the star topology is that the central node is a critical component in the system; if it fails, the whole system fails.

Hierarchy: The system has many of the characteristics of the star, but instead of one central switching node, many of the nodes have to act as switches. Frequently it can closely reflect the actual structure of the application. The addition of new nodes to a hierarchy can be difficult.

Ring: This is probably the most popular method. The ring is typically an active transmission system, that is the ring itself contains regeneration circuits which amplify the signals. The information placed on a ring network continues to circulate until a device removes it from the ring; in some systems the originating device removes the data from the ring. The information is broadcast in the sense that it is available to all devices connected to the ring.

Mesh: The mesh topology allows for random interconnection between the various nodes. It provides a means by which alternative routes between nodes can be found and hence has built into it a form of redundancy. A problem which can arise with the mesh is that there can be a variable delay between the sending and receiving of the message because of the number of nodes through which the message has had to pass. Information is transmitted in the form of 'packets' which may be of fixed or variable length. Early systems used character-oriented packets similar to that illustrated in Figure 3.22. The prevailing standard is now the HDLC protocol (referred to in the previous section) and the format of the packet is shown in Figure 3.24.

The *access* mechanisms used to ensure that only one node of the network is attempting to transmit at any one time divide into two main types:

- synchronous token passing, message slots; and
- asynchronous carrier sense multiple access/collision detection (CSMA/CD).

Ring-based LANs normally use synchronous techniques. The packets of data circulate in one direction around the ring and in the token passing system attached to one, and only one, packet is a token. (If the network is idle a packet containing just the token is circulated.) Each node reads the packet into a buffer and checks the message. There are three actions which can then be taken.

1.   If the message is for that node it is read, marked as accepted and replaced

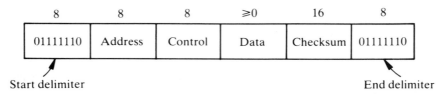

Figure 3.24   Format of HDLC packet.

on the ring — it will be removed from the ring when it reaches the originating node.

2. If the message is not for that node it is simply replaced on the ring.
3. If the packet contains the token and the node wishes to transmit a message, the token is removed from the packet which is then passed on. The node then transmits its own message adding the token to the end of the message.

An alternative to the token passing method is the message slot. A sequence of bits is used to mark a slot and the slots circulate around the ring. If a node detects an empty slot it may insert a message in that slot.

Bus LANs may use token passing (the token is passed from node to node in some predetermined manner) or message slots but asynchronous methods are more common. In the asynchronous systems a node may attempt to transmit at any time. The node listens to the bus and if it is idle begins to transmit. Because of the distances between nodes and the time taken to transmit a message, two (or more) nodes may be transmitting simultaneously. If this happens a collision is said to have occurred; bus systems must therefore have some means of detecting a collision. If a collision is detected the nodes attempting to transmit execute a random delay and then retry.

Several of the major process companies have developed distributed control systems based on the use of LAN technology (mostly rings). These allow a wide range of devices — from individual instruments to large computers — to be connected to a common network.

3.8 STANDARD INTERFACES

Most of the companies which supply computers for real-time control have developed their own 'standard' interfaces, such as the Digital Equipment Corporation's Q-bus for the PDP-11 series, and, typically, they, and independent suppliers, will be able to offer a large range of interface cards for such systems. The difficulty with the standards supported by particular manufacturers is that they are not compatible with each other; hence a change of computer necessitates a redesign of the interface.

An early attempt to produce an independent standard was made by the British Standards Institution (BS 4421, 1969). Unfortunately the standard is limited to the concept of how the devices should interconnect and the standard does not define the hardware. It is not widely used and has been overtaken by more recent developments.

An interface which was originally designed for use in atomic energy research laboratories — the computer automated measurement and control (CAMAC) system — has been widely adopted in laboratories, the nuclear industry and some other industries. There are over 1000 different modules, supported by about 50 manufacturers in eight countries, available for the system. There are also FORTRAN libraries which provide software to support a wide range of the interface modules. One of the attractions of the system is that the CAMAC data highway connects to the computer by a special card; to change to a different computer only requires that the one card be changed.

A general purpose interface bus (GPIB) was developed by the Hewlett Packard Company in the early 1970s for connecting laboratory instruments to a computer. The system was adopted by the IEEE and standardised as the IEEE 488 bus system.

Table 3.1   The ISO seven-layer model

Layer	Description	Standards
Physical	Defines the electrical and mechanical interfacing to a physical medium. Sets up, maintains and disconnects physical links. Includes hardware (I/O ports, modems, communication lines, etc.) and software (device drivers)	RS232-C RS442/443/449 V.24/V.28 V.10/V.11 X.21, X.21 bis, X.26, X.27, X.25 level 1
Data link	Establishes error-free paths over physical channel, frames messages, error detection and correction. Manages access to and use of channels. Ensures proper sequence of transmitted data	ANSI-ADCCP ISO-HDLC LAP DEC DDCMP IBM SDLC, BISYNC X.25 level 2
Network	Addresses and routes messages. Sets up communication paths. Flow control	USA DOD-IP X25, X75 (e.g. Tymnet, Telenet, Transpace, ARPANET, PSS)
Transport	Provides end-to-end control of a communication session. Allows processes to exchange data reliably	USA DOD-TCP IBM SNA DEC DNA
Session	Establishes and controls node-system-dependent aspects. Interfaces transport level to logical functions in node operating system	
Presentation	Allows encoded data transmitted via communications path to be presented in suitable formats for user manipulation	FTP JTMP FAM
Application (user)	Allows a user service to be supported, e.g. resource sharing, file transfers, remote file access, DBM, etc.	

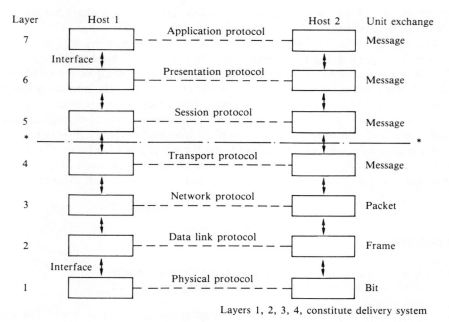

Figure 3.25   ISO seven-layer model.

The bus can connect up to a maximum of 15 devices and is only suited to laboratory or small, simple control applications.

The ISO (International Organisation for Standardisation) have promulgated a standard protocol system in the Open Systems Interconnection (OSI) model. This is a layered (hierarchical) model with seven layers running from the basic physical connection to the highest application protocol. The general structure is illustrated in Figure 3.25. The layers can be described as shown in Table 3.1.

3.9 SUMMARY

This chapter has provided a brief overview of some of the basic hardware ideas that are relevant to using computers in embedded control applications. We have concentrated on the basic ideas and not on particular microprocessors. In order to design a control system involving embedded computers you will need to obtain detailed knowledge about the particular microprocessor, microcomputer or microcontroller that you are going to use. In particular you will need to understand in detail its interface and where appropriate the range of interface support chips available for the particular processor. If you are going to be able to choose an appropriate device you will need to understand the important characteristics of a wide range of devices.

The key to understanding and coping with the complexities of the available hardware is to think in layers or hierarchies. An example of this was given in section 3.2.4 when we described the bus structure: we divided the discussion into physical, electrical and functional characteristics. Another example is the approach adopted by the ISO in its OSI communication model; detail that is not required at a higher level is hidden in the lower levels. As you will see later we will apply this approach to software, and will hide unwanted detail in low-level software modules.

An important aspect of interfacing is the timing of the transfer of data and the synchronisation of transfers. Timing diagrams of the form shown in Figure 3.4 are important and you need to be able to read and understand such diagrams.

EXERCISES

3.1   Why is memory protection important in real-time systems?
What methods can be used to provide memory protection?

3.2   A large valve controlling the flow of steam is operated by a dc motor. The motor controller has two inputs:

      1.   on/off control, 0 V = off, 5 V = on; and
      2.   direction, 0 V = clockwise, 5 V = anti-clockwise;

and two outputs:

      1.   fully open = 5 V;
      2.   fully closed = 5 V.

Show how this valve could be interfaced to a computer controlling the process.

3.3   A turbine flow meter generates pulses proportional to the flow rate of a liquid. What methods can be used to interface the device to a computer?

3.4   There are a number of different types of analog-to-digital converters. List them and discuss typical applications for each type (see, for example, Woolvet (1977) or Barney (1985)).

3.5   The clock on a computer system generates an interrupt every 20 ms. Draw a flowchart for the interrupt service routine. The routine has to keep a 24 hour clock in hours, minutes and seconds.

3.6   Twenty analog signals from a plant have to be processed (sampled and digitised) every 1 s. The analog-to-digital converter and multiplexer which is available can operate in two modes: automatic scan and computer-controlled scan. In the automatic scan mode, on receipt of a 'start' signal the converter cycles through each channel in turn.

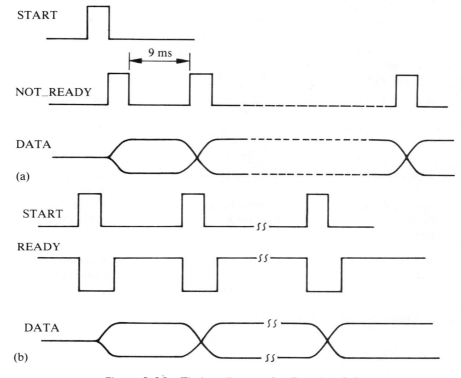

Figure 3.26 Timing diagram for Exercise 3.6.

The data corresponding to the channel sampled is available for 0.9 ms. The signal 'not-ready' is asserted during the conversion period and this indicates that the data is changing and should not be read by the computer. The timing is shown in Figure 3.26. In Mode 2 under computer-controlled scanning, the converter holds the data for each channel sampled until it receives a command from the computer to start the sampling of the next channel. To speed up the operation the multiplexer is switched to the next channel once the current channel has been sampled and before the computer reads the data for the current channel. The converter can be reset to start from Channel 1 by asserting a signal reset. The timing of this mode of operation is shown in Figure 3.26b. Consider the ways in which (a) polling and (b) interrupt methods can be used to interface the converter to a computer. Discuss in detail the advantages and disadvantages of each method.

3.7 We will assume that the simple heat process described in Chapter 1 has, in addition to a temperature sensor and heat controller, some logic signals and control switches. These are:

> *Plant controls*:
> heater on/off;

blower on/off; and
power on/off.
Plant signals:
overtemperature alarm; and
blower failure alarm.
The start-up sequence for the unit is:

1. Turn power on.
2. Turn blower on.
3. Wait 5 seconds.
4. Turn heater on.
5. DDC control action begins.

If at any time the overtemperature alarm becomes true, that is the signal level is set to logic 'high', the heater must be turned off but the blower kept running. If the blower failure alarm is detected, both the blower and the heater must be switched off. Draw a flowchart to show the sequence of operations to be carried out (a) for start-up and (b) in the event of failure.

3.8   The hot-air blower system described in Chapter 1 uses interrupts to indicate: power failure, printer ready, air temperature too high, VDU display ready, blower failure, clock signal and key pressed on the keyboard. Draw up a list of the priority order for the interrupts and explain the reasons for your choice of priority. Should any of the interrupts be connected to a non-maskable interrupt?

4

DDC Algorithms and Their Implementation

The main purpose of this chapter is to consider the methods used to implement simple digital control algorithms and some of the problems that arise in so doing. We shall take as an example a widely used and simple control algorithm: the PID or three-term control algorithm.

We shall consider:

- The digital form of the algorithm.
- The timing requirements.
- Integral wind-up and bumpless transfer.
- Choice of sampling rates.

The last two sections of the chapter introduce some more advanced control ideas concerned with finding the discrete equivalents of continuous controllers and the implementation of control algorithms designed using discrete control system design techniques. These sections can be omitted if you so wish.

4.1 INTRODUCTION

In Chapter 2 we introduced the idea of DDC (Direct Digital Control) and stated the differential equation for a PID controller:

$$m(t) = K_p[e(t) + 1/T_i \int_0^t e(t)dt + T_d de(t)/dt] \tag{4.1}$$

where $e(t) = r(t) - c(t)$, $r(t)$ is the desired value (set point), $c(t)$ the value of the variable being controlled and $m(t)$ the output from the controller. The differential equation is the time domain representation of the controller. The equivalent frequency domain representation is

$$G_c(s) = \frac{M(s)}{E(s)} = K_p\left(1 + \frac{1}{T_i s} + T_d s\right) \tag{4.2}$$

In the frequency domain the overall system of controller and plant can be represented by a block diagram as shown in Figure 4.1.

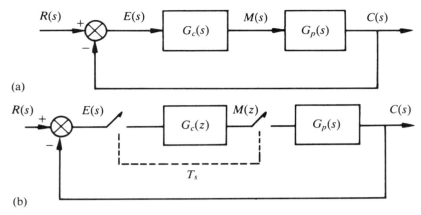

(a)

(b)

Figure 4.1   General form of a control system: (a) continuous form; (b) discrete form.

Both the time domain and frequency domain representations are continuous representations. To implement the controller using a digital algorithm we have to convert from a continuous to a discrete representation of the controller. There are several methods of doing this; the simplest is to use first-order finite differences. Considering the time domain version of the controller (equation 4.1) we replace the differential and integral terms by their discrete equivalents by using the relationships

$$\frac{df}{dt}\bigg|_{k} = \frac{f_k - f_{k-1}}{\Delta t}, \quad \int e(t)dt = \sum_{k=1}^{n} e_k \Delta t \tag{4.3}$$

and hence equation 4.1 becomes

$$m(n) = K_p \left[T_d \left(\frac{e(n) - e(n-1)}{\Delta t} \right) + e(n) + \frac{1}{T_i} \sum_{k=1}^{n} e_k \Delta t \right] \tag{4.4}$$

where $m(n)$ represents the value of m at some time interval $n\Delta t$ where n is an integer.

By introducing new parameters as follows:

$$K_i = K_p(T_s/T_i)$$
$$K_d = K_p(T_d/T_s)$$

where $T_s = \Delta t =$ the sampling interval, equation 4.4 can be expressed as an algorithm of the form

$$s(n) = s(n-1) + e(n)$$
$$m(n) = K_p e(n) + K_i s(n) + K_d[e(n) - e(n-1)] \tag{4.5}$$

where $s(n) =$ the sum of the errors taken over the interval 0 to nT_s.

4.2 IMPLEMENTATION OF THE BASIC PID ALGORITHM

Writing the code to implement the algorithm given in equation 4.5 is a simple job. The basic code statements are:

```
sn :=sn+en;
mn :=Kp*en+Ki*sn+Kd*(en-enOld);
enOld :=en;
```

These statements can be incorporated into a procedure such as:

```
PROCEDURE PIDControl(en: REAL; VAR mn: REAL);
BEGIN
    sn :=sn+en;
    mn :=Kp*en+Ki*sn+Kd*(en-enOld);
    enOld :=en;
END PIDControl;
```

If we assume that the plant output is obtained by using an ADC to sample and convert the output signal, that the actuator control signal is output through a DAC to the actuator and that procedures ADC and DAC are available to read the ADC and to send values to the DAC, then we can write a PID control module as shown in Example 4.1.

EXAMPLE 4.1

```
MODULE PIDcontroller;
(*
Title : PIDcontroller
File : PIDCONTR
Last Edit : 17 Jan 1992
Author : S.Bennett
*)
(*
This is the ideal controller. It ignores all
practical problems and all timing problems.
*)
FROM IOmodule IMPORT ADC, DAC;
FROM IO IMPORT KeyPressed;
CONST
    KpValue=1.0;
    KiValue=0.8;
    KdValue=0.3;
VAR
    sn,Kp,Ki,Kd,en,enOld,mn : REAL;
```

```
PROCEDURE PIDControl(en: REAL; VAR mn: REAL);
BEGIN
    sn := sn+en;
    mn := Kp*en+Ki*sn+Kd*(en-enOld);
    enOld := en;
END PIDControl;

BEGIN (* Main program *)
    sn:=0.0; (* initialise integral action term *)
    Kp:=KpValue;
    Ki:=KiValue;
    Kd:=KdValue;
    enOld:=ADC();
    REPEAT
        en:=ADC();
        PIDControl(en, mn);
        DAC(mn);
    UNTIL KeyPressed();
END PIDcontroller.
```

The above program ignores several important practical problems, for example it does not take into account the need to synchronise the calculation of PIDControl with 'real' time. As written the program makes the sampling interval T_s dependent on the speed of operation of the computer on which the program is run. For correct operation some means of fixing the sampling interval is required since the coefficients K_i and K_d are both calculated assuming a specific value of sampling interval. It is of course possible to change the algorithm to include the sampling interval T_s as a variable.

Other omissions include:

- bumpless transfer – that is, smooth transfer from manual to automatic control;
- actuator limiting and other forms of saturation – these lead to integral wind-up; and
- measurement and process noise.

The program also has the controller parameters built in as program constants; hence modification of the controller settings requires recompilation of the program.

In the next sections we examine some ways of modifying the program to deal with these problems and look at improved forms of the basic algorithm.

4.3 SYNCHRONISATION OF THE CONTROL LOOP

A typical feature of real-time programs is that once they have been started they run continuously until some external event occurs to stop them. We will emphasise this by using the infinite loop programming construct LOOP...END with an EXIT statement to indicate the terminating condition. The general form of a control program will be:

```
MODULE RealTimeControl;
(* declarations *)
BEGIN
    (* initialisation *)
    LOOP
        (* synchronisation *)
        (* get plant data *)
        (* control calculation *)
        (* EXIT condition check *)
        (* put control data to plant *)
    END (* loop *);
END RealTimeControl.
```

Synchronisation can be achieved by several different means such as:

- polling;
- external interrupt signals;
- ballast coding; and
- real-time clock signals.

Two methods, polling and external interrupt signals, were discussed in Chapter 3 (section 3.6). They rely on the plant (or some other unit external to the computer) sending a signal to indicate that it is time for a control action to take place. This signal must be sent at the sampling interval T_s chosen when starting the controller since, as can be seen from equation 4.4, the algorithm is correct only for a particular sampling rate. The difference between the two methods is that in polling the control computer repeatedly reads a value − normally a logical signal − whereas with an external interrupt the computer can be performing other computations; the action of the interrupt is to tell the computer to suspend whatever it is doing and carry out the control task.

4.3.1 Polling

We can write a synchronisation procedure which uses polling as follows:

```
PROCEDURE Synchronisation;
(* Use of polling for synchronisation *)
BEGIN
    LOOP
        WHILE NOT (Digin(SampleTime)) DO
            (* wait until time *)
        END (* while *);
    END (* loop *);
END Synchronisation;
```

In the above example it is assumed that a Boolean function, `Digin(line)`, is available which reads the appropriate logical signal, in this case `SampleTime`, from the plant interface. When the procedure `Synchronisation` is called the computer waits in the `WHILE...DO` loop until the `Digin(SampleTime)` function returns a value true — this wait is referred to as a *busy wait* since no other computation can be carried out while the computer is waiting.

The polling method is simple to program and easy to design and use; however, because of the busy wait its use is restricted to small dedicated systems. An alternative method suitable for simple, dedicated, control systems is to use ballast coding (Hine and Burbridge, 1979).

4.3.2 Ballast Code

The idea of ballast coding is to make the loop time completely dependent on the internal operations of the computer and independent of external timing or synchronisation signals. The method involves finding the time taken to execute each possible path in the control loop of the program and adding code statements — ballast code — to make the execution time for each path equal. If necessary a further block of ballast code is added at the end to make the total execution time for the control loop equal to some desired execution time.

The method can be illustrated by considering the program structure shown in Figure 4.2. For each path (for example, A, $A1$, $A1.1$) the computational time for that particular path is calculated (or measured) and ballast code is added to each so as to make the computational time for each path equal. For path A, $A1$, $A1.1$ ballast $A1.1$ is added. Further ballast code can be added to make the total computational time equal to the sample interval; this is shown as Ballast B.

The method minimises the amount of external hardware required and is thus cost effective for systems that are to be produced in large quantities. An obvious problem is that any change in the code results in the need to adjust the ballast code segments. Also the technique cannot be used if interrupts are being used (why not?) and the code will have to be modified if the CPU clock rate is changed. As is the case with polling, the use of the ballast code technique prevents the computer system being used to carry out any other calculations while it is waiting to carry out the next control calculation.

4.3.3 External Interrupt

For small systems with a limited number of DDC loops (or other actions that require synchronisation), use of an external interrupt for synchronisation can be very effective. The control loop is written as an interrupt which is associated with a particular interrupt line. The interrupt line is activated by some external device — typically a clock. While the control loop is waiting to be activated other programs

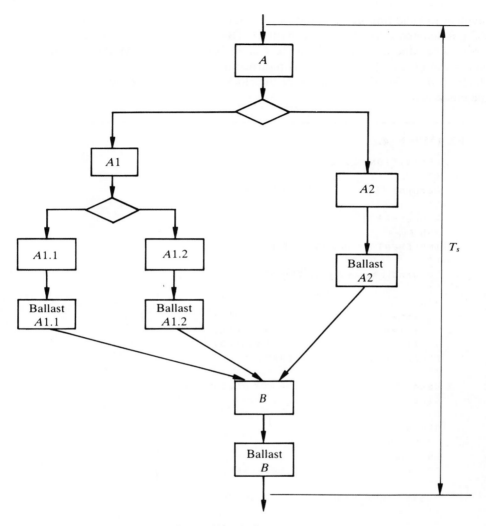

Figure 4.2 Ballast coding.

can be running. This form of operation is typically referred to as a foreground–background operation and is described in more detail later in this chapter.

4.3.4 Real-time Clock

The most general solution to the problem of timing a control loop is provided by adding a real-time clock to the computer system. Provision of a real-time clock involves the addition of some hardware components and some software. The IBM PC (and compatibles) as part of the BIOS (Basic Input Output System) provides a clock. The Module Timer, Example 4.2, shows how, using Modula-2, the clock values can be accessed. The time is returned in terms of the number of *ticks* of the clock, a tick being the resolution of the clock, that is the smallest interval of time the clock can measure.

EXAMPLE 4.2

```
DEFINITION MODULE Timer;
(*
Returns the value of the 'TICK' clock.
*)
PROCEDURE Ticks():LONGCARD;
END Timer.
IMPLEMENTATION MODULE Timer;
(*
Returns the value of the 'TICK' clock.
*)
VAR
    TimerLow  [0:046CH] : CARDINAL;
    TimerHigh [0:046EH] : CARDINAL;
    Time      [0:046CH] : LONGCARD;
PROCEDURE LowTicks():CARDINAL;
(*
Reads TIMER_LOW from the ROM BIOS clock.
(Incremented every 1/18.2 seconds.)
*)
BEGIN
    RETURN TimerLow
END LowTicks;
PROCEDURE HighTicks():CARDINAL;
(*
Reads TIMER_HIGH from the ROM BIOS clock.
(Incremented every hour.)
*)
BEGIN
    RETURN TimerHigh
END HighTicks;
PROCEDURE Ticks():LONGCARD;
(*
Reads the complete ROM BIOS clock. (18.2 ticks/second.)
*)
BEGIN
    RETURN Time
END Ticks;
END Timer.
```

The code fragment, Example 4.3, illustrates how the timer module can be used to synchronise the control loop calculations to real time.

EXAMPLE 4.3

```
FROM Timer IMPORT Ticks;
CONST
    sT=20; (* time between samples in 'TICKS' *)
VAR
    time, NextSampleTime : LONGCARD;
BEGIN   (* Main program *)
    NextSampleTime := Ticks()+sT;
    time:=Ticks()+sT;
    LOOP
        WHILE Ticks() < NextSampleTime DO
            (* nothing *)
        END; (* of WHILE *)
        time:=Ticks();
        (* get plant input *)
        (* control calculation *)
        (* put plant output *)
        NextSampleTime := time+sT;
        IF KeyPressed()THEN EXIT;
        END (* IF *);
    END; (* of LOOP *)
```

This example uses a busy wait in the control loop – the WHILE...DO statement – and during this wait the clock is read continually and checked against the time for the next sample. Immediately on exiting from the wait the current value of time is saved in the variable time. At the end of the control calculation the time for the next sample is updated by adding the sample interval to the variable time. The interval between successive runs of the control loop is thus independent of the time taken in the control calculation. It is good practice to check that at the end of the control loop the value of NextSampleTime is later than the current time and to provide an error indication if this is not the case. A suitable check would be to add the code

```
IF NextSampleTime < Ticks() THEN RaiseError(timing)
END (* IF *);
```

Two examples of PID controllers are given below. You should study both of them carefully and how the method of synchronisation for each differs.

EXAMPLE 4.4

```
MODULE PIDcon2A;
(*
Title : PIDcontroller Example
File : PIDcon2A
*)
(*
This example illustrates a method for synchronising the PID
control calculation but ignores other practical problems.
*)
FROM IOmodule IMPORT ADC, DAC;
FROM Timer IMPORT Ticks;
FROM IO IMPORT KeyPressed;
FROM Error IMPORT RaiseError;
CONST
    KpValue=1.0;
    KiValue=0.8;
    KdValue=0.3;
    sT=20; (* time between samples in 'TICKS' *)
VAR
    sn,Kp,Ki,Kd,en,enOld,mn : REAL;
    time, NextSampleTime : LONGCARD;

BEGIN (* Main program *)
    sn:=0.0; (* initialise integrate action term *)
    Kp:=KpValue;
    Ki:=KiValue;
    Kd:=KdValue;
    en:=ADC();
    NextSampleTime := Ticks()+sT;
    time:=Ticks()+sT;
```

```
    LOOP
        WHILE Ticks() < NextSampleTime DO
            (* nothing *)
        END; (* of WHILE *)
        time:=Ticks();
        enOld:=en;
        en:=ADC();
        sn:=sn+en;
        mn:=Kp*en+Ki*sn+Kd*(en-enOld);
        DAC(mn);
        NextSampleTime := time+sT;
        IF NextSampleTime < Ticks() THEN
            RaiseError(timing);
        END (* IF *);
        IF KeyPressed()THEN EXIT;
        END (* IF *);
    END; (* of LOOP *)
END PIDcon2A.
```

EXAMPLE 4.5

```
MODULE PIDcon2B;
(*
Title : PIDcontroller
File : PIDCON2B
*)
(*
This is an alternative way of providing
synchronisation to that given in MODULE PIDcon2A.
*)
FROM IOmodule IMPORT ADC, DAC;
FROM Timer IMPORT Ticks;
FROM IO IMPORT KeyPressed;
FROM Error IMPORT RaiseError;
CONST
    KpValue=1.0;
    KiValue=0.8;
    KdValue=0.3;
    sT=20; (* time between samples in 'TICKS' *)
VAR
    sn,Kp,Ki,Kd,en,enOld,mn : REAL;
    time:LONGCARD;
```

```
BEGIN (* Main program *)
   sn:=0.0; (* initialise integrate action term *)
   Kp:=KpValue;
   Ki:=KiValue;
   Kd:=KdValue;
   en:=ADC();
   time:=Ticks()+sT;
   LOOP
      WHILE Ticks() < time DO
         (* nothing *)
      END; (* of WHILE *)
      time:=time+sT;
      enOld:=en;
      en:=ADC();
      sn:=sn+en;
      mn:=Kp*en+Ki*sn+Kd*(en-enOld);
      DAC(mn);
      IF Time < Ticks() THEN
         RaiseError(timing);
      END (* IF *);
      IF KeyPressed()THEN EXIT;
      END (* IF *);
   END; (* of LOOP *)
END PIDcon2B.
```

4.4 BUMPLESS TRANSFER

Equation 4.5 implies that in the steady state, with zero error, the controlled variable $m(n)$ is equal to the value of the integral term $K_i s(n)$. Ideally in the steady state with zero error we would like the integral term to be zero, which would mean that $m(n)$ is also zero. In many applications the steady-state operating conditions require that $m(n)$ has some value other than zero. For example, a steam boiler may require the fuel line valves to be half open. In the case of the hot-air blower described in Chapter 1 a non-zero voltage has to be applied to the heater input for the heater to provide heat output. Therefore the normal practice is to modify equation 4.5 by adding a constant term (M) representing the value of the manipulated variable at the steady-state operating point, giving

$$m(n) = K_p e(n) + K_i s(n) + K_d[e(n) - e(n-1)] + M \qquad (4.6)$$

The quantity M can be thought of as setting the operating point for the controller. If it is omitted and integral action is present, the integral action term will compensate for its omission but there will be difficulties in changing smoothly, without disturbance to the plant, from manual to automatic control. There will also be the danger that on change-over, a large change (for example, in a valve position) will be demanded. Plant operating requirements usually demand that manual/automatic change-over be made in the so-called 'bumpless' manner. Bumpless transfer can be achieved by several means.

4.4.1 Method 1 – Preset Change-over Value

The value of M is calculated for a given steady-state operating point and is inserted either as a constant in the program or by the operator prior to the change-over from manual to automatic mode. The transfer to automatic mode is made when the value of the error is zero; at the time of change-over the integral term is set to zero and the output $m(n)$ equal to M. The problem with this technique is obvious: the predetermined value of M is correct only for one specified load. If the load is varying it may not be possible or convenient to make the change-over at the predetermined load value. If the error is not zero on change-over there will be a sudden change in the value of the manipulated variable due to the proportional action.

4.4.2 Method 2 – Tracking of Operator Setting

During operation under operator control the manipulated variable (m) is set from the operator's control panel and the computer system keeps track of the value. This may be done either by obtaining an analog or digital readout from the operator's control panel, or by reading the value of the input to the control actuator on the plant. In both cases it may be necessary to convert and scale the reading obtained to conform to the units being used for m inside the computer. At the point at which change-over is made, the value of m is stored in a variable mc. Two methods of transfer can be used:

1. M is not preset and change-over is made when the error (e) is zero; then $M = mc$. Or
2. M is preset to a value appropriate for the nominal level and change-over is made when the error is not zero. The integral action term needs to be set to an initial value

$$s = mc - K_p ec - M$$

where ec = error value at change-over.

4.4.3 Method 3 – Velocity Algorithm

The PID algorithm given by equation 4.5 is often referred to as the positional algorithm because it is used to calculate the absolute value of the actuator position.

An alternative form of the PID algorithm, the so-called velocity algorithm, is widely used to provide automatic bumpless transfer. The velocity algorithm gives the change in the value of the manipulated variable at each sample time rather than the absolute value of the variable. In continuous terms it can be obtained by differentiating equation 4.1, with respect to time, to give

$$\frac{dm(t)}{dt} = K_p\left(\frac{de(t)}{dt} + \frac{1}{T_i}\,e(t) + T_d\,\frac{d^2e(t)}{dt^2}\right) \tag{4.7}$$

The difference equation can be obtained either by applying backward differences to equation 4.7 or by finding $m(n) - m(n-1)$ using equation 4.4 which gives

$$\Delta m = m(n) - m(n-1)$$

$$= K_p\left([e(n) - e(n-1)] + \frac{\Delta t}{T_i}\,e(n) + \frac{T_d}{\Delta t}\,[e(n) - 2e(n-1) + e(n-2)]\right) \tag{4.8}$$

Rearranging equation 4.8 gives

$$\Delta m(n) = K_p\left[\left(1 + \frac{T_s}{T_i} + \frac{T_d}{T_s}\right)e(n) - \left(1 + 2\,\frac{T_d}{T_s}\right)e(n-1) + \frac{T_d}{T_s}\,e(n-2)\right] \tag{4.9}$$

Writing

$$K_1 = K_p(1 + T_s/T_i + T_d/T_s)$$
$$K_2 = -(1 + 2T_d/T_s)$$
$$K_3 = T_d/T_s$$

equation 4.9 becomes

$$\Delta m(n) = K_1 e(n) + K_2 e(n-1) + K_3 e(n-2) \tag{4.10}$$

which is easily programmed.

Because it outputs only a change in the controller position this algorithm automatically provides bumpless transfer. However, if a large standing error exists on change-over the response of the controller may be slow, particularly if the integral action time is long, that is with a large value of T_i.

4.4.4 Comparison of Position and Velocity Algorithms

Comparing the position algorithm (equation 4.5) and the velocity algorithm (equation 4.10) shows that the latter is simpler to program and is inherently safer in that large changes in demanded actuator position are unlikely to occur. Frequently the maximum value which $m(n)$ can take is limited, thus ensuring that sudden large changes, for example in valve position or motor speed, are avoided. These sudden changes can occur if the measured signal is noisy or if the set point is changed. A method of dealing with noisy measurements is to use a fourth-order difference algorithm to approximate de/dt and this is explained in the next section. The disturbance caused by set point changes can be reduced by modifying the algorithm to use the set point r and the measured output c rather than the error signal e.

In the standard algorithm, based on the use of error e, the value of the set point appears in the derivative term and any change in value is differentiated; hence a sudden step change can cause a large disturbance. If in equation 4.9 we let $e(n) = r - c(n)$ (note that r, the set point, is assumed to be constant), then the equation becomes

$$\Delta m(n) = K_p \Big([c(n-1) - c(n)] + \frac{T_s}{T_i} (r - c(n))$$

$$+ \frac{T_d}{T_s} [2c(n-1) - c(n-2) - c(n)] \Big) \tag{4.11}$$

Changes in the set point are then accommodated by simply changing the value of the constant r.

The set point r appears only in the integral term and hence the controller must always include integral action. For security of operation a check must be included in the program to prevent the T_s/T_i parameter being set to zero or some very small value.

Disturbances can also be caused by on-line parameter changes since with a digital algorithm a large parameter change can be introduced in a single step. Consider the basic form of the PID algorithm

$$\begin{aligned} s(n) &= s(n-1) + e(n) \\ m(n) &= K_p e(n) + K_i s(n) + K_d [e(n) - e(n-1)] \end{aligned} \tag{4.12}$$

If we make a change to the integral action time T_i clearly, unless $s(n)$ is zero, there will be a step change in output since $K_i = K_c(T_s/T_i)$. We can avoid this either by limiting the rate at which a parameter changes or by altering the algorithm as shown below:

$$\begin{aligned} s(n) &= s(n-1) + e(n)(T_s/T_i) \\ m(n) &= K_p e(n) + K_p s(n) + K_d [e(n) - e(n-1)] \end{aligned} \tag{4.13}$$

By writing the algorithm in this form the effects of making a change to T_i are much reduced. Changes to K_p and K_d cause much smaller disturbances unless the error is large and/or the rate of change of error is large.

4.5 SATURATION AND INTEGRAL ACTION WIND-UP

In practical applications the value of the manipulated variable $m(n)$ is limited by physical constraints. A valve cannot be more than fully opened, or more than fully

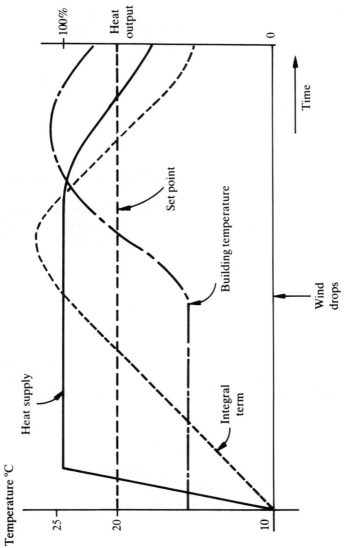

Figure 4.3   Illustration of integral wind-up action.

closed; a thyristor-controlled electric heater can supply only a given maximum amount of heat and cannot supply negative heat. If the value of the manipulated variable exceeds the maximum output of the control actuator effective feedback control is lost: good plant design should ensure that this only occurs in unusual conditions.

A simple example of what can happen is provided by considering a building heating system. The capacity of such a system is usually chosen to cope with an average winter: if extreme low temperature and high winds coincide, the system, even when operating at maximum capacity, will not be able to maintain the desired temperature. Under these conditions a large standing error in temperature will exist. If a PI controller is used, then because there is a standing error, the integral term will continue to grow; that is, the value of $s(n)$ in equation 4.5 will be increased at each sample time. Consequently the value of the manipulated variable will increase and the demanded heat output will continually increase, but since this will already be a maximum, the demand cannot be met. The changes are shown in Figure 4.3. If the wind drops and the outside temperature increases, then the building temperature will increase and eventually reach the desired temperature. The value of $s(n)$, the integral term, will, however, still be large, since it will not be reduced until the building temperature exceeds the demanded temperature. As a consequence the integral term will continue to keep the demanded heat output at its maximum value even though the building temperature is now higher than desired.

The effect is called *integral wind-up* or *integral saturation* and results in the controller having a poor response when it comes out of a constrained condition. Many techniques have been developed for dealing with the problems of integral wind-up and the main ones are:

- fixed limits on integral term;
- stop summation on saturation;
- integral subtraction;
- use of velocity algorithm; and
- analytical method.

4.5.1 Fixed Limits

A maximum and minimum value for the integral summation is fixed and if the term exceeds this value it is reset to the maximum or minimum as appropriate. The value often chosen is the maximum/minimum value of the manipulated variable; thus if $s_{max} = m_{max}$ and $s_{min} = m_{min}$ then the coding in the PROCEDURE PIDControl in Example 4.1 could be modified as shown in Example 4.6.

EXAMPLE 4.6

```
PROCEDURE PIDControl (en: REAL; VAR mn: REAL);
    sn := sn+en; (* integral summation *)
    IF sn > smax THEN
        sn := smax
    ELSE  IF sn < smin THEN
            sn := smin;
        END (* ELSE IF *)
    END (* IF *)
    mn := Kp*en+Ki*sn+Kd*(en-enOld);
    enOld := en;
END PIDControl;
```

4.5.2 Stop Summation

In this method the value of the integrator sum is frozen when the control actuator saturates and the integrator value remains constant while the actuator is in saturation. The scheme can be implemented either by freezing the summation term when the manipulated variable falls outside the range m_{min} to m_{max} or by the use of a digital input signal from the actuator which indicates that it is at a limit.

Both of the above methods stop the integral term building up to large values during saturation but both have the disadvantage that the value of the integral term, when the system emerges from saturation, does not relate to the dynamics of the plant under full power. Consequently the controller offset (provided by the integral term) lags behind the offsets required by the plant and load as the set point is reached.

The stop summation technique gives a better response if the integral term is unfrozen once the sign of the error changes. The sign of the error will change before the actuator comes out of saturation. Assume that a positive error drives the actuator towards its upper limit; then the behaviour required is:

Actuator	Error	Integral summation
upper limit	+	stopped
upper limit	−	active
normal	+	active
normal	−	active
lower limit	+	active
lower limit	−	stopped

The procedure becomes

```
PROCEDURE PIDControl (en: REAL; VAR mn: REAL);
   StopSummation:=((mn > mnmax) AND (en > 0.0))
      OR ((mn < mnmin) AND (en < 0.0));
   IF NOT (StopSummation) THEN sn:=sn+en;
   END (* IF *)
   mn:=Kp*en+Ki*sn+Kd*(en-enOld);
   enOld:=en;
END PIDControl;
```

4.5.3 Integral Subtraction

The idea behind this method is that the integral value is decreased by an amount proportional to the difference between the calculated value of the manipulated variable and the maximum value allowable. The integral summation expression

$$s(n) = s(n - 1) + e(n)$$

is replaced by

$$s(n) = s(n - 1) - K[m(n) - m_{max}] + e(n)$$

The integral sum is thus decreased by the excess actuation and increased by the error. The rate of decrease is dependent on the choice of the parameter K; if it is not properly chosen then a continual saturation/desaturation oscillation can occur.

The method can be modified to stop the addition of the error part during saturation if a logic signal from the actuator indicating saturation or no saturation is available. In this case the value of the integral sum begins to decrease as soon as the actuator enters saturation and continues to decrease until it comes out of saturation, at which point integral summation begins again. The benefit of this method is that the system comes out of saturation as quickly as possible; there is, however, no attempt to match the integral term to the requirements of the plant and the value of K must be chosen by experience rather than by reference to the plant characteristics.

4.5.4 Velocity Algorithm

It is often stated that integral wind-up can be avoided by the use of the velocity algorithm since the integral action is obtained by a summation of the increments in the output device, either at the actuator or at a device connected to the actuator, and it is this device which is subject to limiting. There is therefore an automatic integral limit which prevents a build-up of error. However, as soon as the error changes sign the actuator will come off its limit and hence at desaturation the integral term is lost.

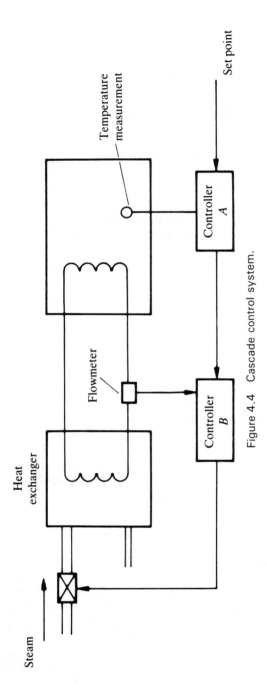

Figure 4.4   Cascade control system.

When controllers are cascaded it must not be assumed that the use of the velocity algorithm will prevent integral wind-up. In the system shown in Figure 4.4, controllers A and B are assumed to use the velocity algorithm and both are assumed to employ PI control. Controller A is used to adjust the steam flow to a heat exchanger in order to maintain a particular water temperature in the hot-water supply used to heat a room. The demanded water temperature is set by controller B. Suppose on a cold day controller B demands a water temperature of $60°C$ but the best the heat exchanger can do is to provide water at $55°C$. The effect will be that the room temperature will remain too low and the integral sum will begin to grow and hence the set point of the controller will continually be increased until it reaches its maximum limit. If the steam valve is fully open there is no action controller A can take. If now the external temperature increases there will be a delay, during which the room temperature might rise well above the desired temperature, before the set point of controller A is reduced to correct the overshoot. In order to avoid this the master controller must know when the subsidiary loop is at a limit.

4.5.5 Analytical Approach

This method has been developed by Thomas, Sandoz and Thomson (1983) and it uses knowledge of the plant to set the integral sum term approximately to the correct value at the point of desaturation such that the normal linear response from steady state is achieved. This is shown in Figure 4.5. For a system of the form shown in Figure 4.6, that is a first-order plant with a PI controller, the integral sum at the time τ when the system desaturates is given by

$$I(\tau) = \frac{1}{K_p K_c} [c(\tau) + sK_p L(s)] \tag{4.14}$$

If it is assumed that the load $L(s)$ is constant, or slowly varying, then

$$L(s) \approx L_0/s \tag{4.15}$$

and hence equation 4.14 becomes

$$I(\tau) = \frac{1}{K_p K_c} [c(\tau) + K_p L_0] \tag{4.16}$$

When the control actuator desaturates the integral value should be set to the value which it would have been holding in the steady state at $c(\tau)$ and then the remaining step $e(\tau)$ will follow as in the linear case. The value of $c(\tau)$ is not known since the time, τ, of desaturation is not known; however, if prior to the control calculation, the integral term $I(\tau)$ is set using equation 4.16 above with the $c(\tau)$ for $t < \tau$, then at the instant of desaturation $I(t) = I(\tau)$. If the actuator is not in saturation then the normal integration of error takes place. This scheme is shown in the program segment in Example 4.7 where the REAL variables load and KPKC are used for $K_p L_0$ and $K_p K_c$ respectively.

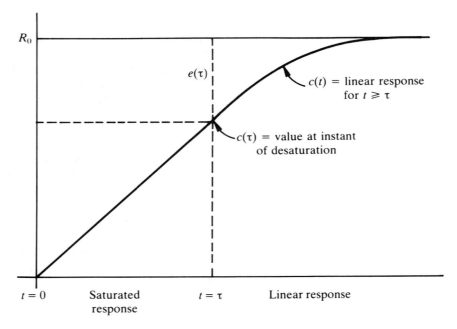

Figure 4.5   Saturated and linear regions of first-order responses.

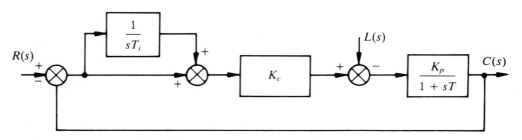

Figure 4.6   PI control of first-order plant.

EXAMPLE 4.7

```
LOOP
   cn :=ADC; (* ADC returns value of plant output
*)
   en :=cn-setpoint;
   IF mn > mnmax THEN
      sn :=(cn+load)/KPKC
   ELSE
      sn :=sn +en; (* integral summation *)
   END (* IF *)
   mn :=Kp*en+Ki*sn+Kd*(en-enOld);
   DAC(mn);
   enOld :=en;
END (* LOOP *)
```

4.6 TUNING

The first papers on methods of tuning analog three-term controllers were those of J.G. Ziegler and N.B. Nichols which were published in 1942 and 1943. Since then there has been extensive work on extending and developing tuning methods. In recent years direct tuning by plant engineers has largely been replaced by autotuning controllers. The basic tuning methods are covered in most texts on control engineering: for adaptation to digital implementations see for example Leigh (1992), Astrom and Wittenmark (1984), Takahashi *et al.* (1970). All of the methods are based on simple plant measurements and on the assumption that the plant can be modelled by the transfer function

$$G(s) = \frac{ke^{-Ls}}{(1 + sT_p)} \tag{4.17}$$

The value of the parameters in the above plant model can be obtained from a simple open-loop step response as shown in Figure 4.7.

For the analog system the tuning problem is, given R, L or T_r, to choose K_p, K_i, K_d so as to minimise (or maximise) some performance criteria (for example, the integral of absolute error, IAE).

The standard Ziegler–Nichols rules are:

	K_p	T_i	T_d
P	$1/RL$		
PI	$0.9/RL$	$3L$	
PID	$1.2/RL$	$2L$	$0.5L$

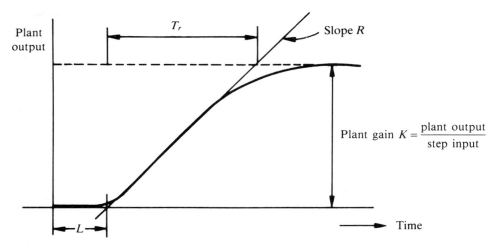

Figure 4.7   Process reaction curve.

The alternative method is to close the feedback loop and use a proportional controller. The gain of the controller is turned up until the system just begins to oscillate with a steady amplitude. The tuning parameters are then calculated in terms of the gain K_m and the period of the oscillation T_p. The settings are

	K_p	T_i	T_d
P	$0.5\,K_m$		
PI	$0.45\,K_m$	$T_p/1.2$	
PID	$0.6\,K_m$	$T_p/2$	$T_p/8$

The digital implementation of the three-term controller involves additional variables: q the unit of quantisation and T_s the sampling time; hence the performance function becomes

$$J = f(q,\ T_s,\ K_p,\ K_i,\ K_d,\ L,\ T_p)$$

The size of the quantisation is not normally a problem in industrial process control since the control computation can be done using either real numbers or fixed point arithmetic. It becomes greater in, for example, aircraft controls and weapons systems where time constants are shorter and, in order to obtain the necessary computational speed, limited wordlength arithmetic has to be used. The problems are discussed extensively by Katz (1981) and Leigh (1992); J.B. Knowles has devised a design method which takes into account quantisation and sample rates (see Bennett and Linkens, 1984).

Smith (1972) has suggested that when using tuning tables based on L values

$(L + T_s/2)$ should be used rather than L, in order to take into account the delay caused by sampling. He also notes that as the value of $T_s/2$ approaches the dead time, L, the performance deteriorates.

If the PID algorithm is expressed in the form given in equation 4.10 but with

$$K_i = K_c T_s/T_i \quad \text{and} \quad K_d = K_c T_d/T_s$$

then Takahashi *et al.* (1970) give the following rule for tuning the parameters:

$$K_p = \frac{1.2}{R(L + T_s)} - \frac{1}{2K_i}$$

$$K_i = \frac{0.6T_s}{R(L + T_s/2)^2}$$

$$K_d = \frac{0.5}{RT_s} \quad \text{to} \quad \frac{0.6}{RT_s}$$

When $T_s = 0$ the above rule converges to the standard Ziegler–Nichols result

$$K_p = 1.2/RL, \quad 1/T_i = 0.5/L, \quad T_d = 0.5L$$

It should be noted that the quality of the control deteriorates when T_s increases relative to the dead time, L, and that the tuning rule fails when L/T_s is very small.

4.7 CHOICE OF SAMPLING INTERVAL

Intuitively we might assume that if we decrease the sampling interval then as T_s tends to zero the system will asymptotically converge towards the performance of the equivalent analog system. However, this is not the case since the digital computation has a finite resolution. Thus as the sample interval decreases the change between successive values becomes less than the resolution of the system and hence information is lost. The interaction between sampling interval and arithmetic resolution is complex and a detailed analysis can be found in Katz (1981) and in Williamson (1991); for a shorter summary of the effects see Leigh (1992).

Conversely increasing the sampling interval can result in destabilising the feedback loop, in loss of information (the sampling effect), and in a loss of accuracy of the control algorithm. Various empirical rules have been given for choosing a sample interval, some of which are listed below:

1. Dominant plant time constant: If the dominant plant time constant is T_p then choose T_s such that

   $$T_s < T_p/10$$

   This is a widely used rule but as Leigh points out it is dangerous under conditions where a high closed-loop performance is forced on a system with a low open-loop performance.

2. Assumption of Ziegler–Nichols plant model: If the plant model is as given in equation 4.17 then the following suggestions are given in the literature:
    (a) general $0.05 < T_s/L < 0.3$,
    (b) large dead time L set $T_i/T_s = 2$,
    (c) small dead time L set $T_i/T_s = 6$,
    (d) choose sampling interval such that $5 < T_d/T_s < 10$.

3. Closed-loop performance requirements: If the closed-loop system is required to have either a settling time of T_{ss} or a natural frequency of ω_n then choose T_s such that
    (a) $T_s < T_{ss}/10$,
    (b) $\omega_s > 10\omega_n$ where $\omega_s = 2\pi/T_s$.

4.8 PLANT INPUT AND OUTPUT

So far we have assumed that the input signal obtained from the plant and used to calculate the error signal e is a clean signal that is in the correct form required for the control algorithm. In practice this is not the case. The signal may be a noisy signal, it may require the application of a calibration factor and it may need scaling. Similarly we have assumed that the output from the control algorithm can be sent directly to the actuator and again this is not generally the case: the value may need scaling and converting to a different form of output, for example in process control applications many actuators require a pulse of varying duration to operate them rather than a signal of varying amplitude.

A way of dealing with the practical details of input and output from the plant is to divide the software into segments as shown in Figure 4.8. The plant input segment is visualised as providing a plant image that is a filtered, scaled and calibrated version of the plant input signal. The control algorithm uses this and calculates the actuator command signal which it places in the output image. The plant output module converts this into a form required to drive the real actuator on the plant.

4.8.1 Noise

In an analog control system, a small amount of high-frequency noise on the measured signal usually does not cause any problems since the dynamic components in the system act as low-pass filters and attenuate the noise. If sampling is involved, high-frequency noise may produce a low-frequency disturbance due to folding or aliasing (see, for example, Leigh, 1992; Kuo, 1980). The low-frequency disturbance has the same sample amplitude as the original noise and its frequency is the

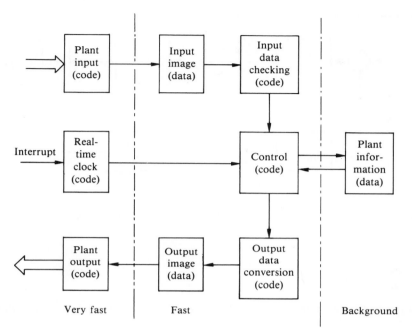

Figure 4.8 Basic division control software.

difference between the original noise frequency and a multiple of the sampling interval. To reduce this effect the measurement signal must be filtered using an analog filter before it is sampled.

For many industrial applications a simple first-order filter

$$G_f(s) = \frac{1}{1 + T_f s}$$

is satisfactory. The choice of $T_f = T_s/2$ (where T_s is the sampling time for the controller) will reduce the aliasing effect by about 90% for white noise. For example, if the bandwidth of the system being controlled is 0 to 2 Hz then from the Nyquist sampling theory we need to sample at a minimum frequency of 4 Hz if we are to be able to reconstruct the signal. In practice we would choose to sample at ten times the bandwidth and hence need to sample at 20 Hz which corresponds to a sampling interval T_s of 0.05 s. To remove high-frequency noise prior to sampling we need to use a filter with a time constant of $T_f = T_s/2$, that is 0.025 s.

Where T_f is small, analog filters can easily be constructed from passive elements. If T_f is greater than a few seconds then combined digital and analog filtering is used. The arrangement is illustrated in Figure 4.9 and the analog filter is used to remove the high-frequency noise and hence reduce the required sampling rate.

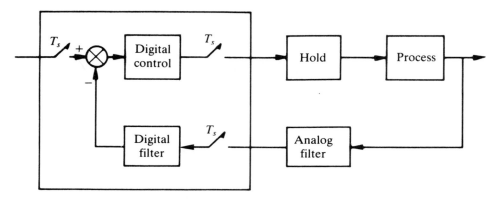

Figure 4.9   General control system with filtering.

For a first-order lag filter

$$T_f \frac{dx}{dt} + x = u$$

with input u, output x and filter sampling interval T_{fs} the numerical approximation is

$$x(n + 1) = [1 - \exp(-T_{fs}/T_f)]x(n) + \exp(-T_{fs}/T_f)u(n)$$

Introducing $a = \exp(-T_{fs}/T_f)$ and applying a backwards time shift gives

$$x(n) = (1 - a)x(n - 1) + au(n - 1)$$

It can be seen that if $a = 1$ there is no smoothing and if $a = 0$ the current measurement (input) is not used.

The sample interval T_{fs} for the input to the digital filter has to be made smaller than the time constant T_f of the filter and hence several samples of the measurement signal are taken for each output of the controller. The time constants of analog pre-filters are usually small and do not significantly degrade the overall performance. Excessively large values of T_f should be avoided.

4.8.2 Actuator Control

In designing and implementing real controllers the characteristics of the actuators used to control the process are important. The normal practice is to take the value of the manipulated variable ($m(n)$ or $\Delta m(n)$) as the input to an actuator control module as shown in Figure 4.10. In some instances there will be local feedback of the actuator position as shown by the dotted line in Figure 4.10. The advantage of this approach is that the basic controller and the detailed control of the actuator are separated. If the actuator is modified or changed then only the actuator controller module is affected.

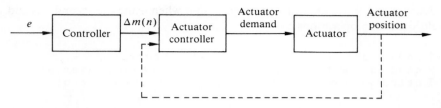

Figure 4.10   Actuator control.

4.8.3 Computational Delay

Analog-to-digital conversions, all computations in the computer – digital filtering, calculation of the control value – and digital-to-analog conversion all take a finite amount of time. There is thus a time difference – the computational delay – between sampling the plant output and changing the value of the actuator. The value of this delay depends on the computations carried out, on the processor and input/output speed of the computer and on the order in which certain operations are done. Intuitively most people will order the digital controller calculations as follows:

```
1 read plant input data
2 calculate control output
3 send output to actuator
```

If the kth sample of the plant input is measured at some time $t(k)$ then the kth sample of the actuator output is sent out at some time $t(k) + h$ where h is the computational delay time (note $h < T_s$). In other words a plant measurement $ct(k)$ gives rise to an output $u[t(k) + h]$. In general h will be variable as it will depend partially on the data values. If the calculations are ordered in this way then h is made as small as possible by performing as few operations as possible between getting the plant input and producing the output.

An alternative way of ordering the computations is as follows:

```
1 send to plant the control value for sample interval (k-1)
2 read plant input data for sample interval k
3 calculate control output for kth sample interval.
```

This approach produces a constant computational delay equal to the sample interval T_s. The advantages are that only statements 1 and 2 above need to be tightly synchronised to the real-time clock; statement 3, the control calculation, can be done at any time providing that the calculation is completed within the sample interval T_s and that the computational delay is constant. Whichever approach is used it is normal to take the computational delay into account by including the delay h or T_s as part of the plant model.

Astrom and Wittenmark (1984) suggest that when using the second method you should order the operations as

```
1 read plant input data for sample interval k
2 send to plant the control value for sample interval (k-1)
3 calculate control output for kth sample interval
```

so as to avoid the risk of electrical cross-coupling. If you do this you should split the processing of the input data so that only the actual read operations are carried out before you do the output to the actuator; any linearisation or other operations on the input data should be done after sending the previous control value to the actuator.

4.9 IMPROVED FORMS OF ALGORITHM FOR INTEGRAL AND DERIVATIVE CALCULATION

The positional and velocity algorithms considered so far use first- and second-order differences respectively to compute the derivative terms. Since differentiation or its numerical equivalent is a 'roughening' process – it accentuates noise and data errors – some form of smoothing, that is filtering, is required. Smoothing can be obtained by using a difference technique which averages the value over several samples. One such technique which has been used is the four-point central difference method (Bibbero, 1977; Takahashi *et al.*, 1970). This gives

$$\frac{de}{dt} = \frac{\Delta e}{T_s} = \frac{1}{6T_s} \left[e(n) + 3e(n-1) - 3e(n-2) - e(n-3) \right] \tag{4.18}$$

Substituting for $T_d/T_s[e(n) - e(n-1)]$ in equation 4.4 gives

$$m(n) = K_p \left(\frac{T_d}{6T_s} \left[e(n) + 3e(n-1) - 3e(n-2) - e(n-3) \right] \right.$$

$$\left. + e(n) + \frac{T_s}{T_i} \sum_{k=1}^{n} e(k) \right) \tag{4.19}$$

The position algorithm using the above technique for the derivative term is thus

$$s(k) = s(k-1) + e(k)$$
$$m(k) = p_1 e(k) + p_2 e(k-1) - p_2 e(k-2) - p_3 e(k-3) + p_4 s(k) \tag{4.20}$$

where

$$p_1 = 1 + K_p T_d/6T_s$$
$$p_2 = 1 + K_p T_d/2T_s$$
$$p_3 = 1 + K_p T_d/6T_s$$
$$p_4 = 1 + K_p T_d/T_i$$

Improvements can also be made to the accuracy of the integration calculation by using the trapezoidal rule instead of the rectangular rule. If this is done, equation 4.4 can be written as

$$m(n) = K_p' \left[e(n) + \left(\frac{T_s}{T_i'} \sum_{k=1}^{n} \frac{e(k) + e(k-1)}{2} \right) + \frac{T_d'}{T_s} [e(n) - e(n-1)] \right] \quad (4.21)$$

and in velocity algorithm form

$$\Delta m(n) = K_p' \left[\left(1 + \frac{T_s}{2T_i'} + \frac{T_d'}{T_s} \right) e(n) + \left(\frac{T_s}{2T_i'} - \frac{2T_d'}{T_s} - 1 \right) e(n-1) \right.$$

$$\left. + \frac{T_d'}{T_s} e(n-2) \right] \quad (4.22)$$

If this is compared with the velocity algorithm using rectangular integration we find that

$$K_p = K_p'(1 - T_s/2T_i')$$
$$T_i = T_i' - T_s/2$$
$$T_d = 2T_d' T_i'/(2T_i' - T_s)$$

Hence if the appropriate values for the coefficients are used the form of the two algorithms is identical.

4.10 IMPLEMENTATION OF CONTROLLER DESIGNS BASED ON PLANT MODELS

In the early days of computer control one of the justifications for using digital control instead of analog control was the ease with which more complex control algorithms could be introduced. In particular control algorithms could be designed on the basis of accurate plant models instead of relying on the Ziegler–Nichols assumption that the plant could be modelled as a first-order lag and a time delay. The performance of the more complex algorithms does not always, in practice, match that suggested by the theory. Leigh (1992, p. 185) summarises the position as follows:

1.   PID algorithms perform surprisingly well in practice. They are robust and difficult to improve on significantly on a day-to-day basis unless very considerable effort is expended.
2.   Feedforward and cascade algorithms perform well in those situations for which they were designed.
3.   More complex algorithms tend to lose their supposed advantages once process parameters drift or noise begins to affect measurements.
4.   In general, the use of a long sampling interval greatly increases the sensitivity of a control loop to process parameter drift. Complex algorithms

tend to use longer sampling intervals and hence are prone to parameter drift and noisy measurement problems.

The use of detailed plant models allows a wide variety of methods to be used in the design of the controller (see, for example, Astrom and Wittenmark, 1984; Franklin and Powell, 1980; Leigh, 1992; Katz, 1981; Kuo, 1980). Use of such methods gives rise to two types of representation for the controller:

- state-space representation of the difference equations; and
- transfer function in z^{-1}.

If the controller is in difference equation form it may be programmed directly; if it is given as a transfer function it has to be realised, that is converted either into an electronic (or other hardware) circuit or into a computer algorithm. There are four techniques for realisation:

- direct method 1;
- direct method 2;
- cascade; and
- parallel.

In terms of computer algorithms it can be shown that for a given quantisation limit (that is, wordlength) the cascade and parallel methods give algorithms in which the numerical errors are smaller than the errors in the algorithms produced by the two direct methods; for details see for example Katz (1981), Leigh (1992). In the next sections we deal briefly with the representation of the PID algorithm using the z-transform notation and with realisation approaches. If you do not have any control systems background you may wish to omit these sections and turn directly to section 4.11. If you want to try reading these sections but the z-transform is unfamiliar to you, you can for the purposes below treat z^{-1} as a delay operator. This means that writing ez^{-1} implies the value of e at the previous sampling interval. Hence

$$M(z) = (a + bz^{-1} + cz^{-2})E(z)$$

implies that, in the time domain, at the nth sample interval

$$m(n) = ae(n) + be(n - 1) + ce(n - 2)$$

or

$$m(nT) = ae(nT) + be[(n - 1)T] + ce[(n - 2)T]$$

where T is the sample interval.

4.10.1 The PID Controller in Z-transform Form

The PID controller can be expressed as a transfer function in z. Consider equation

4.4 and let $d = T_d/T_s$ and $g = T_s/T_i$. Then

$$m(n) = K_p\left(e(n) + g \sum_{k=1}^{n} e(k) + d[e(n) - e(n-1)]\right) \tag{4.23}$$

Since $D(z) = M(z)/E(z)$, $D(z)$ can be found by taking the z-transform of the right-hand side of equation 4.23 term by term to give

$$D(z) = K_p\left(1 + \frac{gz}{z-1} + d - dz^{-1}\right) \tag{4.24}$$

Equation 4.24 represents the parallel realisation of the PID controller and is shown in block form in Figure 4.11. Rewriting $K_p gz/(z-1)$ as $K_p gz/(1-z^{-1})$ the algorithm becomes

$$x_1(i) = K_p(1+d)e(i)$$
$$x_2(i) = K_p g e(i) + x_2(i-1)$$
$$x_3(i) = -K_p d e(i-1)$$
$$m(i) = x_1(i) + x_2(i) + x_3(i)$$

Substituting for d and g gives

$$x_1(i) = K_p(1 + T_d/T_s)e(i)$$
$$x_2(i) = K_p(T_s/T_i)e(i) + x_2(i-1) \tag{4.25}$$
$$x_3(i) = -K_p(T_d/T_s)e(i-1)$$

and it can be seen that equation 4.25 is the integral summation term which in equation 4.5 was expressed in the form $s(n) = s(n) + e(n)$. The algorithm from equation 4.25 is

$$s(n) = K_1 e(n) + s(n-1)$$
$$m(n) = K_2 e(n) + K_3 e(n-1) + s(n)$$

where $K_1 = K_p T_s/T_i$, $K_2 = K_p(1 + T_d/T_s)$ and $K_3 = -K_p T_d/T_s$.

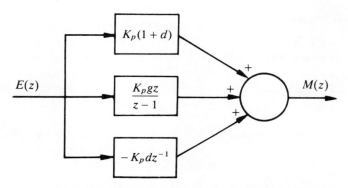

Figure 4.11 *z*-transform function form of PID controller.

Alternatively equation 4.24 can be rearranged to give

$$D(z) = K_p \left(\frac{(1 + g + d)z^2 - (1 + 2d)z - d}{z(z - 1)} \right) \tag{4.26}$$

Dividing the numerator and denominator by z^2 gives

$$D(z) = \frac{a_0 + a_1 z^{-1} + a_2 z^{-2}}{1 + b_1 z^{-1}}$$

where

$a_0 = K_p(1 + g + d)$
$a_1 = -K_p(1 + 2d)$
$a_2 = K_p d$
$b_1 = -1$

Direct implementation gives

$$m(i) = a_0 e(i) + a_1 e(i - 1) + a_2 e(i - 2) - b_1 m(i - 1) \tag{4.27}$$

and substituting for a_0, a_1, a_2, b_1 gives

$$m(i) = K_p \left[\left(1 + \frac{T_s}{T_i} + \frac{T_d}{T_s} \right) e(i) - \left(1 + 2 \frac{T_d}{T_s} \right) e(i - 1) \right.$$

$$\left. + \frac{T_d}{T_s} e(i - 2) + m(i - 1) \right] \tag{4.28}$$

which can easily be rearranged to give the velocity algorithm of equation 4.10.

4.10.2 Direct Method 1

The transfer function can be expressed as the ratio of two polynomials in z^{-1}:

$$\frac{M(z)}{E(z)} = D(z) = \frac{\sum_{j=0}^{n} a_j z^{-j}}{1 + \sum_{j=1}^{n} b_j z^{-j}} \tag{4.29}$$

The transfer function in equation 4.29 is converted directly into the difference equation

$$m_i = \sum_{j=0}^{n} a_j e_{i-j} - \sum_{j=1}^{n} b_j m_{i-j} \tag{4.30}$$

EXAMPLE 4.8

Consider a system with the transfer function

$$\frac{M(z)}{E(z)} = D(z) = \frac{3 + 3.6z^{-1} + 0.6z^{-2}}{1 + 0.1z^{-1} - 0.2z^{-2}} \tag{4.31}$$

Then by direct method 1 the computer algorithm is simply

$$m_i = 3e_i + 3.6e_{i-1} + 0.6e_{i-2} - 0.1m_{i-1} + 0.2m_{i-2}$$

4.10.3 Direct Method 2

Assuming as before that the transfer function can be expressed as in equation 4.29, then in direct method 2 the difference equation is formulated by introducing an auxiliary variable $P(z)$ such that

$$\frac{M(z)}{P(z)} = \sum_{j=1}^{n} a_j z^{-j} \tag{4.32}$$

and

$$\frac{P(z)}{E(z)} = \frac{1}{1 + \sum_{j=1}^{n} b_j z^{-j}} \tag{4.33}$$

From equations 4.32 and 4.33 two different equations are obtained:

$$m_i = \sum_{j=0}^{n} a_j p_{i-j} \tag{4.34}$$

and

$$p_i = e_i - \sum_{j=0}^{n} b_j p_{i-j} \tag{4.35}$$

Using the example above the following algorithm is obtained:

$$p_i = e_i - 0.1p_{i-1} + 0.2p_{i-2}$$
$$m_i = 3p_i + 3.6p_{i-1} + 0.6p_{i-2}$$

4.10.4 Cascade Realisation

If the transfer function is expressed as the product of simple block elements of first and second order as shown in Figure 4.12, then each element can be converted to a difference equation using direct method 1 and the overall algorithm is the set of

Figure 4.12   Cascade realisation.

difference equations. Equation 4.31 when expressed in this form becomes

$$\frac{M(z)}{E(z)} = D(z) = \frac{3(1 + z^{-1})(1 + 0.2z^{-1})}{(1 + 0.5z^{-1})(1 - 0.4z^{-1})} \tag{4.36}$$

Hence

$$D_1 = 3$$
$$D_2 = (1 + z^{-1})$$
$$D_3 = (1 + 0.2z^{-1})$$
$$D_4 = 1/(1 + 0.5z^{-1})$$
$$D_5 = 1/(1 - 0.4z^{-1})$$

and letting x_1, x_2, x_3, x_4, and x_5 be the outputs of blocks D_1, D_2, D_3, D_4 and D_5 respectively, then

$$x_1(i) = 3e(i)$$
$$x_2(i) = x_1(i) + x_1(i - 1)$$
$$x_3(i) = x_2(i) + 0.2x_2(i - 1)$$
$$x_4(i) = x_3(i) - 0.5x_4(i - 1)$$
$$x_5(i) = x_4(i) + 0.4x_5(i - 1)$$

4.10.5 Parallel Realisation

If the transfer function is expressed in fractional form or is expanded into partial fractions then it can be represented as shown in Figure 4.13. In this case each of the transfer functions D_1, D_2, and D_3 is expressed in difference equation form using

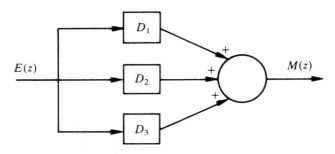

Figure 4.13   Parallel realisation.

direct method 1 and the output is obtained by summing the outputs from each block.

The partial fraction expansion of equation 4.31 is

$$\frac{M(z)}{E(z)} = D(z) = -3 - \frac{1}{1 + 0.5z^{-1}} + \frac{7}{1 - 0.4z^{-1}} \qquad (4.37)$$

Hence $D_1 = -3$, $D_2 = -1/(1 + 0.5z^{-1})$, $D_3 = 7/(1 - 0.4z^{-1})$ and the algorithm is

$$x_1(i) = -3e(i)$$
$$x_2(i) = -e(i) - 0.5x_2(i-1)$$
$$x_3(i) = 7e(i) + 0.4x_3(i-1)$$
$$m(i) = x_1(i) + x_2(i) + x_3(i)$$

4.10.6 Discretisation of Continuous Controllers

In this section we have assumed that a controller designed using a plant model has been designed using discrete design techniques. However, controllers can be designed using continuous system design methods and then discretised for digital implementation.

The problem of discretisation is interesting and considerably more complex than might at first be thought. There are a number of methods which can be used and these are summarised below. However, none of them exactly preserve the characteristics of the continuous system (time response, frequency response, pole-zero locations). The main methods are:

1. impulse invariant transform (z-transform);
2. impulse invariant transform withhold;
3. mapping of differentials;
4. bilinear (or Tustin) transform;
5. bilinear transform with frequency prewarping; and
6. mapping of poles and zeros (matched z-transform).

If it is not possible to give any firm indication of a best method for all applications, however, in general the bilinear transform (4 or 5) and pole-zero matching (6) give the closest approximations to the continuous system. Extensive comparisons of the methods can be found in Astrom and Wittenmark (1984), Franklin and Powell (1980), Leigh (1992) and Katz (1981).

4.11 SUMMARY

We have dealt at length with the implementation of a simple control algorithm in order to illustrate some of the practical problems that have to be overcome in implementing digital controllers. These problems apply whatever form of digital control algorithm is used. An important point to remember is that the more precise and accurate the algorithm the more precisely must the timing requirements be met. The standard PID algorithm, because it is not exactly matched to a particular plant, remains well behaved when there are variations in the sampling interval, and when samples are missed, algorithms that have been designed using discrete control design techniques or discretised forms of continuous algorithms are not always so well behaved and may be sensitive to small variations in sampling interval. For detailed consideration of these issues see Franklin and Powell (1980), Katz (1981) and Williamson (1991).

The reasons for the algorithm's widespread use are that it requires very little knowledge of the plant dynamics and the methods of determining the controller parameters are well known and understood (Auslander and Sagues, 1981; Ahson, 1983; Cohen and Coon, 1953; Leigh, 1992; Smith, 1972; Stephanopoulos, 1985; Takahashi et al., 1970; Ziegler and Nichols, 1942). If knowledge of the plant model is available a wide variety of techniques can be used. The design can be carried out using continuous system design methods to find $G_c(s)$ followed by discretisation of $G_c(s)$ to give $G_c(z)$. A number of methods of mapping $G_c(s)$ to $G_c(z)$ are available (see section 4.10.6) but none give a controller $G_c(z)$ with exactly the same characteristics of $G_c(s)$. Care must be used since, for example, discretisation using first-order finite differences can easily result in changing a stable continuous-time system into an unstable discrete-time system. An alternative is to design $G_c(z)$ on the basis of a discrete-time model of the plant, $G_p(z)$, obtained either by discretisation of the continuous-time model $G_p(s)$ or by determining $G_p(z)$ directly.

In both cases we are concerned with how the controller is programmed, not with how it is designed (information on design techniques can be found in Franklin and Powell (1980), Iserman (1981), Katz (1981), Kuo (1980), Leigh (1992), and Smith (1972)).

Digital microcontrollers for PID control are now widely available from process instrument manufacturers; most of these units contain well-proven PID software and most incorporate some form of self-tuning or expert-system-based automatic tuning. Also available are software packages containing standard implementations of common control algorithms.

In most computer control applications the implementation of the direct digital control algorithms is a minor part of the system: the major efforts and complications arise in communicating with the plant, the operators, and other computers, and in providing a safe system that can handle alarm and fault conditions. In the rest of the book we turn our attention to the tools and techniques that help us in building complex systems in a safe and secure way.

EXERCISES

4.1   Many personal computers have interval timers, that is they have a counter which can be initialised and which is incremented (or decremented) at fixed intervals by an interrupt signal. Using the technique shown in Example 4.2 write a program to output the 'bell' character (07H) at a fixed interval (for example, 2 seconds). If you have access to a personal computer check the accuracy of the timing by using a stopwatch to time a number of rings.

4.2 A person's reaction time can be measured by sending, at random intervals, a character to a VDU screen and asking the subject to press a key when the character appears. If you have access to a personal computer or some other small computer write a program to carry out such an experiment.

4.3 Modify the code of Example 4.1 to incorporate the velocity subtraction method of preventing integral action wind-up. Assume that a logic signal is available to indicate when the control actuator is in saturation.

4.4 (a) Draw a flowchart to show how bumpless transfer (Method 2 – tracking of the manipulated variable) can be incorporated into the standard PID controller.
(b) Based on the flowchart write a program (in any language) for the system.

4.5 Write a program, in any language, to implement the velocity algorithm for the PID controller.

4.6 How would you incorporate (a) into the standard PID digital controller and (b) into the velocity form of the PID controller, the requirement that the manipulated variable should not change by more than 1% between two sample intervals?

4.7 Discuss the problems of testing the computer implementation of a digital control algorithm. Work out a test scheme which would minimise the amount of time required for test purposes on the actual plant. The scheme should show the various stages of testing and should be designed to eliminate coding errors and logic design errors prior to the connection of the controller to the plant.

4.8 The results of an open-loop response to a unit step input for a plant are:

Time (seconds)	Output
0.1	0.01
0.2	0.02
0.3	0.06
0.4	0.14
0.5	0.24
0.6	0.34
0.7	0.44
0.8	0.54
0.9	0.64
1.0	0.71
1.1	0.76
1.2	0.79
1.3	0.80

Find (a) the approximate plant model, (b) a suitable sampling interval for a digital PID controller and (c) estimates of the optimum controller settings for PI and PID control.

4.9   The analog system shown in Figure 4.14 can be discretised using the z-transform plus zero-order hold method. The resulting algorithm is

$$e(n) = r - c(n)$$
$$c(n) = (k/a^2) [Ae(n-1) + Be(n-2)] + [Cc(n-1) - Dc(n-2)]$$

where

$$A = T_s a - 1 + \exp(-aT_s)$$
$$B = 1 - \exp(-aT_s) - T_s a \exp(-aT_s)$$
$$C = 1 + \exp(-aT_s)$$
$$D = \exp(-aT_s)$$
$$T_s = \text{sampling interval}$$

Write a program which will enable you to calculate the change in output of the system, $c(n)$, with time. It is suggested that 50 values are calculated. The program should enable different values of k, a, T_s and r to be entered.

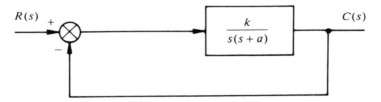

Figure 4.14   Control system for Exercise 4.9.

4.10   Using the program of Exercise 4.9, set $k = 2$, $a = 1$, $r = 1$ and investigate the response of the system for different values of T_s. It is suggested that $T_s = 0.02, 0.05, 0.1, 0.2, 0.5$. Compare the results (for example, in terms of maximum overshoot) with the exact solution for the continuous system (maximum overshoot = 30.5%).

5

Languages for Real-time Applications

Languages are an important implementation tool for all systems that include embedded computers. To understand fully methods for designing software for such systems one needs to have a sound understanding of the range of implementation languages available and of the facilities which they offer. The range of languages with features for real-time use continues to grow, as do the range and type of features offered. In this chapter we concentrate on the fundamental requirements of a good language for real-time applications and will illustrate these with examples drawn largely from Modula-2 and Ada. It is not the purpose of the chapter to compare languages or to offer a complete discussion of all the features of Modula-2 and Ada; such material can be found in the books listed in the Bibliography.

Choosing a language for writing software for embedded real-time systems is an important and serious matter. The choice has implications for the safety, reliability and costs of the final system. The major purpose of this chapter is to provide an understanding of some of the features of programming languages that can influence the choice. This chapter needs to be read in conjunction with the next chapter on Operating systems since some important real-time features − access to the underlying hardware, support for concurrency, intertask communication − are determined by the interface between the language and the operating system.

The aims and objectives of this chapter are to:

- List, explain and prioritise the major requirements for a real-time language.
- Describe the features that assist in the construction of safe, reliable software.
- List and explain techniques used to support the division of code into modules.
- List and explain the major facilities required to support concurrency.
- Assess a language for its suitability for a particular type of application.

5.1 INTRODUCTION

Producing safe real-time software places heavy demands on programming languages. Real-time software must be reliable: the failure of a real-time system can be expensive both in terms of lost production, or in some cases, in the loss of human life (for example, through the failure of an aircraft control system). Real-time systems are frequently large and complex, factors which make development and maintenance costly. Such systems have to respond to external events with a guaranteed response time; they also involve a wide range of interface devices, including non-standard devices. In many applications efficiency in the use of the computer hardware is vital in order to obtain the necessary speed of operation.

Early real-time systems were necessarily programmed using assembly level languages, largely because of the need for efficient use of the CPU, and access interface devices and support interrupts. Assembly coding is still widely used for small systems with very high computing speed requirements, or for small systems which will be used in large numbers. In the latter case the high cost of development is offset by the reduction in unit cost through having a small, efficient, program. Dissatisfaction with assemblers (and with high-level languages such as FORTRAN which began to be used as it was recognised that for many applications the advantages of high-level languages outweighed their disadvantages) led to the development of new languages for programming embedded computers. These languages included CORAL 66, RTL/2 and more recently C. The limitation of all of them is that they are designed essentially for producing sequential programs and hence rely on operating system support for concurrency.

The features that a programmer demands of a real-time language subsume those demanded of a general purpose language and so many of the features described below are also present (or desirable) in languages which do not support real-time operations. Barnes (1976) and Young (1982) divided the requirements that a user looked for in a programming language into six general areas. These are listed below in order of importance for real-time applications:

- Security.
- Readability.
- Flexibility.
- Simplicity.
- Portability.
- Efficiency.

In the following sections we will examine how the basic features of languages meet the requirements of the user as given above. The basic language features examined are:

- Variables and constants: declarations, initialisation.
- Data types – including structured types and pointers.
- Control structures and program layout and syntax.
- Scope and visibility rules.
- Modularity and compilation methods.
- Exception handling.

A language for real-time use must support the construction of programs that exhibit concurrency and this requires support for:

- Construction of modules (software components).
- Creation and management of tasks.
- Handling of interrupts and devices.
- Intertask communication.
- Mutual exclusion.
- Exception handling.

Modern real-time languages differ in how they provide the above facilities. There are two basic approaches: one is to provide a minimum set of language mechanisms with the ability to extend the set; the other is to provide a more extensive set of mechanisms from which the programmer can choose. The first approach gives a simple clean language but at the expense of standardisation as each user creates additional facilities (Modula-2 is an example of this approach), the second approach leads to a more complex, but standardised, language. Ada, CONIC and CUTLASS are examples of this latter approach.

5.1.1 Security

Security of a language is measured in terms of how effective the compiler and the run-time support system is in detecting programming errors automatically. Obviously there are some errors which cannot be detected by the compiler regardless of any features provided by the language: for example, errors in the logical design of the program. The chance of such errors occurring is reduced if the language encourages the programmer to write clear, well-structured, code.

Language features that assist in the detection of errors by the compiler include:

- good modularity support;
- enforced declaration of variables;
- good range of data types, including sub-range types;
- typing of variables; and
- unambiguous syntax.

It is not possible to test software exhaustively and yet a fundamental requirement of real-time systems is that they operate reliably. The intrinsic security of a language is therefore of major importance for the production of reliable programs. In real-time system development the compilation is often performed on a different computer than the one used in the actual system, whereas run-time testing has to be done on the actual hardware and, in the later stages, on the hardware connected to plant. Run-time testing is therefore expensive and can interfere with the hardware development program.

Economically it is important to detect errors at the compilation stage rather than at run-time since the earlier the error is detected the less it costs to correct it. Also checks done at compilation time have no run-time overheads. This is important as a program will be run many more times than it is compiled. Reliance on run-time checking frequently requires additional code to be inserted in the program (normally done by the compiler) and this leads to an increase in program size and a decrease in execution speed. In general strong typing gives good security at compilation time.

5.1.2 Readability

Readability is a measure of the ease with which the operation of a program can be understood without resort to supplementary documentation such as flowcharts or natural language descriptions. The emphasis is on ease of reading because a particular segment of code will be written only once but will be read many times. The benefits of good readability are:

- Reduction in documentation costs: the code itself provides the bulk of the documentation. This is particularly valuable in projects with a long life expectancy in which inevitably there will be a series of modifications. Obtaining up-to-date documentation and keeping documentation up to date can be very difficult and costly.
- Easy error detection: clear readable code makes errors, for example logical errors, easier to detect and hence increases reliability.
- Easy maintenance: it is frequently the case that when modifications to a program are required the person responsible for making the modifications was not involved in the original design − changes can only be made quickly and safely if the operation of the program is clear.

Factors which affect readability are manifold and to some extent readability depends on personal preference. The co-operation of the programmer is also required: it is possible to write unreadable programs in any language. The readability of a program is improved by the adoption of a clear layout which emphasises the structure, and by the careful choice of variable names. The syntax of the language and the layout of the code statements are the two most important factors that affect the readability.

The major disadvantage of using languages that support good readability and of writing readable code is that the source code is longer and it takes longer to write; for all except short-lived software this is a small price to pay for the added security and maintainability of the software.

5.1.3 Flexibility

A language must provide all the features necessary for the expression of all the operations required by the application without requiring the use of complicated constructions and tricks, or resort to assembly level code inserts. The flexibility of a language is a measure of this facility. It is particularly important in real-time systems since frequently non-standard I/O devices will have to be controlled. The achievement of high flexibility can conflict with achieving high security. The compromise that is reached in modern languages is to provide high flexibility and, through the *module* or *package* concept, a means by which the low-level (that is, insecure) operations can be hidden in a limited number of self-contained sections of the program.

5.1.4 Simplicity

In language design, as in other areas of design, the simple is to be preferred to the complex. Simplicity contributes to security. It reduces the cost of training, it reduces the probability of programming errors arising from misinterpretation of the language features, it reduces compiler size and it leads to more efficient object code.

Associated with simplicity is consistency: a good language should not impose arbitrary restrictions (or relaxations) on the use of any feature of the language.

5.1.5 Portability

Portability, while desirable as a means of speeding up development, reducing costs and increasing security, is difficult to achieve in practice. Surface portability has improved with the standardisation agreements on many languages. It is often possible to transfer a program in source code form from one computer to another and find that it will compile and run on the computer to which it has been transferred. There are, however, still problems when the wordlengths of the two machines differ and there may also be problems with the precision with which numbers are represented even on computers with the same wordlength.

Portability is more difficult for real-time systems as they often make use of specific features of the computer hardware and the operating system. A practical solution is to accept that a real-time system will not be directly portable, and to restrict the areas of non-portability to specific modules by restricting the use of low-level features to a restricted range of modules. Portability can be further enhanced by writing the application software to run on a virtual machine, rather than for a specific operating system. A change of computer and operating system then requires the provision of new support software to create a virtual machine on the new system.

5.1.6 Efficiency

In real-time systems, which must provide a guaranteed performance and meet specific time constraints, efficiency is obviously important. In the early computer control systems great emphasis was placed on the efficiency of the coding – both in terms of the size of the object code and in the speed of operation – as computers were both expensive and, by today's standards, very slow. As a consequence programming was carried out using assembly languages and frequently 'tricks' were used to keep the code small and fast. The requirement for generating efficient object code was carried over into the designs of the early real-time languages and in these languages the emphasis was on efficiency rather than security and readability.

The falling costs of hardware and the increase in the computational speed of computers have changed the emphasis. Also in a large number of real-time applications the concept of an efficient language has changed to include considerations of the security and the costs of writing and maintaining the program; speed and compactness of the object code have become, for the majority of applications, of secondary importance. There are, however, still application areas where compactness and speed do matter: in the consumer market where production runs may be 100 000 per year the ability to use a slower, cheaper CPU or to keep down the amount of memory used can make a significant difference to the viability of the product. Other areas in which speed matters are in the control of electromechanical systems, aircraft controls and in the general area of signal processing, for example speech recognition. The efficiency of a language depends much more on the compiler and the run-time support than on the actual language design.

5.2 SYNTAX LAYOUT AND READABILITY

The language syntax and its layout rules have a major impact on the readability of code written in the language. Consider the program fragment given below:

```
BEGIN
NST := TICKS() + ST;
T:=TICKS()+ST;
LOOP
WHILE TICKS()< NST DO (* nothing *) END;
T:=TICKS();
CC;
NST := T+ST;
IF KEYPRESSED()THEN EXIT;
END;
END;
END;
```

Without some explanation and comment the meaning is completely obscure. By using long identifiers instead of, for example NST and ST, it is possible to make the code more readable.

```
BEGIN
NEXTSAMPLETIME := TICKS()+SAMPLETIME;
TIME:=TICKS()+SAMPLETIME;
LOOP
WHILE TICKS()< NEXTSAMPLETIME DO (* NOTHING *)
END;
TIME:=TICKS();
CONTROLCALCULATION;
NEXTSAMPLETIME:=TIME+SAMPLETIME;
IF KEYPRESSED()THEN EXIT;
END;
END;
END;
```

The meaning is now a little clearer, although the code is not easy to read because it is entirely in upper case letters.

We find it much easier to read lower case text than upper case and hence readability is improved if the language permits the use of lower case text. It also helps if we can use a different case (or some form of distinguishing mark) to identify the reserved words of the language. Reserved words are those used to identify particular language constructs, for example repetition statements, variable declarations, etc. In the next version we use upper case for the reserved words and a mixture of upper and lower case for user-defined entities.

```
BEGIN
NextSampleTime := Ticks()+SampleTime;
Time:=Ticks()+SampleTime;
LOOP
WHILE Ticks()< NextSampleTime DO (* nothing *)
END;
Time:=Ticks();
ControlCalculation;
NextSampleTime := Time+SampleTime;
IF KeyPressed()THEN EXIT;
END;
END;
END;
```

The program is now much easier to read in that we can easily and quickly pick out the reserved words. It can be made even easier to read if the language allows embedded spaces and tab characters to be used to improve the layout.

```
BEGIN (* Main program *)
    NextSampleTime := Ticks() + SampleTime;
    Time := Ticks() + SampleTime;
    LOOP
        WHILE Ticks() < NextSampleTime DO
            (* nothing *)
        END (* of WHILE *);
        Time := Ticks();
        ControlCalculation;
        NextSampleTime := Time + SampleTime;
        IF KeyPressed() THEN EXIT;
        END (* IF *);
    END (* of LOOP *);
END (* MAIN *).
```

The exact form of layout adopted is a matter of house style or of personal preference. Modula-2 does not allow names to contain embedded spaces or characters such as the underscore as separators, so it is usual to capitalise the first letter of each word in a compound name.

Modula-2 requires reserved words to be written in upper case. It is also case sensitive; for example, `Time` and `time` would be treated as different entities. Ada is case insensitive but a convention has been established that reserved words are written in lower case and application entities in upper case, with the underscore character used as the separator for compound names. Adopting this approach the code fragment given above would be written as shown below.

```
begin (* Main program *)
    NEXT_SAMPLE_TIME := TICKS() + SAMPLE_TIME;
    TIME := TICKS() + SAMPLE_TIME;
    loop
        while TICKS() < NEXT_SAMPLE_TIME do
            (* nothing *)
        end (* of while *);
        TIME := TICKS();
        CONTROL_CALCULATION;
        NEXT_SAMPLE_TIME := TIME + SAMPLE_TIME;
        if KeyPressed() then exit;
        end (* if *);
    end (* of loop *);
end (* main *).
```

Cooling (1991, p. 269) reports that some Ada programmers consider the adoption of this convention to have been a mistake in that it makes it more difficult to read the code. In the examples illustrating some features of Ada which are given later in this chapter, Modula-2 style is used with upper case for the reserved words.

5.3 DECLARATION AND INITIALISATION OF VARIABLES AND CONSTANTS

5.3.1 Declarations

The purpose of declaring an entity used in a program is to provide the compiler with information on the storage requirements and to inform the system explicitly of the names being used. Languages such as Pascal, Modula-2 and Ada require all objects to be specifically declared and a type to be associated with the entity when it is declared. The provision of type information allows the compiler to check that the entity is used only in operations associated with that type. If, for example, an entity is declared as being of type REAL and then it is used as an operand in logical operation, the compiler should detect the type incompatibility and flag the statement as being incorrect.

Some older languages, for example BASIC and FORTRAN, do not require explicit declarations; the first use of a name is deemed to be its declaration. In FORTRAN explicit declaration is optional and entities can be associated with a type if declared. If entities are not declared then implicit typing takes place: names beginning with the letters I–N are assumed to be integer numbers; names beginning with any other letter are assumed to be real numbers.

Optional declarations are dangerous because they can lead to the construction of syntactically correct but functionally erroneous programs. Consider the following program fragment:

```
100 ERROR=0
    ...
200 IF X=Y THEN GOTO 300
250 EROR=1
300 ...
    ...
400 IF ERROR=0 THEN GOTO 1000
    ...
```

In FORTRAN (or BASIC), ERROR and EROR will be considered as two different variables whereas the programmer's intention was that they should be the same – the variable EROR in line 250 has been mistyped. FORTRAN compilers cannot detect this type of error and it is a *characteristic* error of FORTRAN. Many organisations which use FORTRAN extensively avoid such errors by insisting that all entities are declared and the code is processed by a preprocessor which checks that all names used are mentioned in declaration statements.

The implicit typing which takes place in FORTRAN can also lead to confusion and misinterpretation. Consider the program fragment

```
REAL KP,KD,KI
INTEGER DACV,ADCV
...
100     DACV=KP*(KD/KI)
...
END
```

A programmer reading the statement with the label 100, without reference to the declaration statement, might assume that the variables on the right-hand side of the statement are all integers and that the resultant value is then floated and assigned to a real variable, whereas the statement is doing just the opposite: the variables KP, KD and KI are real and the resultant is truncated and assigned to an integer variable.

A common method of avoiding this problem is to insist that all variable names conform to the implicit typing rules. Meaningful names that do not conform are prefixed with an appropriate letter. Therefore a normally implicit real name used for an integer is prefixed I and a normally implicit integer name used for a real number is prefixed X. The above fragment is written

```
REAL XKP, XKD, XKI
INTEGER IDACV, IADCV
100 IDACV = XKP*(XKD/XKI)
```

The most notorious example of the lack of security in programs written in FORTRAN is the error that caused the misdirection of the Voyager spacecraft. The FORTRAN statement intended was

```
DO 20 I = 1,100
```

which is a loop construct. What was typed was

```
DO 20 I = 1.100
```

Because embedded spaces in names are ignored and variables need not be declared, the FORTRAN compiler treated the characters to the left of the assignment operator as a variable name DO20I, and because the name begins with the character D treated it as a real number and assigned the value 1.100 to it.

5.3.2 Initialisation

It is useful if a variable can be given an initial value when it is declared. It is bad practice to rely on the compiler to initialise variables to zero or some other value.

This is not, of course, strictly necessary as a value can always be assigned to a variable. In terms of the security of a language it is important that the compiler checks that a variable is not used before it has had a value assigned to it. The security of languages such as Modula-2 is enhanced by the compiler checking that all variables have been given an initial value. However, a weakness of Modula-2 is that variables cannot be given an initial value when they are declared but have to be initialised explicitly using an assignment statement.

5.3.3 Constants

Some of the entities referenced in a program will have constant values either because they are physical or mathematical entities such as the speed of light or π, or because they are a parameter which is fixed for that particular implementation of the program, for example the number of control loops being used or the bus address of an input or output device. It is always possible to provide constants by initialising a variable to the appropriate quantity, but this has the disadvantage that it is insecure in that the compiler cannot detect if a further assignment is made which changes the value of the constant. It is also confusing to the reader since there is no indication which entities are constants and which are variables (unless the initial assignment is carefully documented).

Pascal provides a mechanism for declaring constants, but since the constant declarations must precede the type declarations, only constants of the predefined types can be declared. This is a severe restriction on the constant mechanism. For example, it is not possible to do the following:

```
TYPE
    AMotorState = (OFF,LOW,MEDIUM,HIGH);
CONST
    motorStop = AMotorState(OFF);
```

A further restriction in the constant declaration mechanism in Pascal is that the value of the constant must be known at compilation time and expressions are not permitted in constant declarations. The restriction on the use of expressions in constant declarations is removed in Modula-2 (experienced assembler programmers will know the usefulness of being able to use expressions in constant declarations). For example, in Modula-2 the following are valid constant declarations:

```
CONST
    message = 'a string of characters';
    length = 1.6;
    breadth = 0.5;
    area = length * breadth;
```

In Ada the value of the constant can be assigned at run-time and hence it is more appropriate to consider constants in Ada as special variables which become read only following the first assignment to them.

5.4 MODULARITY AND VARIABLES

5.4.1 Scope and Visibility

The scope of a variable is defined as the region of a program in which the variable is potentially accessible or modifiable. The regions in which it may actually be accessed or modified are the regions in which it is said to be visible.

Most languages provide mechanisms for controlling scope and visibility. There are two general approaches: languages such as FORTRAN provide a single level of locality whereas the block-structured languages such as Modula-2 provide multi-level locality.

In the block-structured languages entities which are declared within a block may only be referenced inside that block. Blocks can be nested and the scope extends throughout any nested blocks. This is illustrated in Example 5.1 which shows the scope for a nested PROCEDURE in Modula-2.

EXAMPLE 5.1

```
MODULE  ScopeExample1;
VAR
    A,B: INTEGER;
PROCEDURE LevelOne;
VAR
    B,C: INTEGER;
BEGIN
    (*
        LevelOne.B and LevelOne.C visible here
    *)
END (* LevelOne *);
BEGIN
    (*
        A and B visible here but not LevelOne.B and
LevelOne.C
    *)
END ScopeExample1.
```

The *scope* of variables A and B declared in the main module ScopeExample1 extends throughout the program, that is they are global variables. However, because the variable name B is reused in PROCEDURE LevelOne it hides the global variable B while procedure LevelOne is executing. The scope of variables LevelOne.B and LevelOne.C extends over PROCEDURE LevelOne and both are visible within the procedure.

As in this example, in a block-structured language an entity declared in an inner block may have the same name as an entity declared in an outer block. This does not cause any confusion to the compiler which simply provides new storage for the entity in the inner block and the entity in the outer block temporarily *disappears*, to reappear when the inner block is left.

In Modula-2, as in most block-structured languages, variables declared inside a procedure have storage allocated to them only while that procedure is being executed. When the procedure is entered storage in RAM is allocated from available memory (often referred to as *heap*); if no memory is available a run-time error is generated. When an exit is made from the procedure the memory is released and can be reused.

This dynamic allocation and release of memory leads to two problems:

● variables declared within a procedure cannot be used to hold values for reuse on the next entry to the procedure; and
● if a procedure is called recursively it is possible that the program may fail because there is no more memory available.

Modula-2 overcomes the first of these problems through its use of a program unit MODULE which has some specific properties. It is because of the second problem that recursively called procedures should not be used in real-time systems, particularly if failure will compromise the safety of people or equipment (even if the procedures do not declare any variables each call makes demands on the stack space for storage of its volatile environment). To construct a safe system it must be possible to predict the maximum memory requirements.

5.4.2 Global and Local Variables

Although the compiler can easily handle the reuse of names, it is not as easy for the programmer and the use of deeply nested PROCEDURE blocks with the reuse of names can compromise the security of a Pascal or Modula-2 program. As the program shown in Example 5.2 illustrates the reuse of names can cause confusion as to which entity is being referenced.

EXAMPLE 5.2

Loss of Visibility in Nested Procedures

```
MODULE ScopeL2;
VAR X, Y, Z : INTEGER;
PROCEDURE L1;
    VAR Y : INTEGER;
    PROCEDURE L2;
        VAR X : INTEGER;
        PROCEDURE L3;
            VAR Z : INTEGER;
            PROCEDURE L4;
            BEGIN
                Y : = 25; (* L1.Y NOT L0.Y*)
            END L4;
            BEGIN
                (* L1.Y, L2.X, L3.Z visible *)
            END L3;
        BEGIN
            (* L1.Y, L2.X, L0.Z visible *)
        END L2;
    BEGIN
        (* L0.X, L1.Y, L0.Z visible *)
    END L1;
    BEGIN
    (* ... *)
    END ScopeL2.
```

It is very easy to assume in assigning the value 25 to Y in PROCEDURE L4 that the global variable Y is being referenced, when in fact it is the variable Y declared in PROCEDURE L1 that is being referenced.

The argument over global or local declaration of entities has been almost as fierce as that over the use of GOTO statements. The proponents of local declarations argue that it is good practice to introduce and name entities close to where they are to be used and thus to limit the scope of the entity and its visibility. Those arguing in favour of global visibility of names claim that it is the only way in which consistency and control of the naming of entities can be achieved for large systems being developed by a team of programmers. They argue that local declaration leads to duplication of names and difficulties in subsequent maintenance of programs. A sensible compromise position is probably to declare globally the names of all entities which directly relate to the outside world, that is to the system being modelled or controlled, and to use local declaration for the names of all internal entities. Many of the difficulties disappear if the language permits explicit control over the scope and visibility of entities and does not rely on default rules.

5.4.3 Control of Visibility and Scope

Modern approaches to software design place a lot of emphasis on modularity and information hiding, both of which contribute to security. For effective use of modularisation in program design a language must provide mechanisms that enable the programmer to control explicitly the scope and visibility of all entities. This includes the ability to extend the scope of entities declared within a program unit to areas outside that unit. These problems have been addressed by a number of language designers. Program units which provide the necessary facilities have been devised; they are known by several different names, for example a *class* in SIMULA, a *module* in Modula-2, a *segment* in ALGOL and a *package* in Ada.

5.4.4 Modularity

In Modula-2 the main program unit is a `MODULE` and local modules can be nested within the main module. The nesting of modules is illustrated in Example 5.3.

EXAMPLE 5.3

Scope and Visibility Control in Modula-2

```
MODULE ImportExport;
(*
Title : Example of Import and Export of objects
File : sb1 Importex.mod
LastEdit:
Author : S. Bennett
*)
```

```
    VAR a,b,c : INTEGER;
(* ... *)
    MODULE L1;
        IMPORT a;
        EXPORT d,g;
        VAR d,e,f : INTEGER;
            MODULE L2;
                IMPORT e,f;
                EXPORT g,h;
                VAR g,h,i : INTEGER;
                (* ...
                    e,f,g,h,i visible here
                *)
            END L2;
        (* ...
            a,d,e,f,g,h are visible here
        *)
    END L1;
(* ...
a,b,c,d,g are visible here
*)
END ImportExport.
```

To allow entities which are declared within the body of a module to be visible outside the module they must be listed in an EXPORT list. Similarly entities which are declared outside the module must be specifically imported into the module by naming them in an IMPORT list. It should be noted that all entities can be imported and exported, that is variables, constants, types and procedures. Entities which are declared in a module are created at the initialisation of the program and remain in existence throughout the existence of the program, that is they are not like entities created inside a procedure which cease to exist when the procedure body is left.

The more general concept of a module allows a program to be split into many separate units, each of which can be compiled separately. The facilities for separation also allow for the construction of program libraries in which the library segments are held in compiled form rather than as source code. In order to do this the module is split into two parts:

DEFINITION MODULE — this contains information about entities which are exported from the module;
IMPLEMENTATION MODULE — this is the body of the module which contains the code which carries out the functions of the module.

The DEFINITION part is made available to the client program in source form, but the IMPLEMENTATION part is only provided in object form and its source code remains private to the module designer. The separation provides an excellent method of hiding implementation details from a user, and the actual implementation can be changed without informing the user, providing that the DEFINITION part does not change. An example of a DEFINITION MODULE is given in Example 5.4.

EXAMPLE 5.4

DEFINITION MODULE in Modula-2

```
DEFINITION MODULE Buffer;
(*
Title : Example of definition module
File : sb1 Buffer.mod
*)

EXPORT QUALIFIED
    put, get, nonEmpty, nonFull;
VAR
    nonEmpty, nonFull : BOOLEAN; (* used to test the
status of buffer *)

PROCEDURE put (x : CARDINAL);
    (* used to add items to the buffer *)
PROCEDURE get (VAR x: CARDINAL)
    (* used to remove items from the buffer *)
END Buffer.
```

5.5 COMPILATION OF MODULAR PROGRAMS

If we have to use a modular approach in designing software how do we compile the modules to obtain executable object code? There are two basic approaches: either combine at the source code level to form a single unit which is then compiled, or compile the individual modules separately and then in some way link the compiled version of each module to form the executable program code. Using the second approach a special piece of software called a *linker* has to be provided as part of the compilation support to do the linking of the modules.

A reason for the popularity and widespread use of FORTRAN for engineering and scientific work is that subroutines can be compiled independently from the main program, and from each other. The ability to carry out compilation independently arises from the single-level scope rules of FORTRAN; the compiler makes the assumption that any entity which is referenced in a subroutine, but not declared within that subroutine, will be declared externally and hence it simply inserts the necessary external linkage to enable the linker to attach the appropriate code. It must be stressed that the compilation is *independent*, that is when a main program is compiled the compiler has no information available which will enable it to check that the reference to the subroutine is correct. For example, a subroutine may expect three real variables as parameters, but if the user supplies four integer variables in the call statement the error will not be detected by the compiler. Independent compilation of most block-structured languages is even more difficult and prone to errors in that arbitrary restrictions on the use of variables have to be imposed. Many errors can be detected at the linking stage. However, because linking comes later in the implementation process errors discovered at this stage are more costly to correct. It is preferable to design the language and compilation system in such a way as to be able to detect as many errors as possible during compilation instead of when linking.

Both Modula-2 and Ada have introduced the idea of *separate* compilation units. Separate compilation implies that the compiler is provided with some information about the previously or separately compiled units which are to be incorporated into a program. In the case of Modula-2 the source code of the DEFINITION part of a separately compiled module must be made available to the user, and hence the compiler. This enables the compiler to carry out the normal type checking and procedure parameter matching checks. Thus in Modula-2 type mismatches and procedure parameter errors are detectable by the compiler. It also makes available the scope control features of Modula-2.

The provision of independent compilation of the type introduced in FORTRAN represented a major advance in supporting software development because it enabled the development of extensive object code libraries. Languages which support separate compilation represent a further advance in that they add greater security and easy error checking to library use.

5.6 DATA TYPES

As we have seen above, the allocation of types is closely associated with the declaration of entities. The allocation of a type defines the set of values that can be taken by an entity of that type and the set of operations that can be performed on the entity. The richness of types supported by a language and the degree of rigour with which type compatibility is enforced by the language are important influences on the security of programs written in the language. Languages which rigorously enforce type compatibility are said to be *strongly* typed; languages which do not enforce type compatibility are said to be *weakly* typed.

FORTRAN and BASIC are weakly typed languages: they enforce some type checking; for example, the statements A$=25 or A=X$+Y are not allowed in BASIC, but they allow mixed integer and real arithmetic and provide implicit type changing in arithmetic statements. Both languages support only a limited number of types.

An example of a language which is strongly typed is Modula-2. In addition to enforcing type checking on standard types, Modula-2 also supports enumerated types. The enumerated type allows programmers to define their own types in addition to using the predefined types. Consider a simple motor speed control system which has four settings OFF, LOW, MEDIUM, HIGH and which is controlled from a computer system. Using Modula-2 the programmer could make the declarations:

```
TYPE
    AMotorState = (OFF,LOW,MEDIUM,HIGH);
VAR
    motorSpeed : AMotorState;
```

The variable motorSpeed can be assigned only one of the values enumerated in the TYPE definition statement. An attempt to assign any other value will be trapped by the compiler, for example the statement

```
motorSpeed := 150;
```

will be flagged as an error.

If we contrast this with the way in which the system could be programmed using FORTRAN we can see some of the protection which strong typing provides. In ANSI FORTRAN integers must be used to represent the four states of the motor control:

```
INTEGER OFF,LOW,MEDIUM,HIGH
DATA OFF/0/,LOW/1/,MEDIUM/2/,HIGH/3/
```

If the programmer is disciplined and only uses the defined integers to set MSPEED then the program is clear and readable, but there is no mechanism to prevent direct assignment of any value to MSPEED. Hence the statements

```
MSPEED = 24
MSPEED = 150
```

would be considered as valid and would not be flagged as errors either by the compiler or by the run-time system. The only way in which they could be detected is if the programmer inserted some code to check the range of values before sending them to the controller. In FORTRAN a programmer-inserted check would be necessary since the output of a value outside the range 0 to 3 may have an unpredictable effect on the motor speed.

5.6.1 Sub-range Types

Another valuable feature which enhances security is the ability to declare a sub-range of a type. In Modula-2 sub-ranges of ordinal types (that is, INTEGER, CHAR and ENUMERATED types) can be defined. The following statements define sub-range types:

```
TYPE
    ADACValue = 0..255;
    ALowerCaseChar = 'a'..'z';
```

and if the variables are defined as

```
VAR
    output : ADACValue;
    character : ALowerCaseChar;
```

then the assignments

```
output := -25;
character := 'A';
```

will be flagged as errors by the compiler. The compiler will also insert run-time checks on all assignment statements involving sub-range types and any assignment which violates the permitted values will generate a run-time error. The use of sub-range types can increase the security of a program, but in a real-time system full use of sub-range types may not be appropriate. They can be used if the run-time system permits the transfer of control to a user-supplied error analysis segment on detection of a run-time error; if it does not and it terminates execution of the program then the security of the system can be jeopardised by their use. Sub-range types can be useful during the development stages of the system as violations of correctly set sub-ranges can indicate logical errors in the code. For this reason many compilers provide an option switch to control the inclusion of sub-range checking.

Sub-range types have been extended in Ada to include sub-ranges of REAL. Again the usefulness is limited because of the extra code introduced in order to check for violations. In applications involving a large amount of computation the use of sub-ranges of REAL can significantly slow down the computation and hence in many applications the efficiency requirements will necessitate the use of explicit range checks at appropriate points rather than the use of compiler-supplied checks through the use of sub-range types.

5.6.2 Derived Types

In many languages new types can be created from the implicit types: these are known as derived types and they inherit all the characteristics of the parent type. The use of derived types can make the meaning of the code clearer to the reader, for example

```
TYPE
    AVoltage = REAL;
    AResistance = REAL;
    ACurrent = REAL;
VAR
    V1 : AVoltage;
    R3 : AResistance;
    I2 : ACurrent;
BEGIN
    I2 := V1/R3;
END;
```

In the above code the reader can easily see what is implied by the calculation: Ohm's law is being used to calculate the current flowing through a resistance. If the statement read

```
I2 := V1*R3;
```

then because I2 has been declared as of type ACurrent the reader would be suspicious that there was an error in the line.

In a very strongly typed language such as Ada neither of the above statements would be accepted by the compiler for although derived types inherit the properties of the parent type they are treated as distinct types and hence are not compatible. Strong typing of this form has the advantage that the compiler can detect errors such as assigning V1 which represents a voltage to the variable I2, which represents a current, as is shown in the first statement in the program fragment below, but it would also flag as an error the perfectly legitimate operation shown in the second statement:

```
I2 := V1;
I2 := V1/R3
```

In Ada this problem is overcome by a mechanism that permits operators on types to be *overloaded*, that is the operator can be redefined to be compatible with a different set of types. In this case the operator / would have to be defined as performing the division operation with a variable of type AVoltage as the dividend, a variable of type AResistance as the divisor and a variable of type ACurrent as the resultant. It is arguable whether the benefits of the strong typing outweigh the disadvantage of the complexity of overloading.

5.6.3 Structured Types

Many programming languages provide only one structured type, the array. Arrays, though powerful, are limited in that all the elements of the array must be of the same type. There are many applications in which it would be useful to be able to declare entities which are made up of elements of different types.

EXAMPLE 5.5

```
TYPE
    AController = RECORD
        inputAddress : INTEGER;
        outputAddress : INTEGER;
        maxOutput : REAL;
        name : ARRAY [0..16] OF CHAR;
        status : (OFF,ON);
    END (* RECORD *);
VAR
    airFlow : AController;
    fuelFlow : AController;
    reactant : AController;

PROCEDURE Control(VAR loop : AController);
(* controller that is used to control several loops
*)
END (* CONTROL *);
BEGIN
    LOOP
        Control(fuelFlow);
        Control(reactant);
    END (* LOOP *);
END.
```

In Example 5.5 the whole of the information relevant to a particular control loop for the plant is contained in one variable of type RECORD. The advantages of using a structure type such as the record structure is that the programmer does not continually have to consider the details of the way in which the information relevant to the variable is stored: it contributes to the process known as abstraction. In addition to RECORD many of the modern languages support types such as FILE and SET.

5.6.4 Pointers

Pointers provide a mechanism for indirect reference to variables. They are widely used in systems programming and in some data processing applications. They can be used, for example, to create linked lists and tree structures. The unrestricted

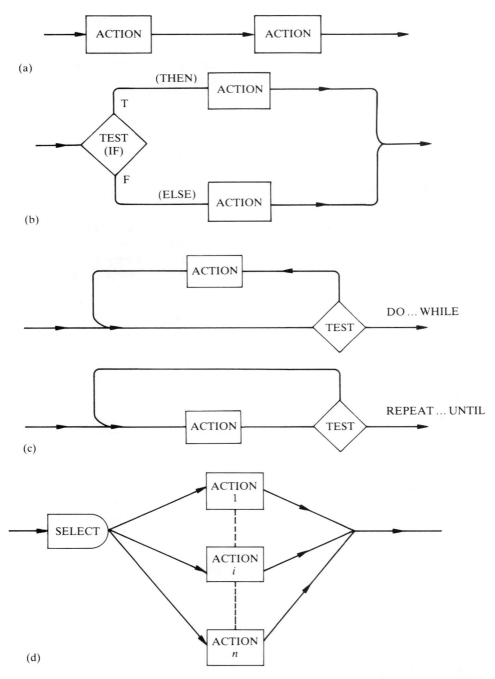

Figure 5.1 Standard structured program constructs: (a) sequence; (b) decision; (c) repetition; (d) selection.

availability of pointers, for example permitting pointers to entities of different types to be interchanged, can give rise to insecurity. Implicit in the use of pointers is the dynamic allocation and consequent deallocation of storage; the overheads involved in the required mechanisms can be considerable and care in the use of pointer types is necessary in real-time applications.

5.7 CONTROL STRUCTURES

There have been extensive arguments in the past about the use of both conditional and unconditional GOTO statements in high-level languages. It is argued that the use of GOTOs makes a program difficult to read and it has been shown that any program can be expressed without the use of GOTOs as long as the language supports the WHILE statement, the IF...THEN...ELSE conditional and BOOLEAN variables. Most modern languages support such statements. The standard structured programming constructs are shown in Figure 5.1.

From a theoretical point of view the avoidance of GOTOs is attractive. There are, however, some practical situations in which the judicious use of a GOTO can avoid complicated and confusing coding. An example of such a situation is when it is required to exit from a loop in order to avoid further processing when a particular condition occurs.

EXAMPLE 5.6

Consider the following scenario. A stream of data in character form is received from a remote station over a serial link. The data has to be processed character by character by a routine ProcessItem until the end-of-transmission character (EOT — ASCII code = 4) is received. The EOT character must not be processed.

A simple loop structure of the form

```
REPEAT
    get(character);
    ProcessItem(character);
UNTIL character = EOT;
```

cannot be used since the EOT character would be processed. Possible solutions are:
1.

```
Get(Character);
WHILE Character <> EOT DO
    ProcessItem(Character);
    Get(Character)
END (* WHILE *);
```

2.
```
Finished := FALSE;
REPEAT
    Get(Character);
    IF Character = EOT THEN
        Finished := TRUE
    ELSE
        ProcessItem(Character)
    END (* IF *);
UNTIL Finished;
```

A much cleaner solution is provided by the use of an EXIT statement as in the fragment below:

3.
```
LOOP
    Get(Character);
    IF Character=EOT THEN
        EXIT
    END (* IF *);
    ProcessItem(Character);
    END (* LOOP *);
(* EXIT causes a jump to statement here if EOT is
detected *)
```

The solution shown in (3) which is possible in Modula-2 is much clearer because all the operations are shown within the LOOP statement, whereas in solutions (1) and (2) some operation has to be performed outside the loop. Either the first character has to be obtained before entering the loop or a Boolean variable has to be set before the loop is entered. The general LOOP...END statement becomes particularly valuable if several different exceptions require an exit from the loop.

A general non-terminating loop statement which can contain one or more EXIT statements that result in control passing to the statement immediately following the end of the loop is useful as is a RETURN statement (for use in PROCEDUREs) that results in an immediate exit from the PROCEDURE. Both are useful additions to control structures. Although not strictly required they can lead to clearer and simpler coding. A program illustrating their use is given in Example 5.7.

EXAMPLE 5.7

Behaviour of RETURN and EXIT statements.
```
MODULE ReturnExitLoop ;
(*
Title : Example of different methods of leaving LOOP
statement
*)
```

```
FROM InOut IMPORT
    WriteString, WriteLn;
IMPORT Terminal;
VAR
    ch: CHAR;
    return : BOOLEAN;

PROCEDURE ReadKey ;
BEGIN
    WriteString ('ReadKey called  ');
    LOOP
        WHILE NOT (Terminal.KeyPressed()) DO
            (* do nothing *)
        END (* while *);
        Terminal.Read(ch);
        IF ch='r' THEN
            RETURN
        END (* if *);
        IF ch='e' THEN
            EXIT
        END (* if *);
    END (* loop *);
    WriteString ('this was EXIT not RETURN');
    WriteLn;
    return:=FALSE
END ReadKey;
BEGIN
    LOOP
        return:=TRUE;
        ReadKey;
        IF return THEN
            WriteString ('this was RETURN not EXIT');
            WriteLn
        END (* if *);
        WriteString ('To repeat type c, to stop any
other character');
        WHILE NOT (Terminal.KeyPressed()) DO
            (* do nothing *)
        END (* while *);
        Terminal.Read(ch);
        IF ch<>'c' THEN
            EXIT
        END (* if *);
        WriteLn
    END (* LOOP *)
END ReturnExitLoop.
```

5.8 EXCEPTION HANDLING

One of the most difficult areas of program design and implementation is the handling of errors, unexpected events (in the sense of not being anticipated and hence catered for at the design stage) and exceptions which make the processing of data by the subsequent segments superfluous, or possibly dangerous. The designer has to make decisions on such questions as what errors are to be detected? What sort of mechanism is to be used to do the detection? And what should be done when an error is detected?

Most languages provide some sort of automatic error detection mechanisms as part of their run-time support system. Typically they trap errors such as an attempt to divide by zero, arithmetic overflow, array bound violations, and sub-range violations; they may also include traps for input/output errors. For many of the checks the compiler has to add code to the program; hence the checks increase the size of the code and reduce the speed at which it executes. In most languages the normal response when an error is detected is to halt the program and display an error message on the user's terminal.

In a development environment it may be acceptable for a program to halt following an error; in a real-time system halting the program is not acceptable as it may compromise the safety of the system. Every attempt must be made to keep the system running.

EXAMPLE 5.8

Error Checking

Consider a boiler control system (similar to the one described in Chapter 2). One control loop uses the ratio of fuel flow and air flow to calculate the set point for the controller. Assume that the values of fuel flow and air flow are read from the measuring instruments and stored in REAL variables FuelFlow and AirFlow respectively. The instruments provide values which are in the range 0.0 to 4096.0 corresponding to the flow ranges 0 to 100%. The control setting is to be held in a real variable, RatioSetPoint. The programmer might write the statement:

```
RatioSetPoint := FuelFlow / AirFlow
```

A program using this statement may function correctly for a long period of time, but suppose a fault either on the air flow measuring instrument or in the interface unit results in AirFlow being set equal to zero. The program, and hence the system, would halt with an error − attempt to divide by zero.

One method of dealing with this problem is to validate the data prior to executing the statement by adding code to give

```
IF AirFlow > 0 THEN
   RatioSetPoint := FuelFlow / AirFlow
ELSE
   AirFlowAlarm := TRUE;
   RatioSetPoint := DefaultValue
END (* IF *);
```

An alternative method, if the language supports sub-ranges of REAL variables, is to declare AirFlow to be in the range 0.0 to 4096.0 and to allow the compiler to insert the necessary sub-range checks; but what happens when the compiler detects a sub-range violation – does it simply halt the program? If it does, then nothing has been gained.

In Example 5.8 it is easy for the programmer to put in the necessary checks. However, checking for all possible data errors can become very complex, can obscure the general flow of the code, and can slow down execution.

A better solution for error handling in real-time systems is for the language to be designed so as to allow the application software to deal with errors when they are detected, and to pass the error on to the built-in error handling mechanism only if the application software fails to deal with the error. We need the language run-time support software to detect as many types of error as possible, to inform the application software that an error exists, and to contain error handling routines that will deal with the error if the application software does not do so.

One of the first languages to provide error handling in this way was BASIC with its ON ERROR GOTO and ON ERROR GOSUB statements. Use of this type of statement enables error trap routines to be inserted in the application program and can simplify the flow of the code as it permits grouping of the error handling and analysis into one place. The run-time error can be checked and action taken to keep the system running; if it cannot be kept running then at least there may be an opportunity to close it down safely and warn the operator.

EXAMPLE 5.9

Using BASIC the problem given in Example 5.8 can be dealt with as follows:

```
1000 ON ERROR GOSUB 9500;
     (All errors detected in the following lines of
     code will be
     (trapped and referred to the subroutine at line
     9500 until a
     (further ON ERROR statement is executed.
1040
1050 RatioSetPoint = FuelFlow / AirFlow
1060
1070 REM A return from the ERROR subroutine at
     9500 is made here.
     (...
     (Rest of code inserted here
     (...
9500 REM Start of error handler
9510 IF ERRORNO=DivideByZero THEN RatioSetPoint =
     DefaultValue
9520 ELSE GOTO ERRORTRAP
9530 RETURN
```

The error detection mechanism is assumed to return an error number to the program; if this error number matches the attempt to divide by zero error number then `RatioSetPoint` is set to a default value, otherwise the error was not expected at this point and has to be either passed to some more general error analysis routine or returned to the underlying default error handler which will normally halt the running of the program.

The obvious advantage of this approach is that the designer can consider the normal function and the error actions as two separate problems. It also has the advantage that execution speed is improved and in more complex cases the readability of the code is improved. The disadvantage is that it encourages reliance on the run-time system to catch errors rather than careful consideration of the possible errors.

The use of error trapping or exception statements requires careful consideration of the action which follows the execution of the error handling routine. In Example 5.9 it is assumed that execution resumes at the next statement after the statement which generated the error. Alternatively we could have written the example as follows:

```
1000 ON ERROR GOTO 9500;
1040
1050 RatioSetPoint = FuelFlow / AirFlow
1060
    (...
    (Rest of code inserted here
    (...
9500 REM Start of error handler
9510 IF AirFlow < 0 THEN AirFlow = 100.0
9520 GOTO 1050
```

In the above code, return is made to the statement that caused the error, the cause of the error having been corrected (hopefully!). (A potential problem with returning to the statement that caused the error to be raised is that if the correction is incorrect the same error will be raised again and the software will cycle endlessly round the same loop.)

The facility to trap and return information on potentially fatal errors to the application program is only one aspect of error handling. There will be abnormal or error conditions which the underlying language support system will not detect. For example, if in Example 5.9 the valid range of `AirFlow` was 100.0 to 4096.0 then there would be an error if `AirFlow` was set to 50.0. This would not result in system error (unless sub-range violations were being checked); however, the application program designer would be helped if, having dealt with the immediate problem, there was some means by which some further error analysis or error handling could easily be invoked. It is always possible to write such error handling mechanisms explicitly as Example 5.10 illustrates.

EXAMPLE 5.10

Suppose that the measured values of air flow and fuel flow obtained from the instruments have to be converted to actual flow rates by taking the square root of the input value. An instrument error causing a negative input value results in an error return from the square root function. We wish to protect against this error and also want to notify other parts of the system that an instrument error has been detected. Hence we might code the conversion as follows:

```
BEGIN
    IF RawFuelFlow < 0 THEN
        RawFuelFlow := MinValueRawFuelFlow;
        RaiseError('RawFuelFlow')
    END (* IF *);
    FuelFlow := SQRT(RawFuelFlow);
    IF RawAirFlow <= 0 THEN
        RawAirFlow := MinValueRawAirFlow;
        RaiseError('RawAirFlow')
    END (* IF *);
    AirFlow : = SQRT(RawAirFlow);
    RatioSetPoint := FuelFlow/AirFlow;
END

PROCEDURE RaiseError(VAR ErrorMessage: STRING);
(*
    Procedure to raise alarms according to error
message value
*)
END RaiseError;
```

In Ada there is explicit language provision for handling errors. The above could be written as follows:

```
PROCEDURE Convert_Flows;
BEGIN
    IF Raw_Fuel_Flow < 0 THEN
    RAISE Fuel_Error;
    Fuel_Flow := SQRT(Raw_Fuel_Flow);
    IF Raw_Air_Flow <= 0 THEN
    RAISE Air_Error;
    Air_Flow : = SQRT(Raw_Air_Flow);
    Ratio_Set_Point := Fuel_Flow/Air_Flow;
    EXCEPTION
        WHEN Fuel_Error =>
            ALARM(Fuel);
        WHEN Air_Error =>
            ALARM(Air);
    END Convert_Flows;
```

The major problem in exception handling occurs when procedures are nested and a report of the error has to be passed from one procedure to another.

EXAMPLE 5.11

Consider the following system:

```
PROCEDURE A;
BEGIN
...
B;
           PROCEDURE
           B;
           BEGIN

           C;
                          PROCEDURE C;
                          BEGIN
                          IF RawFuelFlow < 0 THEN
                            RawFuelFlow :=
                          MinValueRawFuelFlow;
                            RaiseError('RawFuelFlow')
                          END (* IF *);
                          FuelFlow :=
                          SQRT(RawFuelFlow);
                          IF RawAirFlow <= 0 THEN
                            RawAirFlow :=
                          MinValueRawAirFlow;
                            RaiseError('RawAirFlow')
                          END (* IF *);
                          AirFlow : =
                          SQRT(RawAirFlow);
                          FuelFlow :=
                          SQRT(RawFuelFlow);
                          END C;

           ...
           END B;

...
(* value of AirFlow is to be used here *)
END A;
```

In Example 5.11, if the value of `RawAirFlow` is found to be in error in `PROCEDURE C` then the code in `PROCEDURE B` and in `PROCEDURE A` may not be relevant if it requires a correct value for RawAirFlow. The standard means of passing back the information that an incorrect value of RawAirFlow has been used is to pass an error flag value in the procedure calls, that is to change the simple procedure calls to ones with the form PROCEDURE ((* arguments *), VAR ErrorMessage:STRING).

Ada offers an alternative solution in that when an exception is raised it is passed from block to block until the appropriate exception handler is found. Thus in the above example, if the exception was raised in PROCEDURE C and there was no exception handler in C, then C would be terminated and control would pass to B; if there were no handler in B it would also be terminated and control passed to A.

One special purpose language, CUTLASS, which was developed for real-time use, offers an interesting approach to this problem. It marks data as *bad* by using one bit in the storage used for data values as a *tag* bit. When an error is detected the data value is marked as bad and then allowed to propagate normally through the system. Any action needed to deal with bad data is then taken at the point at which it is used. The rules used in the propagation of bad data produce a result which is marked as bad if any of the operands in an arithmetic operation are bad. An example illustrating the technique is given in Section 5.16.2.

5.9 LOW-LEVEL FACILITIES

In programming real-time systems we frequently need to manipulate directly data in specific registers in the computer system, for example in memory registers, CPU registers and registers in an input/output device. In the older, high-level languages, assembly-coded routines are used to do this. Some languages provide extensions to avoid the use of assembly routines and these typically are of the type found in many versions of BASIC. These take the following form:

> PEEK(address) – returns as INTEGER variable contents of the location address.
>
> POKE(address, value) – puts the INTEGER value in the location address.

It should be noted that on eight-bit computers the integer values must be in the range 0 to 255 and on 16 bit machines they can be in the range 0 to 65 535. For computer systems in which the input/output devices are not memory mapped, for example Z80 systems, additional functions are usually provided such as

> INP(address) and
> OUT(address, value).

A slightly different approach has been adopted in BBC BASIC which uses an 'indirection' operator. The indirection operator indicates that the variable which follows it is to be treated as a pointer which contains the address of the operand rather than the operand itself (the term indirection is derived from the indirect addressing mode in assembly languages). Thus in BBC BASIC the following code

```
100 DACAddress=&FE60
120 ?DACAddress=&34
```

results in the hexadecimal number 34 being loaded into location FE60H; the indirection operator is '?'.

In some of the so-called Process FORTRAN languages and in CORAL and RTL/2 additional features which allow manipulation of the bits in an integer variable are provided, for example

```
SET BIT J(I),
IF BIT J(I) n1,n2 (where I refers to the bit in
variable J).
```

Also available are operations such as AND, OR, SLA, SRA, etc., which mimic the operations available at assembly level. The weakness of implementing low-level facilities in this way is that all type checking is lost and it is very easy to make mistakes. A much more secure method is to allow the programmer to declare the address of the register or memory location and to be able to associate a type with the declaration, for example

```
VAR charout AT 0FE60H :CHAR;
```

which declares a variable of type CHAR located at memory location 0FE60H. Characters can then be written to this location by simple assignment

```
charout := 'a';
```

Note that the compiler would detect and flag as an error an attempt to make the assignment

```
charout := 45;
```

since the variable is typed. Both Modula-2 and Ada permit declarations of the above type.

Modula-2 provides a low-level support mechanism through a simple set of primitives which have to be encapsulated in a small nucleus coded in the assembly language of the computer on which the system is to run. Access to the primitives is through a module SYSTEM which is known to the compiler. SYSTEM can be thought of as the software bus linking the nucleus to the rest of the software modules. SYSTEM makes available three data types, WORD, ADDRESS, PROCESS, and six procedures, ADR, SIZE, TSIZE, NEWPROCESS, TRANSFER, IOTRANSFER. WORD is the data type which specifies a variable which maps onto one unit of the specific computer storage. As such the number of bits in a WORD will vary from implementation to implementation; for example, on a PDP-11 implementation a WORD is 16 bits, but on a 68000 it would be 32 bits. ADDRESS corresponds to the definition TYPE ADDRESS = POINTER TO WORD, that is objects of type ADDRESS are pointers to memory units and can be used to compute the addresses of memory words. Objects of type PROCESS have associated with them storage for the volatile environment of the particular computer on which Modula-2 is implemented; they make it possible to create easily process (task) descriptors (see Chapter 6 for detailed information on task descriptors).

Three of the procedures provided by SYSTEM are for address manipulation:

```
ADR(v)    returns the ADDRESS of variable v
SIZE(v)   returns the SIZE of variable v in WORDs
TSIZE(t) returns the SIZE of any variable of type t
          in WORDs.
```

In addition variables can be mapped onto specific memory locations. This facility can be used for writing device driver modules in Modula-2. A combination of the low-level access facilities and the module concept allows details of the hardware device to be hidden within a module with only the procedures for accessing the module being made available to the end user.

Example 5.12 shows the DEFINITION MODULE for the analog input and output module for an 11/23 computer system. For normal use the two procedures ReadAnalog and WriteAnalog are all that the user requires. Additional information and procedures are made available to the expert user, including information on the actual hardware addresses (these may vary from system to system).

EXAMPLE 5.12

DEFINITION MODULE for a Device Handler

```
DEFINITION MODULE AnalogIO;
(*-----------------------------------------------------------
*)
(* Analog Input/Output
*)
(* Version 0, S. White, 2-Jul-85, adapted from ADC and DAC *)
(*              modules.                        *)
(*-----------------------------------------------------------
*)
```

```
FROM SYSTEM IMPORT
   ADDRESS;
EXPORT QUALIFIED
   moduleName, moduleVersion,
   ReadAnalog, WriteAnalog;
CONST
   moduleName = 'AnalogIO';
   moduleVersion = 1;
PROCEDURE ReadAnalog( adcNum: CARDINAL; VAR val: CARDINAL);
PROCEDURE WriteAnalog( dacNum; CARDINAL; val: CARDINAL);
(* Hardware configuration-expert use only *)
TYPE
AReadProc=PROCEDURE (CARDINAL, VAR CARDINAL );
AWriteProc=PROCEDURE (CARDINAL, CARDINAL );
PROCEDURE InitAnalog(
   initProc: PROC; readProc: AReadProc;
   writeProc: AWriteProc; doneProc: PROC);
(*-------------------MNCAIO--------------------*)
PROCEDURE InitMNCAIO;
PROCEDURE ReadMNCAD( adcNum: CARDINAL; VAR val: CARDINAL );
PROCEDURE WriteMNCAA( dacNum: CARDINAL; val: CARDINAL );
PROCEDURE DoneMNCAIO;
VAR
   MNCADReg: POINTER TO
      RECORD
         CSR: BITSET; (* Control/Status *)
         DBE: CARDINAL (* BufferPreset *)
      END;
   MNCADVec;
      RECORD
         ConversionComplete: ADDRESS;
         Error: ADDRESS;
      END;
CONST
   MNCADBaseReg = 171000B;
   MNCADBaseVec = 400B;
VAR
   MNCAAReg: POINTER TO
      RECORD
         DAC0: CARDINAL;
         DAC1: CARDINAL;
         DAC2: CARDINAL;
         DAC3: CARDINAL;
      END;
CONST
   MNCAABaseReg=171060B;
END AnalogIO.
```

5.10 COROUTINES

In Modula-2 the basic form of concurrency is provided by coroutines. The two procedures NEWPROCESS and TRANSFER exported by SYSTEM are defined as follows:

```
PROCEDURE NEWPROCESS(ParameterlessProcedure:PROC;
               workspaceAddress: ADDRESS;
               workspaceSize: CARDINAL;
               VAR coroutine: ADDRESS (* PROCESS *));

PROCEDURE TRANSFER(VAR source, destination : ADDRESS
               (*PROCESS*));
```

Any parameterless procedure can be declared as a PROCESS. The procedure NEWPROCESS associates with the procedure storage for the process parameters (the process descriptor or task control block — see Chapter 6) and some storage to act as workspace for the process. It is the programmer's responsibility to allocate sufficient workspace. The amount to be allocated depends on the number and size of the variables local to the procedure forming the coroutine, and to the procedures which it calls. Failure to allocate sufficient space will usually result in a stack overflow error at run-time.

The variable coroutine is initialised to the address which identifies the newly created coroutine and is used as a parameter in calls to TRANSFER. The transfer of control between coroutines is made using a standard procedure TRANSFER which has two arguments of type ADDRESS (PROCESS). The first is the calling coroutine and the second is the coroutine to which control is to be transferred. The mechanism is illustrated in Example 5.13. In this example the two parameterless procedures Coroutine1 and Coroutine2 form the two coroutines which pass control to each other so that the message

```
coroutine one coroutine two
```

is printed out 25 times. At the end of the loop, Coroutine2 passes control back to MainProgram.

EXAMPLE 5.13

Example Showing the Use of Coroutines

```
MODULE CoroutinesExample ;
(*
Title : Example of use of coroutines
*)
FROM SYSTEM IMPORT
    ADDRESS, WORD, NEWPROCESS, TRANSFER, ADR, SIZE, PROCESS;
FROM InOut IMPORT
    WriteString, WriteLn;
```

```
VAR
    coroutine1Id, coroutine2Id, MainProgram :
    PROCESS;
    worksp1, worksp2 : ARRAY [1..600] OF WORD;
PROCEDURE Coroutine1;
BEGIN
    LOOP
        WriteString('coroutine one');
        TRANSFER(coroutine1Id, coroutine2Id);
    END (* loop *);
END Coroutine1;
PROCEDURE Coroutine2 ;
VAR
    count : CARDINAL;
BEGIN
    count : = 0;
    LOOP
        WriteString (' coroutine two');
        WriteLn;
        IF count=25 THEN
            TRANSFER(coroutine2Id, MainProgram);
        ELSE
            INC(count);
            TRANSFER(coroutine2Id, coroutine1Id)
        END (* if *);
    END (* loop *);
END Coroutine2;
BEGIN
    NEWPROCESS(Coroutine1, ADR(worksp1),
SIZE(worksp1), coroutine1Id);
    NEWPROCESS(Coroutine2, ADR(worksp2),
SIZE(worksp2), coroutine2Id);
    TRANSFER(MainProgram, coroutine1Id)
END CoroutinesExample.
```

The fact that the concurrent programming implementation is based on coroutines has led some commentators to state that Modula-2 cannot be used for real-time systems. It cannot be used directly, but it is possible to use the primitive coroutine operations to create a real-time executive as is described in Chapter 6.

5.11 INTERRUPTS AND DEVICE HANDLING

If full I/O device support is to be provided by a high-level language then the language must provide support for the handling of interrupts. This demands support for some minimum form of concurrent operation since an interrupt causes a suspension of the running program (task) and the execution of some other code. There are two basic approaches in language design to doing this: one is to provide a set of low-level primitive operations which can be used either directly or to build higher-level constructs; the other is to provide a high-level set of primitive operations which must be used. Modula-2 is typical of the first method and Ada of the second. In the following sections Modula-2 is used to illustrate the various requirements.

Hardware interrupts can be handled from within a Modula-2 program. A device handling process can enable the external device and then suspend itself by a call to a procedure `IOTRANSFER`. This procedure is similar to `TRANSFER` but has an additional parameter which allows the hardware interrupt belonging to the device to be identified. When an interrupt occurs control is passed back to the device routine by a return from `IOTRANSFER`.

The procedure, `IOTRANSFER`, has the format

```
IOTRANSFER (VAR interruptHandler : PROCESS;
            interruptedProcess : PROCESS;
            interruptVector : CARDINAL)
```

The action of `IOTRANSFER` is to save the current status of `interruptHandler` and to resume execution of `interruptedProcess`, that is to wait for an interrupt. When an interrupt occurs the equivalent of

```
TRANSFER (interruptedProcess, interruptHandler)
```

occurs. A skeleton interrupt handler would thus take the form

```
BEGIN
   LOOP
...
      IOTRANSFER (interruptHandler, interruptedProcess,
interruptVector);
(* interrupt handler waits at this point for interrupt *)
...
      END Loop
END;
```

The interrupt handler code is placed inside the `LOOP ... END` construct, and is initiated by an explicit `TRANSFER` operation; it then waits for an interrupt at the `IOTRANSFER` statement. The first time this statement is executed control is returned to the initiating task. Subsequent executions will be after an interrupt has occurred and a return will be made to whichever task was interrupted.

An example of a Modula-2 program which uses the low-level facilities is given in Example 5.14. This program illustrates a further requirement for low-level support from the high-level language, namely the ability to handle interrupts. In the example shown `SuspendUntilInterrupt` is a high-level procedure provided by the module `Processes` which is part of a real-time support library (further information about this is given in Chapter 6).

EXAMPLE 5.14

```
MODULE TermOut[4 (* interrupt priority of device *)];
FROM PROCESSES IMPORT
    SuspendUntilInterrupt;
EXPORT
    PutC;

CONST
    readyBit = 7;
    interruptEnableBit = 6;
    interruptVector = 64B;
VAR
    ttyReg[177564B] : BITSET;
    ttyBuf[177566B] :CHAR;
PROCEDURE PutC(c: CHAR);
BEGIN
    IF NOT(readyBit IN ttyReg) THEN
        INCL(ttyReg, InterruptEnableBit);
        (* high processor priority will fend off
        the interrupt until ...*)
        SuspendUntilInterrupt(interruptVector);
        EXCL(ttyReg, interruptEnableBit);
        END; (* IF *)
    ttyBuf := c;
END PutC;
BEGIN
END TermOut;
```

This example also shows how Modula-2 handles bit-level manipulation. It is possible to declare a register as of type `BITSET` and perform set operations on the register. The operators `INCL` and `EXCL` are respectively the operations of including a bit in the set, that is setting a bit in the register, and excluding a bit from the set, that is resetting a bit in the register.

5.12 CONCURRENCY

Wirth (1982) defined a standard module `Processes` which provides a higher-level mechanism than coroutines for concurrent programming. The module makes no assumption as to how the processes (tasks) will be implemented; in particular it does not assume that the processes will be implemented on a single processor. If they are so implemented then the underlying mechanism is that of coroutines. The DEFINITION MODULE for Processes is as follows:

```
DEFINITION MODULE Processes ;
TYPE
SIGNAL;
    (* opaque type: variables of this type are used to
    provide synchronisation between processes. The
    variable must be initialised by a call to Init before use
    *)
PROCEDURE StartProcess(P:PROC;
    workSpaceSize:CARDINAL);
    (* start a new process. P is parameterless procedure
    which will form the process, workSpaceSize is the number
    of bytes of storage which will be allocated to the
    process
    *)
PROCEDURE Send(VAR s: SIGNAL);
    (* If no processes are waiting for s, then SEND has no
    effect. If some process is waiting for s then that
    process is given control and is allowed to proceed
    *)
PROCEDURE WAIT(VAR s: SIGNAL);
    (* The current process waits for the signal s. If at some
    later time a SEND(s) is issued by another process then
    this process will return from wait. Note if all
    processes are waiting the program terminates
    *)
PROCEDURE Awaited(s:SIGNAL):BOOLEAN;
    (* Test to see if process is waiting on s, if one or more
    processes are waiting then TRUE is returned
    *)
PROCEDURE Init(s: SIGNAL);
    (* The variable s is initialised, after initialisation
    Awaited(s) returns FALSE
    *)
END Processes.
```

Concurrency and multi-tasking will be dealt with in more detail in Chapter 6.

5.13 RUN-TIME SUPPORT

An important contributor both to the efficiency of implementation of a language and to the run-time security are the run-time support mechanisms that are provided by the particular implementation of the language. A problem with run-time support, however, is that security and efficiency frequently come into conflict. One way of trying to resolve the conflict is to provide optional run-time error checking and trapping mechanisms. For test purposes the error traps are switched in but once the software has been tested the error traps are switched out and the software runs faster.

A typical example is the checking of array bounds. Many compilers allow the insertion of check code as an option which is selected when the program is compiled. This is an acceptable approach for standard programming: during the initial testing the array bound check is selected but once the program appears to function correctly it is omitted and hence the program runs faster. Although the facility is often used in this way for real-time systems it is not a reliable solution. (Why not?)

There are two ways to do the array checking. Consider an array of 20 numbers denoted by *array*[i] where *i* is an integer variable. For security the system must ensure that whenever *array*[i] is used the condition $1 \leqslant i \leqslant 20$ is satisfied. The first way of checking is to insert a check before every array access to ensure that the condition is not violated. An alternative technique, used by the more modern compilers, is to check the condition whenever an assignment is made to *i*. The use of the second method requires the co-operation of the programmer who must specify the permitted range of variable *i* when it is declared. Provided that the checking of the assignment of values to *i* is done then there is no need to check access to *array*[i]. A further reduction in run-time checking is also found with this technique in that a large number of the assignments to *i* can be tested during compilation. The second method relies on the language being strongly typed. For a secure, real-time system, the second method of array bound checking is obviously preferable as it allows checking to remain in the final version of the software without imposing large run-time overheads on the execution time.

For real-time systems the other important feature of the run-time support software is whether it allows the application software to intercept the error traps. This was discussed in the section above where we dealt with exception handling.

5.14 OVERVIEW OF REAL-TIME LANGUAGES

The best way to start an argument among a group of computer scientists, software engineers or systems engineers is to ask them which is the best language to use for writing software. Rational arguments about the merits and demerits of any particular language are likely to be submerged and lost in a sea of prejudice.

Since 1970 high-level languages for the programming and construction of real-time systems have become widely available. Early languages include: CORAL (Woodward *et al.*, 1970) and RTL/2 (Barnes, 1976) as well as modifications to FORTRAN and BASIC. More recently the interest in concurrency and multi-processing has resulted in many languages with the potential for use with real-time systems. These include Ada (see Young, 1982; Burns and Wellings, 1990), ARGUS (Liskov and Scheifler, 1983), CONIC (Kramer *et al.*, 1983), CSP (Hoare, 1978), CUTLASS (CEGB, see Bennett and Linkens, 1984), FORTH (Brodie, 1986),

Modula-2 (Wirth, 1982), occam (Burns, 1988), PEARL and SR (Andrews, 1981, 1982). The most widely used language for programming embedded real-time systems is probably C. These languages range from comprehensive, general purpose languages such as Ada and Modula-2 to more restrictive, special purpose languages such as CONIC and CUTLASS. The advantage of the general purpose languages is that they provide flexibility that can support the building of virtual machines; for example, Budgen (1985) has shown how Modula-2 can be used to create a MASCOT virtual machine. However, general purpose, flexible languages can be too complex and permit operations that can compromise the security of a real-time system.

A language suitable for programming real-time and distributed systems must have all the characteristics of a good, modern, non-real-time language; that is, it should have a clean syntax, a rational procedure for declarations, initialisation and typing of variables, simple and consistent control structures, clear scope and visibility rules, and should provide support for modular construction. The additions required for real-time use include support for concurrency or multi-tasking and mechanisms to permit access to the basic computer functions (usually referred to as low-level constructs).

Modula-2 and Ada represent two very different approaches to language design: Modula-2 is based on a small set of mandatory facilities which can be extended using library modules, whereas in Ada all facilities considered necessary are mandatory and extensions are prohibited. The benefits of Ada are that it is standardised and hence application software should be highly portable; the disadvantages are size and complexity. The reverse is true of Modula-2: although the core is standardised there are many different sets of libraries supplied by different compiler implementers. For example, input and output routines are not provided as part of the language but have to be supplied as a library module and consequently the range and type of routines differ according to the supplier. The advantages of the Modula-2 approach are that the language is kept simple and the user can choose to add the features necessary and appropriate to the type of application. The major disadvantage is the loss of standardisation and hence portability.

5.15 APPLICATION-ORIENTED SOFTWARE

A large number of software packages and languages have been developed with the intention of providing a means by which the end user can easily write or modify the software for a particular problem. The major reason for wanting the end user, rather than a specialist programmer, to write the software is to avoid the communication problem. A large proportion of 'errors' in a system arise from a misunderstanding of the operation or structure of a plant by the programmer. This is not always the programmer's fault; often the engineers and managers responsible for the plant do not communicate their requirements clearly and precisely.

The misunderstanding can largely be avoided if the engineers responsible for the plant can themselves write the software.

The engineers are not, however, expected to be specialists in computer programming and hence they must be provided with simple programming tools which reflect the particular application. Because for a given type of application, for example process control, the range of facilities required is small and predictable, it is not too difficult to devise special application software.

Three main approaches are used:

1. table-driven;
2. block-structured; and
3. specialised languages.

These are considered in more detail in the following sections.

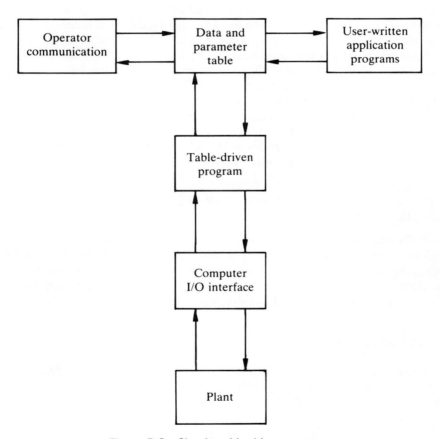

Figure 5.2 Simple table-driven system.

5.15.1 Table-driven Approach

The table-driven approach arose out of systems programmed in assembly code. It was soon realised that many code segments were used again and again in different applications as well as in the particular application. For example, the code segment used for PID control need only be written once if it contains only pure code and all references to parameters and data are made indirectly.

A simple table-driven system is illustrated in Figure 5.2. As well as allowing the control program to communicate with the data and parameter table, provision is also made for the operator to obtain information from the table (and in some instances to change values in the table). In addition some systems allow the user to write application programs in a normal computer language (usually FORTRAN or BASIC) which can interact with the table-driven software. The actual table-driven program is supplied as part of the system and cannot be modified by the user.

An alternative approach is shown in Figure 5.3 in which a database manager program is inserted between the data table and all users; access to the data table is now controlled by the database manager. The use of a database manager to control access places additional overheads on the system and can slow it down. It has the advantage of improving the security of the system in that checks can be built into it, for example to limit the items which can be changed from the operator's console, or to restrict the access of user-written application programs to certain areas of the data table. Typically a database manager program would provide, by means of a password or a key-operated switch on the console, different access rights to the operator and the plant engineer.

A crucial factor in the usefulness of table-driven software is the method of setting up and modifying the tables. There are three main methods:

1. direct entry into the data tables;
2. use of language DATA statements; and
3. the filling in of forms.

In the simplest systems the data has to be entered into specified memory locations, but this method is rarely used nowadays. It is more normal for the entry program to allow names to be used for the locations. Thus, for example, in a system with eight analog inputs the conversion values for each channel may be set by entering from a keyboard statements such as

```
ANCONV (1) = 950.2
ANCONV (6) = 0.328
```

which would set the conversion factors for the signals coming in on channels 1 and 6. On some systems it is possible to specify signal names rather than use the table index numbers. For example,

```
ANINP (1) = FEEDFL
ANINP (2) = FEEDTP
```

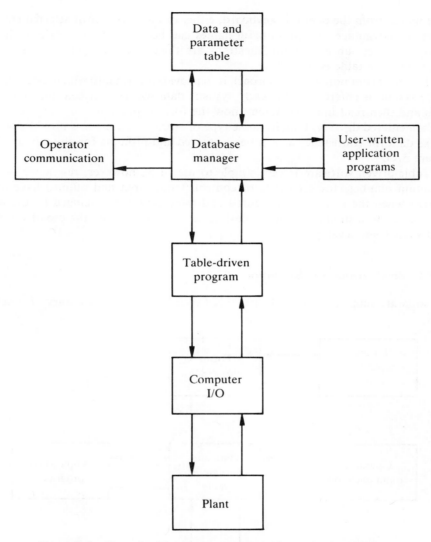

Figure 5.3   Use of table-driven system with database manager.

would specify that the signals on analog inputs 1 and 2 are to be known as FEEDFL (representing, say, feedwater flow rate) and FEEDTP (representing feedwater temperature). Once such names have been declared they can be used elsewhere to reference the particular signals. This is a feature which is useful when user-written application software is used, say, to produce specialised displays.

The use of language statements to set up the data tables is useful in large systems for initially setting up the system. It is normally supplemented by allowing changes to be made, from the operator keyboard, either of all entries or of selected entries. It has the advantage that meaningful names can be given to the entries at initial setting-up time; subsequent modifications are then made using the plant names rather than the table entry index.

The most commonly used approach is to provide a form into which data relating to the system is entered. In the early systems data was transcribed onto punched cards and then read into the system; now the form is presented on the screen and the data entered directly. Usually some type of 'forms' processor is provided which checks data for consistency as it is being entered and prompts for any missing data (Figure 5.4).

Table-driven software is very simple to use; it is, however, restrictive in that maximum numbers for each type of control (loop, input and output) have to be inserted when the system is configured and these cannot be changed by the user. Because of this a similar, but more flexible, approach based on the use of function blocks has been developed.

5.15.2 Block-structured Software

The software supplied in a block-structured system consists of a library of function

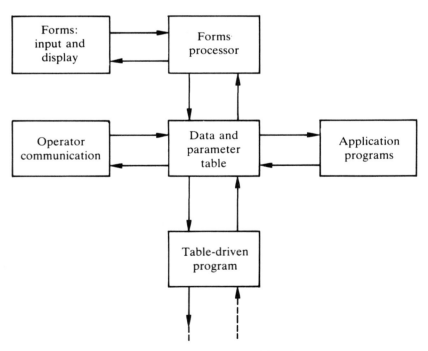

Figure 5.4   Forms processor for table-driven system.

```
┌─────────────────────────────────────────────────────────────────────────────┐
│ If the input is from another block, enter the block number.              5)  │
│                               or                                             │
│ If the input is from the PCM, enter the register number. (0-59)          5)  │
│                               or                                             │
│ If the input is from the Analog Input Module, enter the following:           │
│       Analog Input Module Type (1=contact, 2=fixed gain,                     │
│                                 3=programmable gain,                         │
│                                 4=Interspec)                             5)  │
│                                                                              │
│ If type 1, enter the following information:                                  │
│       Multiplexer address. (0-1023)                                      5)  │
│       Gain code. (0=1V, 2=50V, 3=10MV)                                       │
│                                                                              │
│ If type 2, enter the following information:                                  │
│       Next address. (0-15)                                               5)  │
│       Card address. (0-13)                                                   │
│       Point address. (0-7)                                                   │
│       Gain code. (0=x₁, 1=x₂, 2=x₃, 3=x₄)                                    │
│       (Note: x₁-x₄ are defined at SYSGEN time)                               │
│                                                                              │
│ If type 3, enter the following information:                                  │
│       Next address. (0-15)                                               5)  │
│       Card address. (0-13)                                                   │
│       Point address. (0-7)                                                   │
│       Gain code. (3=1V, 4=500MV, 5=200MV, 6=100MV, 7=50MV,                   │
│       8=20MV, 9=10MV)                                                         │
│       Bandwidth. (0=1KH, 1=3KH, 2=10KH, 3=100KH)                             │
│                                                                              │
│ If type 4, enter the following information:                                  │
│       ISCM number. (1-3)                                                 5)  │
│       CCM number. (1-16)                                                     │
│       Type of input. (M-Measurement, S-Setpoint, O-Output)                  │
│       Point number. (1-16)                                                   │
└─────────────────────────────────────────────────────────────────────────────┘

┌─────────────────────────────────────────────────────────────────────────────┐
│ Range of the input in engineering units:                                     │
│       Lowest value. (-32767. to +32767.)                                 6)  │
│       Highest value. (-32767. to +99999.)                                    │
│       Units. (As specified by user at System Generation.)                    │
└─────────────────────────────────────────────────────────────────────────────┘

┌─────────────────────────────────────────────────────────────────────────────┐
│ Signal conditioning index. (0-7)                                         7) -│
│ Thermocouple type if thermocouple input is through the Analog                │
│ Input Module. (J, K, T, R) [Otherwise enter 11.]                             │
│ Linearization polynomial index. (0-511) [For signal conditioning indexes     │
│ 0 or 5 only Enter 5 only if input is from PCM.]                               │
└─────────────────────────────────────────────────────────────────────────────┘

┌─────────────────────────────────────────────────────────────────────────────┐
│ Is digital integration required (Y or N)                                 8)  │
│       If Y, enter integration multiplier K1, (1-32767)                       │
│       and integration divisor K2. (1-32767)                                  │
│       If N, enter the smoothing index. (0-63)                                │
└─────────────────────────────────────────────────────────────────────────────┘

┌─────────────────────────────────────────────────────────────────────────────┐
│ Operator Console Number (1, 2, or 3)                                    11)  │
└─────────────────────────────────────────────────────────────────────────────┘

┌─────────────────────────────────────────────────────────────────────────────┐
│ Process unit number (1-127;0=none)                                           │
│                                                                         12)  │
└─────────────────────────────────────────────────────────────────────────────┘

┌─────────────────────────────────────────────────────────────────────────────┐
│ Block description for alarm messages, Leading and imbedded blanks will be included.  13)│
└─────────────────────────────────────────────────────────────────────────────┘

┌─────────────────────────────────────────────────────────────────────────────┐
│ Is a supervisory program called when an alarm occurs? (Y or N)          15)  │
│ If Y, enter program call number. (0-2047)                                    │
└─────────────────────────────────────────────────────────────────────────────┘

┌─────────────────────────────────────────────────────────────────────────────┐
│ Should this block inhibit the passing of initialization requests?            │
│       (Y or N)                                                          18)  │
└─────────────────────────────────────────────────────────────────────────────┘

┌─────────────────────────────────────────────────────────────────────────────┐
│ If an input fails, should this block continue control using the last good value?  │
│       (Y or N)                                                          19)  │
└─────────────────────────────────────────────────────────────────────────────┘
```

Figure 5.5   Typical layout sheet for forms entry (adapted from Mellichamp, *Real-time Computing*, Van Nostrand Reinhold (1983)).

blocks (scanner routines, PID control, output routines, arithmetic functions, scaling blocks, alarm routines and display routines), a range of supervisory programs and programs for manipulating the block functions. The engineer programs a control scheme by connecting together the various function blocks which he or she requires and entering the parameters for each block. This is typically done using a VDU with a graphical representation of the block connections. An example of the information which has to be supplied is shown in Figure 5.5.

The block-structured approach is used in a wide range of systems, from large process control systems with several hundred loops and multiple operator display stations to simple programmable controllers used for sequence control. The range of controllers available is shown in Figure 5.6.

5.15.3 Application Languages

Application languages range from simple interpreters which allow for interaction with table-driven or functional block systems to complex high-level languages which have to be compiled. The major feature of such languages is that they provide a syntax which reflects the nature of the application. In the following section a large, complex application language is described in outline.

Software functions:	Basic	Advanced	Process
	Boolean	Block transfer	Signalling
	Timers	Jump	Monitoring
	Counters	File	PID control
	Data move	Shift registers	Communication
	Comparison	Sequencers	Logging
	Arithmetic	Floating point	Display
Hardware functions:	Small PCs		Large PCs
Inputs	16		4096
Outputs	16		4096
Timers	8		256
Counters	8		256
User program	2K		48K
Cycle time (per 1K)	100 ms		1 ms

Figure 5.6   Block functions in programmable controllers.

5.16 CUTLASS

CUTLASS is a high-level language which is oriented towards use by the engineer rather than the professional programmer. It has been developed by the UK Central Electricity Generating Board with the aim of enabling engineering staff to develop, modify and maintain application software independently of professional software support staff.

The major requirements which CUTLASS had to meet are:

1. It should be suitable for a wide range of applications within power stations. All applications packages should operate within the same general framework so that future developments can easily be incorporated without rendering obsolete previous work. To achieve these aims the language was developed in the form of a number of compatible subsets which cover the following functions:
    (a) modulating control;
    (b) data logging;
    (c) data analysis;
    (d) sequence control;
    (e) alarm handling;
    (f) visual display; and
    (g) history recording.
    These subsets operate within a framework which provides:
    (a) a real-time executive – TOPSY;
    (b) communications network management;
    (c) support facilities; and
    (d) integrated I/O and file handling.
2. The software should have a high degree of independence from any particular computer type. This is a particularly important requirement for software which is expected to have a long lifetime – 20 to 30 years – during which period the actual control computers may have to be replaced and it is important that this can be done without having either to use obsolete technology or to incur high program modification costs. In order to achieve this the CUTLASS language has been written in CORAL.
3. The software should be simple and safe to use so that engineers based in the power station can produce and modify programs. This has been achieved by providing within the language an extensive range of subroutines and by hiding the detailed operational and security features from the user.

5.16.1 General Features of CUTLASS

The basic unit of a CUTLASS program is a SCHEME which is an independently compilable unit. A SCHEME is defined as

```
<subset type> SCHEME <name>
GLOBAL <data>
COMMON <data>
TASK <qualifying data>
...
TASK <qualifying data>
ENDSCHEME
```

A scheme may contain any number of tasks and a program may contain any number of schemes. The schemes may run on different computers in a distributed network. The software is developed on a host machine and downloaded to the target machine(s); a typical system is shown in Figure 5.7. During the operation of the overall system the host may remain connected to the target systems.

In addition to the schemes generated by the programmer, support software − including the TOPSY executive − is loaded into the target machine. The schemes can be enabled and disabled by the user from a keyboard connected to the target machine or from the host machine or from within a supervisory task.

As an example of the division of the software for an application into schemes and tasks, consider the hot-air blower system described in Chapter 1. For this system it is required that the temperature measurement be filtered by taking a running average over four samples at 10 ms intervals. The actual control is to be a PID

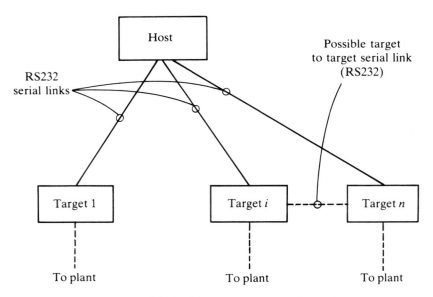

Figure 5.7   CUTLASS host−target configuration.

controller running at 40 ms intervals. The display of the input, output and error is to be updated at 5 s intervals.

EXAMPLE 5.15

Scheme and Task Outline

```
1. TEMPERATURE CONTROL SCHEME
    1A TASK FILTER RUN EVERY 10 MSECS
        Reads temperature from heater and computes the
        running average.
    1B TASK TEMPCON RUN EVERY 40 MSECS
        Uses the average temperature obtained from
        filter and computes, using a PID control
        algorithm, the output for the heater.
2. DISPLAY SCHEME
    2A TASK DISPLAY UPDATE RUN EVERY 5 SECS
        Update the display with the values of
        temperature, error and heater output.
```

A possible arrangement is shown in Example 5.15 in which the program is split into two schemes: TEMPCON, used for the control of the temperature in which there are two tasks (FILTER and TMPCON), and DISPLAY in which there is one task, UPDATE.

The timing of the tasks is controlled by TOPSY and the timing requirement is specified as part of the task qualifying data. The syntax is

```
TASK <identifier> PRIORITY = <priority level> RUN
EVERY <integer> <time interval>
identifier = name
priority level = 1..250
time interval = (MSECS:SECS:MINS:HOURS:DAYS)
```

The outline of the TEMPCON scheme would be as follows:

```
DDC SCHEME TEMPCON
;
; variable declarations placed here
;
TASK FILTER PRIORITY=240 RUN EVERY 10 MSEC
;
; task local declarations placed here
;
START
;
; task body
;
ENDTASK
```

```
;
TASK TMPCON PRIORITY=225 RUN EVERY 40 MSECS
;
; task declarations
;
START
;
; task body
;
ENDTASK
ENDSCHEME
```

As is seen from the above statements, the TOPSY executive supports CYCLIC- or CLOCK-based tasks; it also supports a DELAY timing function, but the DELAY function can be used only within the GEN subject, and then only for tasks which are non-repetitive.

5.16.2 Data Typing and Bad Data

As in Pascal, CUTLASS requires all variables to be associated with a data type when they are declared. The types supported are: logical, integer, real, text, string, array and record.

An additional feature which is used in CUTLASS is that variables can be flagged as good or bad. The concept of 'bad' data is useful in an environment in which a large number of values are derived from plant measurements. Instruments on the plant can become faulty and supply data which is incorrect. It is often easy to detect when an instrument is supplying false information as when, for example, typical plant transducers supply signals in the range 4–20 mA; if the signal falls below 4 mA it is an indication that the transducer is faulty. The difficulty arises in transmitting through the system an indication that the reading from that particular instrument is faulty.

The ultra-safe approach would be to put the whole scheme into manual mode and allow the operator to take over until the instrument is repaired or replaced. In many circumstances such extreme action may be unnecessary; the instrument may only be supplying an operator display, or there may be an alternative measurement available.

The approach adopted is to mark the data value as bad and allow the bad data to propagate through the system, action, if necessary, being taken elsewhere in the system. The rules used in the propagation of bad data produce a result which is marked as bad if any of the operands in an arithmetic operation are bad. For example,

```
A := B + C
E := D / C
```

would result in E being marked as bad if B were bad. The rules for propagation through logic operations are slightly more complex in that a logic variable which is bad is treated as a 'don't know' condition. For example,

```
MODEA := (TEMP > 30.0) OR FANRUN
```

would produce a valid result if TEMP was bad and FANRUN was true, but a bad value if FANRUN was false. An advantage of the bad data flagging is that large sections of the code can be written on the assumption that the data is good: if it is not good then the fact that it is bad will be automatically propagated through the system to the point at which action must be taken.

5.16.3 Language Subsets

The language is divided into four subsets which share some common features but which perform essentially different functions. A scheme may use only instructions from one subset and the subset being used is declared as part of the scheme heading. The subsets are:

1. GEN A general purpose subset which is used to support text input and output; a scheme which uses only the features common to all subsets is also referred to as a GEN scheme.
2. DDC This subset provides the set of instructions used for direct digital control algorithms.
3. SEQ A subset which supports the construction of sequential control algorithms; there is some restriction on the use of the common language features within an SEQ scheme.
4. VDU A subset which provides graphical and text support for a range of VDUs including both colour and monochrome devices.

Some indication of the power of the DDC subset can be obtained from the list of instructions supported which is shown in Figure 5.8.

5.16.4 Scope and Visibility

The CUTLASS system has very simple scope and visibility rules. Objects declared within a task are local to that task: the scope extends throughout the task and the object is visible within the body of the task. The object is not visible within a subroutine called by the task.

Variables may be shared between tasks by declaring them in a COMMON block in the scheme declarations. The compiler restricts variables declared within the COMMON block to the sharing of information between tasks and prevents their use

Function Output
Non-history-dependent functions

AVE   average of all good inputs
EAVE   average of all inputs; if any input is bad then output is bad
MIN   minimum of good inputs
AMIN   absolute minimum of good inputs
EMIN   minimum of inputs; if any input is bad then output is bad
EAMIN   absolute minimum of inputs; if any is bad then output is bad
MAX   maximum of good inputs
AMAX   absolute maximum of good inputs
EMAX   maximum of inputs; if any is bad then output is bad
EAMAX   absolute maximum of inputs; if any bad then output bad
INHIBIT   depends on condition of inhibit raise and lower flags
INCS   converts integer value to incremental units
LIMIT   limit range of real variable
BRAND   deadband function

History-dependent functions

BUCKET   counts increments supplied and used
FLIPFLOP logical flipflop triggered at twice the task repeat interval
CHANGE   logic function $A_n = A_n$ FOR A_{n-1}
RISING   logic function $A_n = A_n$ AND HCT (A_{n-1})
FALLING   logic function A_n NOT (A_n) AND A_{n-1}
ACC   sets accumulator to an initial condition
ACCLIMIT sets hard limits on an accumulator

History- and time-dependent functions

INT   approximate integrator
DELTA   approximate differentiator
FIRST   first-order filter
IMCPID   incremental PID control algorithm, includes roll-off filter
PID   absolute PID algorithm including roll-off filter
EFORM   controller expressed as a polynomial in z
RAMP   limits rate of change of a variable.

Figure 5.8 Example of a CUTLASS DDC scheme (reproduced with permission from
Bennett and Linkens, *Real-time Computer Control*, Peter Peregrinus (1984)).

as local variables as well by enforcing the following rules:

1. Only one task may write to a COMMON variable. For this purpose arrays
    are treated as indivisible objects so that if a task writes to an individual
    element of an array then all the elements become write only for that task.
2. Tasks are not allowed read access to variables to which they write.

Communication between tasks in different schemes is by means of GLOBAL
variables. These variables can be created only by use of a special utility which is run

by a privileged user. GLOBAL variables are owned by the user who created them and may be removed only by the owner. The same access rules as for common variables are applied to GLOBALs with the added restriction that only a task belonging to the owner of a GLOBAL variable may write to that variable.

The rules are relaxed for tasks which belong to users with privileged status such that:

1. GLOBALs may be declared as private to the user.
2. GLOBALs may be written to by more than one task (subject to the ownership rule).
3. Tasks may have both read and write access to the variable.

5.16.5 Summary

CUTLASS represents an attempt to resolve many of the problems which arise in real-time systems. It does not purport to be a general purpose language and hence the solutions which are adopted are not always particularly elegant. The emphasis in the system is on enabling inexperienced computer users to write reliable, secure, programs. In this the language is successful.

However, the overall system is complex and setting up and maintaining it requires the support of experienced computer staff. This does not detract from one of the aims of the system, which was to allow engineers to write their own application programs; this has been achieved. It would be difficult, however, to operate the CUTLASS system in an organisation which did not have expert computer staff.

Although the language syntax is an improvement on CORAL 66, readability is not high in that identifiers are restricted to nine characters and must be in upper case. A further restriction is that user subroutine libraries cannot be developed; all user (as opposed to system) subroutines must be included within the TASK declaration area in source code form. The language supports constants in a useful form, variables can be preset and such variables are then treated as read-only variables.

5.17 A NOTE ON BASIC

BASIC (Beginners' All-purpose Symbolic Instruction Code) was developed as a language which would provide a reasonably powerful range of facilities but which would be easy for the novice to learn and use. In particular the use of an interpretative mode of implementation was intended to make the writing, running and debugging of programs as quick and easy as possible. There is little doubt that BASIC has achieved the designer's aims and its widespread availability on microprocessor systems has also contributed to its use.

There are now many versions of BASIC with numerous extensions to the original language, some of which make them suitable for certain types of real-time systems, for example experimental, prototype and development work where the speed and ease with which BASIC programs can be written and debugged is a great advantage. These BASICs can also be used for small systems where safety is not a major issue.

The simplest extensions to BASIC which provide support for embedded system use are the provisions of low-level access mechanisms. Other extensions include functions for handling input and output to analog-to-digital and digital-to-analog converters (see Mellichamp, 1983), event handling (interrupts) and multi-tasking.

The simplest form of event handling allows the use of statements of the form

```
ON EVENT GOSUB <n>
```

where <n> is a specified line number. This may be extended in some BASICs to allow several different events:

```
ON INTERRUPT0 GOSUB <n1>
ON INTERRUPT1 GOSUB <n2>
ON TIMEOUT GOSUB <n3>
```

The GOSUB is used so that a RETURN statement can be used to indicate the end of the section of code for the particular event. Further extensions provide support for multiple tasks.

Each task body is indicate by statements

```
TASK <name>
...
EXIT
```

which are used to bracket the statements which form the task. The task remains dormant (in terms of the task states described in Chapter 6, existent but not active) until an ENABLE statement is executed:

```
ENABLE <name> EVERY 10cs
```

or

```
ENABLE <name> WHEN event
```

The tasks can be disabled by the statement DISABLE <name>.

The major advantage of BASIC is the ease with which the language can be learnt. If well-trained programmers with experience of other languages are not available then BASIC may be a very sound choice and result in better, more secure software than could be achieved by using an inherently more secure language.

5.18 SUMMARY

In this chapter we have reviewed some of the important language features that might influence our choice of a language for writing real-time software. Particular attention has been paid to elements of languages that contribute to the security of the resulting software. We have not attempted to compare real-time languages; if you are interested in such comparisons you will find a brief survey in Cooling (1991) and a more extensive survey in Tucker (1985). For a greater in-depth study of real-time languages see Young (1982) and Burns and Wellings (1990).

EXERCISES

5.1   Define the scope and visibility of the variables and parameters in the following code:

```
MODULE MyProgram;
VAR A,B:REAL;
    C,D:INTEGER;
PROCEDURE Pone ( A1:REAL;VAR A2:REAL);
  VAR M,N:INTEGER
  BEGIN (* Pone *)
    ..
  END Pone
PROCEDURE Ptwo;
  VAR P,D:INTEGER;
      Q,R:REAL;
  BEGIN (* TWO *)
    ...
    Pone (Q,R);
    ...
  END Ptwo;
  BEGIN (*MyProgram*)
    ...
  END MyProgram.
```

5.2   In the computer science literature you will find lots of arguments about 'global' and 'local' variables. What guidance would you give to somebody who asked for advice on how to decide on the use of global or local variables?

5.3   How does strong data typing contribute to the security of a programming language?

5.4   Why is it useful to have available a predefined data type BITSET in Modula-2? Give an example to illustrate how, and under what circumstances, BITSET would be used.

6

Operating Systems

This chapter is not a complete discussion of operating systems. In it we concentrate on the aspects of operating systems that are particularly relevant to real-time control applications. We first look at what they are, how they differ from non-real-time operating systems and why we use them. We will then examine in some detail how they handle the management of tasks. Finally, we will look briefly at some ways of implementing real-time operating systems.

The aims of the chapter are to:

- Explain why we use a real-time operating system (RTOS).
- Explain what an RTOS does.
- Explain how an RTOS works.
- Describe the benefits and drawbacks of an RTOS.
- List the minimum language primitives required for creating an RTOS.
- Describe the problem of sharing resources and explain several techniques for providing mutual exclusion.
- Explain what a binary semaphore does and write a program in Modula-2 to demonstrate its use.
- Describe and explain the basic task synchronisation mechanisms.

6.1 INTRODUCTION

Software design is simplified if details of the lower levels of implementation on a specific computer using a particular language can be hidden from the designer. An operating system for a given computer converts the hardware of the system into a virtual machine with characteristics defined by the operating system. Operating systems were developed, as their name implies, to assist the operator in running a batch processing computer; they then developed to support both real-time systems and multi-access on-line systems.

The traditional approach is to incorporate all the requirements inside a general purpose operating system as illustrated in Figure 6.1. Access to the hardware of the system and to the I/O devices is through the operating system. In many real-time

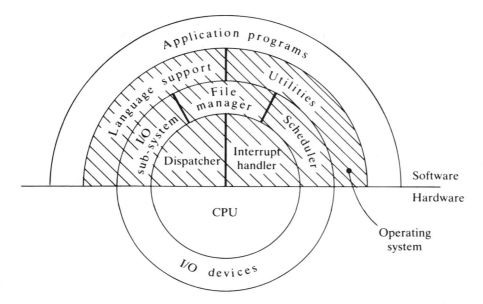

Figure 6.1   General purpose operating system.

and multi-programming systems restriction of access is enforced by hardware and software traps. The operating system is constructed, in these cases, as a monolithic monitor. In single-job operating systems access through the operating system is not usually enforced; however, it is good programming practice and it facilitates portability since the operating system entry points remain constant across different implementations. In addition to supporting and controlling the basic activities, operating systems provide various utility programs, for example loaders, linkers, assemblers and debuggers, as well as run-time support for high-level languages.

A general purpose operating system will provide some facilities that are not required in a particular application, and to be forced to include them adds unnecessarily to the system overheads. Usually during the installation of an operating system certain features can be selected or omitted. A general purpose operating system can thus be 'tailored' to meet a specific application requirement.

Recently operating systems which provide only a minimum kernel or nucleus have become popular; additional features can be added by the applications programmer writing in a high-level language. This structure is shown in Figure 6.2. In this type of operating system the distinction between the operating system and the application software becomes blurred. The approach has many advantages for applications that involve small, embedded systems.

The relationship between the various sections of a simple operating system, the computer hardware and the user is illustrated in Figure 6.3. The command processor provides a means by which the user can communicate with the operating system from the computer console device. Through it the user issues commands to the

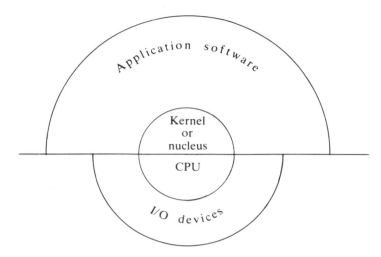

Figure 6.2   Minimal operating system.

operating system and it provides the user with information about the actions being performed by the operating system. The actual processing of the commands issued by the user is done by the BDOS (Basic Disk Operating System) which also handles the input and output and the file operations on the disks. The BDOS makes the actual management of the file and input/output operations transparent to the user. Application programs will normally communicate with the hardware of the system through *system calls* which are processed by the BDOS.

The BIOS (Basic Input Output System) contains the various device drivers which manipulate the physical devices and this section of the operating system may vary from implementation to implementation as it has to operate directly with the underlying hardware of the computer. For example, the physical addresses of the peripherals may vary according to the manufacturer; these differences will be accommodated in the coding of the BIOS.

Devices are treated as *logical* or *physical* units. Logical devices are software constructs used to simplify the user interface; user programs perform input and output to logical devices and the BDOS connects the logical device to the physical device. The actual operation of the physical device is performed by software in the BIOS.

Access to the operating system functions is by means of subroutine calls and information is passed in the CPU registers of the machine. Functions cannot be called directly from most high-level languages and this provides isolation between the operating system and a programmer using a high-level language. The isolation is deliberate; it is an example of *information hiding*. The connection between the high-level language and the operating system is made by the compiler writer through the provision of run-time support routines which convert the operating system into the virtual machine described by the high-level language.

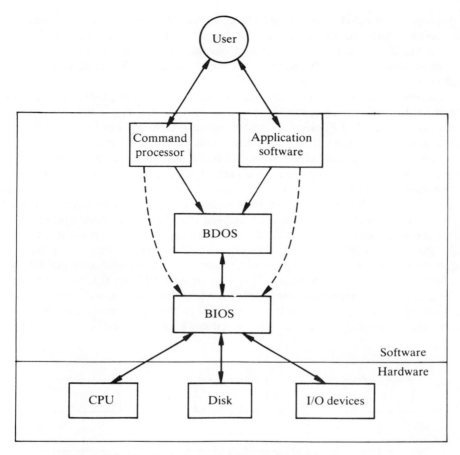

Figure 6.3   General structure of a simple operating system.

The isolation is not complete in that it is possible to call assembly-coded routines from high-level languages and to pass parameters between the high-level language code and the assembly code; this does, however, require detailed knowledge of the system. Again information hiding is used in that the details of the physical implementation on the CPU and of the I/O devices are hidden within the operating system and hence operations are performed on the operating system virtual machine.

6.2 REAL-TIME MULTI-TASKING OPERATING SYSTEMS

There are many different types of operating systems and until the early 1980s there was a clear distinction between operating systems designed for use in real-time applications and other types of operating system. In recent years the dividing line has become blurred. For example, languages such as Modula-2 enable us to construct multi-tasking real-time applications that run on top of single-user, single-task operating systems. And operating systems such as UNIX and OS/2 support multi-user, multi-tasking applications.

Confusion can arise between multi-user or multi-programming operating systems and multi-tasking operating systems. The function of a multi-user operating system is illustrated in Figure 6.4: the operating system ensures that each user can run a single program as if they had the whole of the computer system for their program. Although at any given instance it is not possible to predict which user will have the use of the CPU, or even if the user's code is in the memory, the operating system ensures that one user program cannot interfere with the operation of another user program. Each user program runs in its own protected environment. A primary concern of the operating system is to prevent one program, either deliberately or through error, corrupting another. In a multi-tasking operating system it is assumed that there is a single user and that the various tasks co-operate to serve the requirements of the user. Co-operation requires that the tasks communicate with each other and share common data. This is illustrated in Figure 6.5. In a good multi-tasking operating system task communication and data sharing will be regulated so that the operating system is able to prevent inadvertent communication or data access (that is, arising through an error in the coding of one task) and hence protect

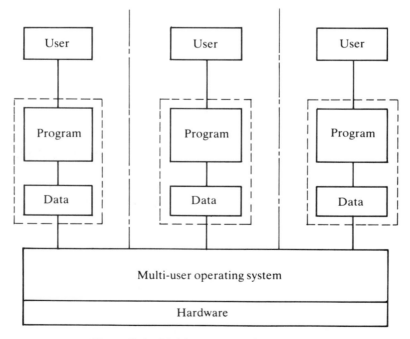

Figure 6.4 Multi-user operating system.

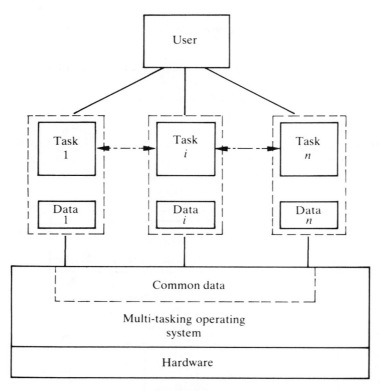

Figure 6.5 Multi-tasking operating system.

data which is private to a task (note that deliberate interference cannot be prevented – the tasks are assumed to be co-operating).

A fundamental requirement of an operating system is to allocate the resources of the computer to the various activities which have to be performed. In a real-time operating system this allocation procedure is complicated by the fact that some of the activities are time critical and hence have a higher priority than others. Therefore there must be some means of allocating priorities to tasks and of scheduling allocation of CPU time to the tasks according to some priority scheme.

A task may use another task, that is it may require certain activities which are contained in another task to be performed and it may itself be used by another task. Thus tasks may need to communicate with each other. The operating system must have some means of enabling tasks either to share memory for the exchange of data, or to provide a mechanism by which tasks can send messages to each other. Also tasks may need to be invoked by external events and hence the operating system must support the use of interrupts. Similarly tasks may need to share data and they may require access to various hardware and software components; hence there has to be a mechanism for preventing two tasks from attempting to use the same resource at the same time.

In summary a real-time multi-tasking operating system has to support the resource sharing and the timing requirements of the tasks and the functions can be divided as follows:

Task management: the allocation of memory and processor time (scheduling) to tasks.

Memory management: control of memory allocation.

Resource control: control of all shared resources other than memory and CPU time.

Intertask communication and synchronisation: provision of support mechanisms to provide safe communication between tasks and to enable tasks to synchronise their activities.

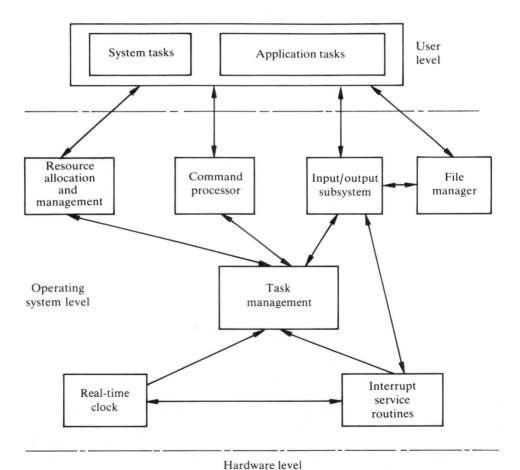

Figure 6.6   Typical structure of a real-time operating system.

In addition to the above the system has to provide the standard features such as support for disk files, basic input/output device drivers and utility programs. The typical structure is illustrated in Figure 6.6. The overall control of the system is provided by the *task management* module which is responsible for allocating the use of the CPU. This module is often referred to as the *monitor* or as the *executive control program* (or more simply the executive). At the user level, in addition to application tasks, a box labelled 'system tasks' is also shown since in many operating systems some operations performed by the operating system and the utility programs run in the memory space allocated to the user or applications – this space is sometimes called 'working memory'.

From the user's viewpoint the two most important features of task management are how to create a task, that is make its existence known to the RTOS, and what scheduling strategy or strategies the RTOS supports. Task creation is largely a function of the interface between the operating system and high-level programming language and we gave some examples of the mechanisms involved in the previous chapter.

6.3 SCHEDULING STRATEGIES

If we consider the scheduling of time allocation on a single CPU there are two basic strategies:

1. Cyclic.
2. Pre-emptive.

6.3.1 Cyclic

The first of these, cyclic, allocates the CPU to a task in turn. The task uses the CPU for as long as it wishes. When it no longer requires it the scheduler allocates it to the next task in the list. This is a very simple strategy which is highly efficient in that it minimises the time lost in switching between tasks. It is an effective strategy for small embedded systems for which the execution times for each task run are carefully calculated (often by counting the number of machine instruction cycles for the task) and for which the software is carefully divided into appropriate task segments. In general this approach is too restrictive since it requires that the task units have similar execution times. It is also difficult to deal with random events using this method.

6.3.2 Pre-emptive

There are many pre-emptive strategies. All involve the possibility that a task will be interrupted – hence the term pre-emptive – before it has completed a particular invocation. A consequence of this is that the executive has to make provision to save the volatile environment for each task, since at some later time it will be allocated CPU time and will want to continue from the exact point at which it was interrupted. This process is called *context switching* and a mechanism for supporting it is described below.

The simplest form of pre-emptive scheduling is to use a time slicing approach (sometimes called a round-robin method). Using this strategy each task is allocated a fixed amount of CPU time – a specified number of *ticks* of the clock – and at the end of this time it is stopped and the next task in the list is run. Thus each task in turn is allocated an equal share of the CPU time. If a task completes before the end of its time slice the next task in the list is run immediately.

The majority of existing RTOSs use a priority scheduling mechanism. Tasks are allocated a priority level and at the end of a predetermined time slice the task with the highest priority of those ready to run is chosen and is given control of the CPU. Note that this may mean that the task which is currently running continues to run.

Task priorities may be fixed – a *static* priority system – or may be changed during system execution – a *dynamic* priority system. Dynamic priority schemes can increase the flexibility of the system, for example they can be used to increase the priority of particular tasks under alarm conditions. Changing priorities is, however, risky as it makes it much harder to predict the behaviour of the system and to test it. There is the risk of locking out certain tasks for long periods of time. If the software is well designed and there is adequate computing power there should be no need to change priorities – all the necessary constraints will be met. If it is badly designed and/or there are inadequate computing resources then dynamic allocation of priorities will not produce a viable, reliable system.

Whatever scheduling strategy is adopted the task management system has to deal with the handling of interrupts. These may be hardware interrupts caused by external events, or software interrupts generated by a running task. An interrupt forces a context switch. The running task is suspended and an interrupt handler is run. The interrupt handler should only contain a small amount of code and should execute very quickly. When the handler terminates either the task that was interrupted is restored or the scheduler is entered and it determines which task should run. The RTOS designer has to decide which approach to adopt.

EXAMPLE 6.1

Interrupt Handling and Scheduling

A system receives an alarm signal interrupt from a plant and in response to the alarm it is to run an alarm alert task which is a high-priority, base level task. The interrupt service routine for the alarm signal will, by some mechanism, cause the alarm alert task to be placed in the runnable queue, and there are then two actions which it can take: (a) return to the interrupted task, (b) jump to the scheduler. If a return to the interrupted task is made then the alarm alert task will not be run until the system reschedules either at the system rescheduling interval or because the running task terminates or becomes suspended waiting for a system resource. However, if a jump is made directly to the dispatcher from the interrupt service routine, then if the alarm alert task is of higher priority than the interrupted task it will be run immediately and the interrupted task will have been pre-empted. The argument for entering the scheduler and rescheduling is that the occurrence of an interrupt is likely to have changed the state of a task and hence the task that was running may no longer be the highest-priority task. The argument for returning to the running task is that it involves less time loss since performing only the context switch requires less CPU time than running the scheduler as well. A rational decision depends on the executive having some knowledge of the application: if it is known that the majority of interrupts are generated by events that have high priority – alarms, or important changes in plant conditions – then entering the scheduler after an interrupt is the best choice; on the other hand, if a large number of interrupts are from serial-based input and output and communications devices that simply involve placing a character in a buffer then to enter the scheduler every time would be a waste of CPU time.

6.4 PRIORITY STRUCTURES

In a real-time system the designer has to assign priorities to the tasks in the system. The priority will depend on how quickly a task will have to respond to a particular event. An event may be some activity of the process or may be the elapsing of a specified amount of time. Most RTOSs provide facilities such that tasks can be divided into three broad levels of priority as shown in Figure 6.7.

1. Interrupt level: at this level are the service routines for the tasks and devices which require very fast response – measured in milliseconds. One of these tasks will be the real-time clock task and clock level dispatcher.
2. Clock level: at this level are the tasks which require repetitive processing, such as the sampling and control tasks, and tasks which require accurate timing. The lowest-priority task at this level is the base level scheduler.
3. Base level: tasks at this level are of low priority and either have no deadlines to meet or are allowed a wide margin of error in their timing. Tasks at this level may be allocated priorities or may all run at a single priority level – that of the base level scheduler.

6.4.1 Interrupt Level

As we have already seen an interrupt forces a rescheduling of the work of the CPU and the system has no control over the timing of the rescheduling. Because an interrupt-generated rescheduling is outside the control of the system it is necessary to keep the amount of processing to be done by the interrupt handling routine to a minimum. Usually the interrupt handling routine does sufficient processing to preserve the necessary information and to pass this information to a further handling routine which operates at a lower-priority level, either clock level or base level. Interrupt handling routines have to provide a mechanism for task swapping, that is they have to save the volatile environment. On completion the routine either

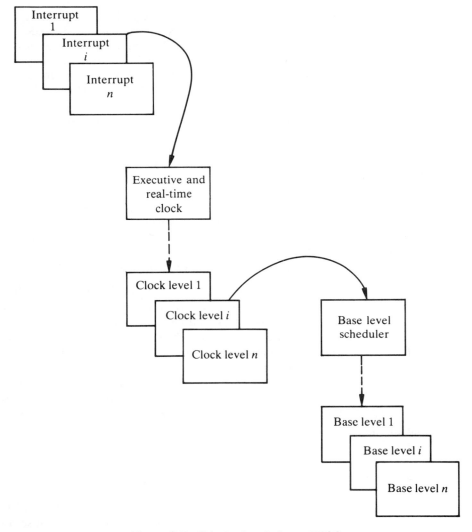

Figure 6.7   Priority levels in an RTOS.

will simply restore the volatile environment and hence will return to the interrupted task, or it may exit to the scheduler.

Within the interrupt level of tasks there will be different priorities and there will have to be provision for preventing interrupts of lower priority interrupting higher-priority interrupt tasks. On most modern computer systems there will be hardware to assist in this operation (see Chapter 3).

6.4.2 Clock Level

One interrupt level task will be the real-time clock handling routine which will be entered at some interval, usually determined by the required activation rate for the most frequently required task. Typical values are 1 to 200 ms. Each clock interrupt is known as a *tick* and represents the smallest time interval known to the system. The function of the clock interrupt handling routine is to update the time-of-day clock in the system and to transfer control to the dispatcher. The scheduler selects which task is to run at a particular clock tick.

Clock level tasks divide into two categories:

1. *CYCLIC*: these are tasks which require accurate synchronisation with the outside world.
2. *DELAY*: these tasks simply wish to have a fixed delay between successive repetitions or to delay their activities for a given period of time.

6.4.3 Cyclic Tasks

The *cyclic* tasks are ordered in a priority which reflects the accuracy of timing required for the task, those which require high accuracy being given the highest priority. Tasks of lower priority within the clock level will have some jitter since they will have to await completion of the higher-level tasks.

EXAMPLE 6.2

Cyclic Tasks

Three tasks *A*, *B* and *C* are required to run at 20 ms, 40 ms and 80 ms intervals (corresponding to 1 tick, 2 ticks and 4 ticks, if the clock interrupt rate is set at 20 ms). If the task priority order is set as *A*, *B* and *C* with *A* as the highest priority then the processing will proceed as shown in Figure 6.8a with the result that the tasks will be run at constant intervals. It should be noted that using a single CPU it is not possible to have all the tasks starting in synchronism with the clock tick. All but one of the tasks will be delayed relative to the clock tick; however, the interval between

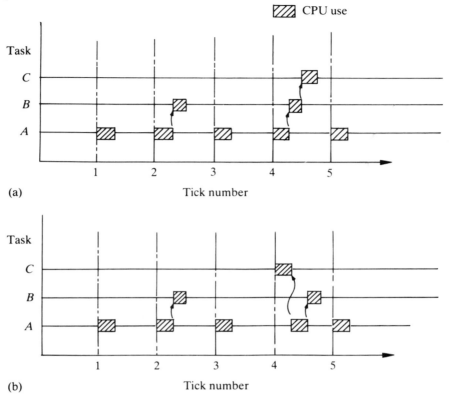

Figure 6.8   Task activation diagram for Example 6.2: (a) task priorities A,B,C;
(b) task priorities C,A,B.

successive invocations of the task will be constant (if the execution time for each task
is a constant value). If the priority order is now rearranged so that it is *C, A* and
B then the activation diagram is as shown in Figure 6.8b and every fourth tick of
the clock there will be a delay in the timing of tasks *A* and *B*. In practice there
is unlikely to be any justification for choosing a priority order *C, A* and *B* rather
than *A, B* and *C*. Usually the task with the highest repetition rate will have
the most stringent timing requirements and hence will be assigned the highest
priority.

A further problem which can arise is that a clock level task may require a longer
time than the interval between clock interrupts to complete its processing (note that
for overall satisfactory operation of the system such a task cannot run at a high
repetition rate).

EXAMPLE 6.3

Timing of Cyclic Tasks

Assume that in Example 6.2 task *C* takes 25 ms to complete, task *A* takes 1 ms and task *B* takes 6 ms. If task *C* is allowed to run to completion then the activity diagram will be as shown in Figure 6.9 and task *A* will be delayed by 11 ms at every fourth invocation. It is normal therefore to divide the *cyclic* tasks into high-priority tasks which are guaranteed to complete within the clock interval and lower-priority tasks which can be interrupted by the next clock tick.

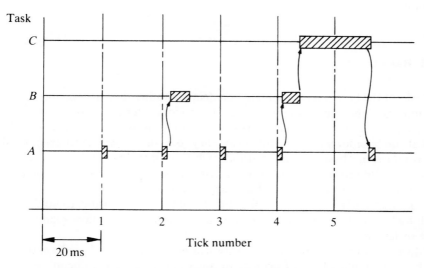

Figure 6.9   Task activation diagram for Example 6.3.

6.4.4 Delay Tasks

The tasks which wish to delay their activities for a fixed period of time, either to allow some external event to complete (for example, a relay may take 20 ms to close) or because they only need to run at certain intervals (for example, to update the operator display), usually run at the base level. When a task requests a delay its status is changed from runnable to suspended and remains suspended until the delay period has elapsed.

One method of implementing the delay function is to use a queue of task descriptors, say identified by the name DELAYED. This queue is an ordered list of task descriptors, the task at the front of the queue being that whose next running time is nearest to the current time. When a task delays itself it calls an executive task which calculates the time when the task is next due to run and inserts the task descriptor in the appropriate place in the queue.

A task running at the clock level checks the first task in the DELAYED queue to see if it is time for that task to run. If the task is due to run it is removed from the DELAYED queue and placed in the runnable queue. The task which checks the DELAYED queue may be either the dispatcher which is entered every time the real-time clock interrupts or another clock level task which runs at less frequent intervals, say every 10 ticks, in which case it is then frequently part of the base level scheduler. Many real-time operating systems do not support the cycle operation and the user has to create an accurate repetitive timing for the task by using the delay function.

6.4.5 Base Level

The tasks at the base level are initiated on demand rather than at some predetermined time interval. The demand may be user input from a terminal, some process event or some particular requirement of the data being processed. The way in which the tasks at the base level are scheduled can vary; one simple way is to use time slicing on a round-robin basis. In this method each task in the runnable queue is selected in turn and allowed to run until either it suspends or the base level scheduler is again entered. For real-time work in which there is usually some element of priority this is not a particularly satisfactory solution. It would not be sensible to hold up a task, which had been delayed waiting for a relay to close but was now ready to run, in order to let the logging task run.

Most real-time systems use a priority strategy even for the base level tasks. This may be either a fixed level of priority or a variable level. The difficulty with a fixed level of priority is in determining the correct priorities for satisfactory operation. The ability to change priorities dynamically allows the system to adapt to particular circumstances. Dynamic allocation of priorities can be carried out using a high-level scheduler or can be done on an *ad hoc* basis from within specific tasks. The high-level scheduler is an operating system task which is able to examine the use of the system resources; it may for example check how long tasks have been waiting and increase the priority of the tasks which have been waiting a long time. The difficulty with the high-level scheduler is that the algorithms used can become complicated and hence the overhead in running can become significant.

Alternatively priorities can be adjusted in response to particular events or under the control of the operator. For example, alarm tasks will usually have a high priority and during an alarm condition tasks such as the log of plant data may be delayed with the consequence that the output of the log lags behind real time (note that the data will be stored in buffer areas inside the computer). So that the log can catch up with real time quickly it may be advisable to increase, temporarily, the priority of the printer output task.

6.5 TASK MANAGEMENT

The basic functions of the task management module or executive are:

1. to keep a record of the state of each task;
2. to schedule an allocation of CPU time to each task; and
3. to perform the context switch, that is to save the status of the task that is currently using the CPU and restore the status of the task that is being allocated CPU time.

In most real-time operating systems the executive dealing with the task management functions is split into two parts: a scheduler which determines which task is to run next and which keeps a record of the state of the tasks, and a dispatcher which performs the context switch.

6.5.1 Task States

With one processor only one task can be running at any given time and hence the other tasks must be in some other state. The number of other states, the names

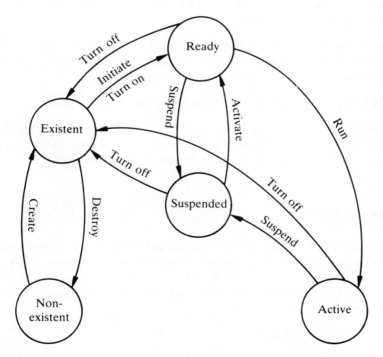

Figure 6.10 Example of a typical task state diagram.

given to the states, and the transition paths between the different states vary from operating system to operating system. A typical state diagram is given in Figure 6.10 and the various states are as follows (names in parentheses are commonly used alternatives):

- *Active (running)*: this is the task which has control of the CPU. It will normally be the task with the highest priority of the tasks which are ready to run.
- *Ready (runnable, on)*: there may be several tasks in this state. The attributes of the task and the resources required to run the task must be available for the task to be placed in the *Ready* state.
- *Suspended (waiting, locked out, delayed)*: the execution of tasks placed in this state has been suspended because the task requires some resource which is not available or because the task is waiting for some signal from the plant, for example input from the analog-to-digital converter, or because the task is waiting for the elapse of time.
- *Existent (dormant, off)*: the operating system is aware of the existence of this task, but the task has not been allocated a priority and has not been made runnable.
- *Non-existent (terminated)*: the operating system has not as yet been made aware of the existence of this task, although it may be resident in the memory of the computer.

The status of the various tasks may be changed by actions within the operating system – a resource becoming available or unavailable – or by commands from the application tasks. A typical command is:

TURN ON (ID) – transfer a task from *existent* to *ready* state,

where ID is the name by which the task is known to the operating system. A typical set of commands is given in Table 6.1. It should be noted that the transition from *ready* to *active* can only be made at the behest of the dispatcher.

6.5.2 Task Descriptor

Information about the status of each task is held in a block of memory by the RTOS. This block is referred to by various names: *task descriptor* (TD), *process descriptor* (PD), *task control block* (TCB) or *task data block* (TDB). The information held in the TD will vary from system to system, but will typically consist of the following:

- task identification (ID);
- task priority (P);
- current state of task;

Table 6.1   RTOS task state transition commands

OFFC01	Turn off the task leaving the memory marked as occupied
OFFC02	Turn off the task leaving the memory marked as unoccupied
DELC01	Delay the task leaving the memory marked as occupied; delay is calculated using current value of time
DELC02	Delay the task leaving the memory marked as unoccupied; delay is calculated as for DELC01
DELC03	Delay the task leaving the memory marked as occupied; delay is calculated by adding the delay to the value of time stored in the task descriptor
TPNC01	Turn the task on; will be accepted if the task is ON, OFF or DELAYED; either the ON constant can be placed in the task descriptor or a specified turn-on time
TPNC02	Turn on the task; will only be accepted if the task is in the OFF state
TPNC03	Run the task immediately regardless of priority; will be accepted if the task is ON, OFF or DELAYED

- area to store volatile environment (or a pointer to an area for storing the volatile environment); and
- pointer to next task in a list.

The reason for including the last item in the list above is that the task descriptors are usually held in a linked list structure. The executive keeps a set of lists, one for each task state as shown in Figure 6.11. There is one active task (task ID = 10) and three tasks that are ready to run (IDs = 20, 9 and 6). The entry held in the executive for the ready queue head points to task 20, which in turn points to task 9 and so on.

The advantage of the list structure is that the actual task descriptor can be located anywhere in the memory and hence the operating system is not restricted to a fixed number of tasks as was often the case in the older operating systems which used fixed length tables to hold task state information. With the list structure moving tasks between lists, reordering the lists, creating and deleting tasks can all be achieved simply by changing pointers. There is no need to copy or move the task descriptors themselves.

The information that has to be stored in order to continue running a task that has been suspended by the scheduler for some reason or other comprises:

Housekeeping information:	CPU register contents;
	stack pointer;
	program counter.
Task data:	task stack;
	task general work area (heap).

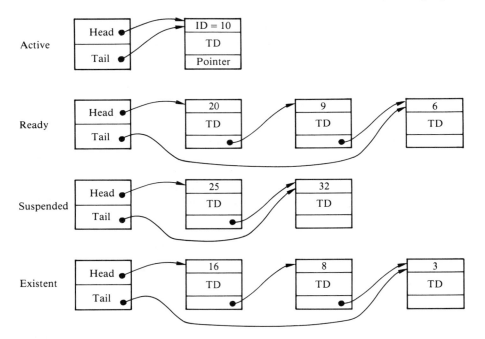

Figure 6.11   List structure for holding task state information.

6.6 SCHEDULER AND REAL-TIME CLOCK INTERRUPT HANDLER

The real-time clock handler and the scheduler for the clock level tasks must be carefully designed as they run at frequent intervals. Particular attention has to be paid to the method of selecting the tasks to be run at each clock interval. If a check of all tasks were to be carried out then the overheads involved could become significant.

6.6.1 System Commands Which Change Task Status

The range of system commands affecting task status varies with the operating system. Typical states and commands are shown in Figure 6.12 and fuller details of the commands are given in Table 6.1. Note that this system distinguishes between tasks which are suspended awaiting the passage of time − these tasks are marked as delayed − and those tasks which are waiting for an event or a system resource − these are marked as locked out.

The system does not explicitly support base level tasks; however, the lowest four priority levels of the clock level tasks can be used to create a base level system. A

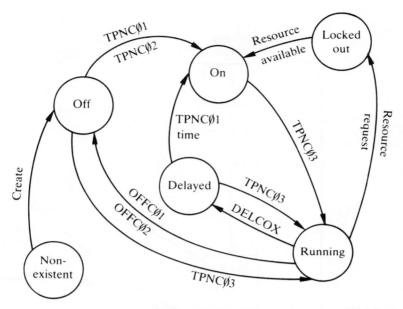

Figure 6.12 RTOS task state diagram.

so-called free time executive (FTX) is provided which if used runs at priority level $n - 3$ (see Figure 6.13) where n is the lowest-priority task number. The FTX is used to run tasks at priority levels $n - 2$, $n - 1$ and n; it also provides support for the chaining of tasks. The dispatcher is unaware of the fact that tasks at these three priority levels are being changed; it simply treats whichever tasks are in the lowest three priority levels as low-priority tasks. Tasks run under the FTX do not have access to the system commands (except OFFC01, that is turn task off).

6.6.2 Dispatcher – Search for Work

The dispatcher/scheduler has two entry conditions:

1. the real-time clock interrupt and any interrupt which signals the completion of an input/output request;
2. a task suspension due to a task delaying, completing or requesting an input/output transfer.

In response to the first condition the scheduler searches for work starting with the highest-priority task and checking each task in priority order (see Figure 6.14). Thus if tasks with a high repetition rate are given a high priority they will be treated as if they were clock level tasks, that is they will be run first during each system clock period. In response to the second condition a search for work is started at the task

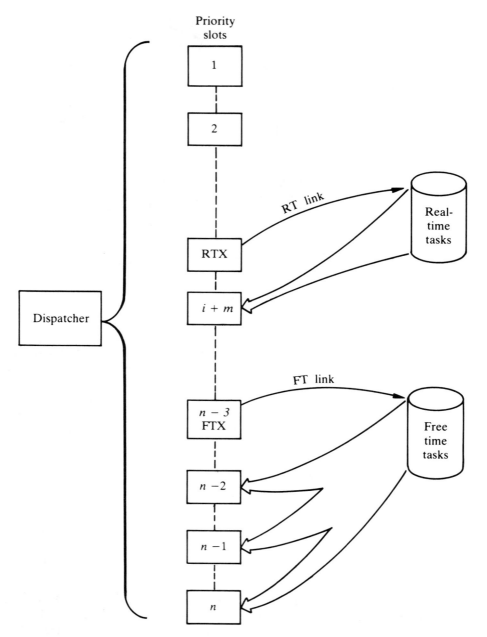

Figure 6.13   RTOS task structure diagram.

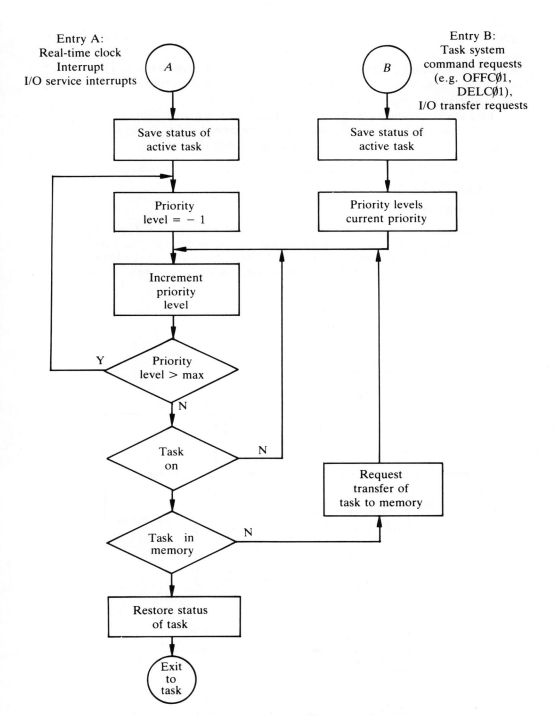

Figure 6.14   RTOS search for work by the dispatcher.

with the next lowest priority to the task which has just been running. There cannot be another higher-priority task ready to run since a higher-priority task becoming ready always pre-empts a lower-priority-running task.

The system commands for task management are issued as calls from the assembly level language and the parameters are passed either in the CPU registers or as a control word immediately following the call statement.

EXAMPLE 6.4

Use of RTOS System Calls

As an example consider the system whose outline design is given in Figure 6.15. It is assumed that the Control, Display and Operator input programs are to be run as separate tasks with priorities 1, 10, 20, respectively. The Control task has to run at 40 ms intervals and the Display update task at 5 s intervals. The system

Foreground

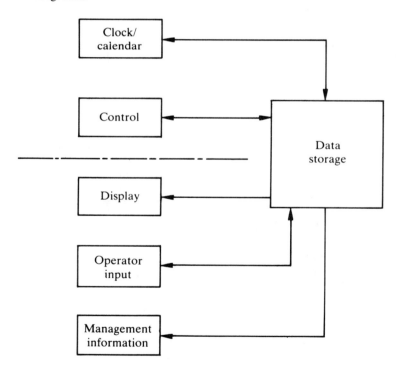

Background

Figure 6.15   Software modules for foreground/background system showing data storage.

clock is set at 20 ms and hence the `Control` task has to run every 2 system ticks. The outline structure of the system is given below:

```
TASK MAIN
;
; Starts up the system by creating the various
; tasks and setting them to the ON condition
;
  CREATE(CONTROL,1,STCTRL)
  CREATE(DISPLAY,10,STDISP)
  CREATE(OPERATOR,20,STOPR)
;
; STCTRL, STDISP, STOPR are common symbols which
; define the starting locations for each task,
; the values will be inserted by the linker/loader.
;
  LDA TIME ; TIME is system variable which
      ; gives current time
  CALL TPNC01
  FCB 1 ; turn on control task
;
  LDA 0
  CALL TPNC02
  FCB 10    ; turn on display task
;
  LDA 0
  CALL TPNC02
  FCB 20 ; turn on operator input task
;
  CALL OFFC02 ; terminate main task
;
  END
```

In the above code by using `TPNC01` to turn on the control task the current value of time is placed in the task descriptor and hence the task can be synchronised to the clock.

```
TASK CONTROL
;
  .....
  main body of task
  .....
;
  CALL DELC03
  FCB 0,2 ; set next time for running
        ; to previous time plus two ticks
;
  END
```

```
    TASK DISPLAY
;
    ....
    main body of task
    ....
;
    CALL DELCO2
    FCB 5,0 ; delay task by 5 seconds
;
    END
```

The difference between using DELCO3 and DELCO2 in the above task segments is that DELCO3 adds the delay increment to the value of time stored in the task descriptor; this time is the time at which the task was last due to run. The use of DELCO3 therefore provides a means of running tasks in a cyclic mode at clock level. The DELCO2 command adds the delay value to the current time and stores the result in the task descriptor; hence the delay is calculated not from when the task was last due to run, but from the time at which the delay command is issued.

In designing a real-time system it is important to know how the scheduler searches for work. In the system described in Example 6.4, the scheduler searches in strict priority order and hence the overheads in terms of the time spent searching for work will be increased if some of the high-priority tasks rarely run. A careful assessment of task priority is required and particular attention will have to be paid to alarm action tasks. Such tasks are normally accorded high priority; however, it is hoped that they will rarely be required. One solution with the above system which avoids having a group of high-priority but rarely run alarm tasks is to make use of the TPNCO3 command. The alarm action tasks are given low priority, but can be made to run immediately if the alarm scanning routine uses the TPNCO3 call to invoke the appropriate task.

6.7 MEMORY MANAGEMENT

Since the majority of control application software is static — the software is not dynamically created or eliminated at run-time — the problem of memory management is simpler than for multi-programming, on-line systems. Indeed with the cost of computer hardware, both processors and memory, reducing many control applications use programs which are permanently resident in fast access memory.

With permanently resident software the memory can be divided as shown in Figure 6.16. The user space is treated as one unit and the software is linked and loaded as a single program into the user area. The information about the various

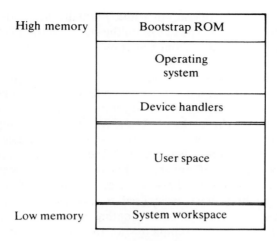

High memory

Low memory

Figure 6.16   Non-partitioned memory.

tasks is conveyed to the operating system by means of a create task statement. Such a statement may be of the form

```
Create(TaskID, Priority, StartAddress, WorkSpace)
```

The exact form of the statement will depend on the interface between the high-level language and the operating system.

An alternative arrangement is shown in Figure 6.17. The available memory is divided into predetermined segments and the tasks are loaded individually into the

| Bootstrap ROM |
| Operating system |
| Device handlers |
| User task area 1 |
| User task area 2 |
| User task area 3 |
| User task area 4 |
| System workspace |

Figure 6.17   Partitioned memory.

various segments. The load operation would normally be carried out using the command processor. With this type of system the entries in the TD (or the operating system tables) have to be made from the console using a memory examine and change facility.

Divided (partitioned) memory was widely used in many early real-time operating systems and it was frequently extended to allow several tasks to share one partition; the tasks were kept on the backing store and loaded into the appropriate partition when required. There was of course a need to keep any tasks in which timing was crucial (hard time constraint tasks) in fast access memory permanently. Other tasks could be swapped between fast memory and backing store. The difficulty with this method is, of course, in choosing the best mix of partition sizes. The partition size and boundaries have to be determined at system generation.

A number of methods have been used to overcome the problem of fixed partitions. One method, referred to as floating memory, divides the available memory into small blocks, for example 64 words. The tasks are installed on the backing store and when a task is required to run the operating system examines a map of memory and finds a contiguous area of memory which will hold the task. The task is loaded into the memory and the memory blocks occupied by the tasks are marked as occupied in the memory map. The area occupied by a task is closely related to the actual size of the task and not to some predetermined fixed partition size. A task which for some reason becomes suspended or delayed will have the memory area it occupies marked in the memory map as occupied but available; hence if another task becomes ready then the suspended task can be returned to backing store and the ready task loaded into its area. In this type of system information must be held in the task descriptor to indicate if the task can be swapped, since, for example, a control task which has to run every 40 ms would have to be held permanently in memory in order to guarantee the sampling rate. A problem which is generated by this system is fragmentation of the available memory. Small areas of free memory become spread about the memory address space; none of the individual areas are large enough to take a task but the combined areas could if they could be brought together. Some form of *garbage* collection is necessary to bring dispersed areas into contiguous blocks.

Other systems which permit dynamic allocation of memory allow the tasks themselves to initiate program segment transfers, either by chaining or by overlaying. In chaining the task is divided into several segments which run sequentially. On completion of one segment the next segment is loaded from memory into the area occupied by the previous segment; any data required to be passed is held either on the disk or in a common area of memory.

Task swapping involves one task invoking another task: the first task is transferred to backing store and the second task brought into memory and made available to run. The procedure is shown in Figure 6.18. Task 1 invokes task 5 by swapping it into priority level 41 and in turn task 5 chains task 6 into level 41. Task 6 swaps task 7 into level 42. When task 7 terminates the operating system returns control to task 6 and when it terminates control is returned to task 1. It should be noted that task 1 remains suspended until task 6 terminates and similarly task 6 is suspended until task 7 terminates.

The difference between chaining and overlaying is that in overlaying a part of the task, the root task, remains in memory and the various segments are brought into an overlay area of memory. In a multi-tasking system there may be several

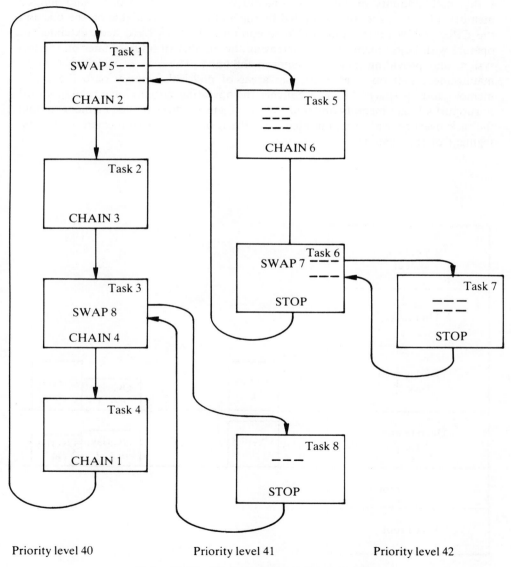

Priority level 40   Priority level 41   Priority level 42

Figure 6.18   Task chaining and swapping.

different overlay areas each of which may be shared by several tasks. A typical arrangement is shown in Figure 6.19 in which it is assumed that two tasks (1 and 15) have overlay segments; each task maintains a root segment and overlay area and the various overlays are loaded into and out of the overlay areas.

Many of the real-time operating systems which provide facilities to swap tasks between fast access memory and backing store were designed for computer systems with small amounts of fast access memory, typically 32K words, and limited memory address space (this is limited by the address lines available on the bus and the CPU architecture). Some of these computer systems have been extended to operate with larger memories by increasing the number of address lines on the bus system and providing memory management units. The way in which a memory management unit operates is to map areas of physical memory onto the actual memory address space. As a consequence the operating systems have been modified to support a larger memory and many do this by using the extended memory to hold the task overlays with a requirement that the root task still be located in the first segment of the memory.

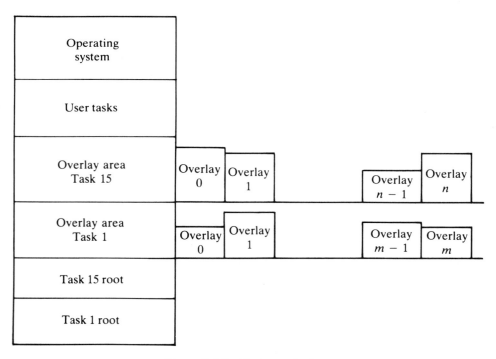

Figure 6.19   Task overlaying.

Dynamic memory allocation is complex to handle and should be avoided wherever possible in embedded real-time systems. RAM is now so cheap that the cost of adding extra memory is usually much less than the cost of programming to provide dynamic memory allocation. It should **never** be used in safety-critical applications.

6.8 CODE SHARING

In many applications the same actions have to be carried out in several different tasks. In a conventional program the actions would be coded as a subroutine and one copy of the subroutine would be included in the program. In a multi-tasking system each task must have its own copy of the subroutine or some mechanism must be provided to prevent one task interfering with the use of the code by another task. The problems which can arise are illustrated in Figure 6.20. Two tasks share the

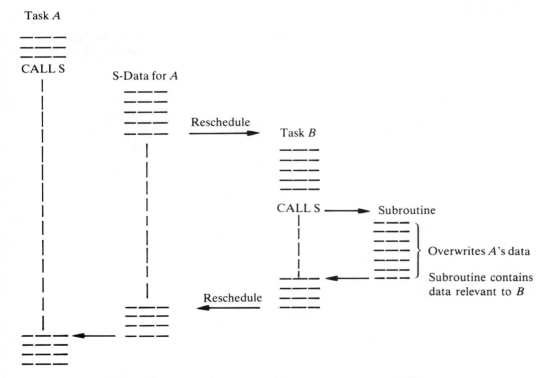

Figure 6.20   Sharing a subroutine in a multi-tasking system.

subroutine s. If task *A* is using the subroutine but before it finishes some event occurs which causes a rescheduling of the tasks and task *B* runs and uses the subroutine, then when a return is made to task *A*, although it will begin to use subroutine *S* again at the correct place, the values of locally held data will have been changed and will reflect the information processed within the subroutine by task *B*.

Two methods can be used to overcome this problem:

- serially reusable code; and
- re-entrant code.

6.8.1 Serially Reusable Code

As shown in Figure 6.21, some form of lock mechanism is placed at the beginning of the routine such that if any task is already using the routine the calling task will not be allowed entry until the task which is using the routine unlocks it. The use of a lock mechanism to protect a subroutine is an example of the need for mechanisms to support mutual exclusion when constructing an operating system.

6.8.2 Re-entrant Code

If the subroutine can be coded such that it does not hold within it any data, that is it is purely code − any intermediate results are stored in the calling task or in a

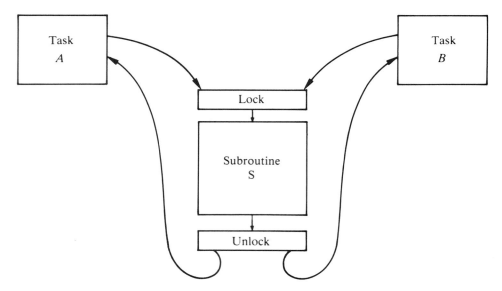

Figure 6.21   Serially reusable code.

stack associated with the task – then the subroutine is said to be re-entrant. Figure 6.22 shows an arrangement which can be used: the task descriptor for each task contains a pointer to a data area – usually a stack area – which is used for the storage of all information relevant to that task when using the subroutine. Swapping between tasks while they are using the subroutine will not now cause any problems since the contents of the stack pointer will be saved with the volatile environment of the task and will be restored when the task resumes. All accesses to data by the subroutine will be through the stack and hence it will automatically manipulate the correct data.

Re-entrant routines can be shared between several tasks since they contain no data relevant to a particular task and hence can be stopped and restarted at a different point in the routine without any loss of information. The data held in the working registers of the CPU is stored in the relevant task descriptor when task swapping takes place.

Device drivers in conventional operating systems are frequently implemented using re-entrant code. Another application could be for the actual three-term (PID)

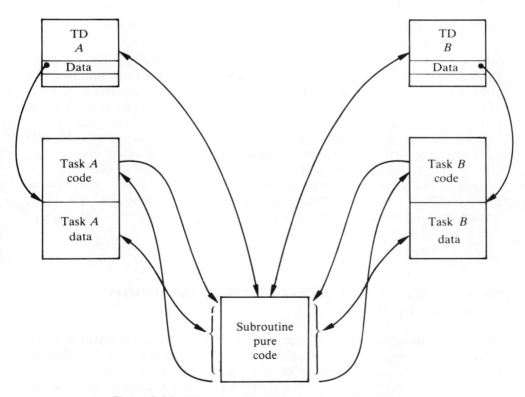

Figure 6.22 Use of re-entrant code for code sharing.

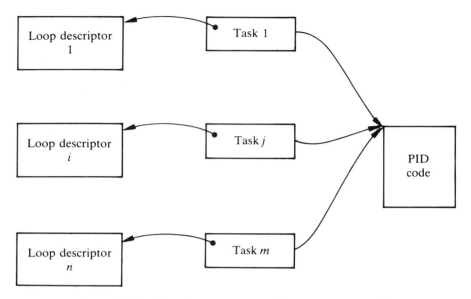

Figure 6.23   Use of re-entrant code in process control.

control algorithm in a process control system with a large number of control loops. The mechanism is illustrated in Figure 6.23; associated with each control loop is a LOOP descriptor as well as a TASK descriptor. The LOOP descriptor contains information about the measuring and actuation devices for the particular loop, for example the scaling of the measuring instrument, the actuator limits, the physical addresses of the input and output devices, and the parameters for the PID controller. The PID controller code segment uses the information in the LOOP descriptor and the TASK to calculate the control value and to send it to the controller. The actual task is made up of the LOOP descriptor, the TASK segment and the PID control code segment. The addition of another loop to the system requires the provision of new loop descriptors; the actual PID control code remains unchanged.

6.9 RESOURCE CONTROL: AN EXAMPLE OF AN INPUT/OUTPUT SUBSYSTEM (IOSS)

One of the most difficult areas of programming is the transfer of information to and from external devices. The availability of a well-designed and implemented input/output subsystem (IOSS) in an operating system is essential for efficient programming. The presence of such a system enables the application programmer to perform input or output by means of system calls either from a high-level language or from the assembler. The IOSS handles all the details of the devices. In a multi-tasking system the IOSS should also deal with all the problems of several tasks attempting to access the same device.

A typical IOSS will be divided into two levels as shown in Figure 6.24. The I/O manager accepts the system calls from the user tasks and transfers the information contained in the calls to the *device control block* (DCB) for the particular device. The information supplied in the call by the user task will be, for example, the

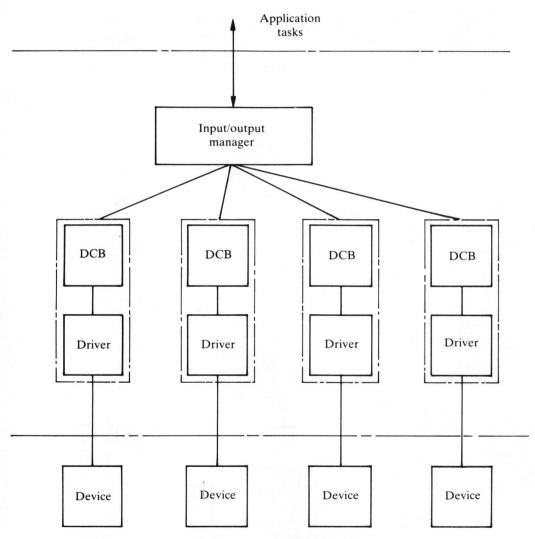

Figure 6.24   General structure of IOSS.

location of a buffer area in which the data to be transferred is stored (output) or is to be stored (input); the amount of data to be transferred; type of data, for example binary or ASCII; the direction of transfer; and the device to be used.

The actual transfer of the data between the user task and the device will be carried out by the device driver and this segment of code will make use of other information stored in the DCB. A separate device driver may be provided for each device or, as is shown in Figure 6.25, a single driver may be shared between several devices; however, each device will require its own DCB. The actual data transfer will usually be carried out under interrupt control.

Typically a DCB will contain the information shown in Table 6.2. The physical device name is the name by which the operating system recognises the device and the type of device is usually given in the form of a code recognised by the operating system. The operating system will normally be supplied with DCBs for the more

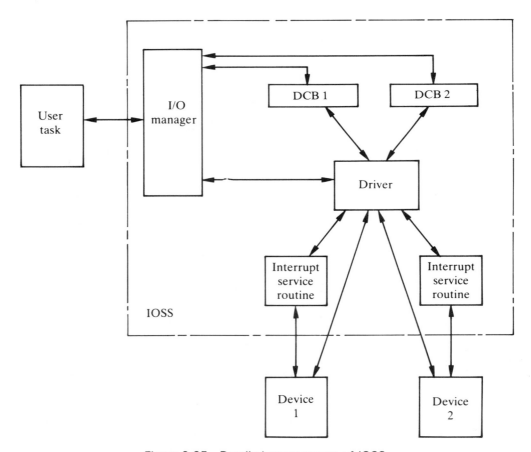

Figure 6.25 Detailed arrangement of IOSS.

Table 6.2   Device control block

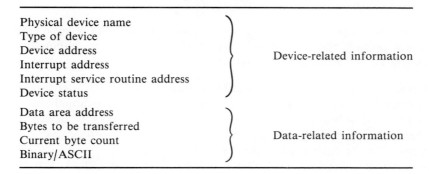

Physical device name	
Type of device	
Device address	Device-related information
Interrupt address	
Interrupt service routine address	
Device status	
Data area address	
Bytes to be transferred	Data-related information
Current byte count	
Binary/ASCII	

common devices. The DCBs may require modifying to reflect the addresses used in a particular system, although many suppliers adopt the policy of using standard addresses both for the physical address of the device on the bus and for the interrupt locations and interrupt handling routines. The addition of non-standard devices will require the user to provide appropriate DCBs. This task is usually made reasonably simple by providing source code for sample DCBs which can be modified to meet particular needs.

In a multi-tasking system provision has to be made to deal with overlapping requests for a particular device, for example several tasks may wish to send information to the log device – typically a printer. The normal way of handling output to a printer in a single-user environment is to send a record, that is one line, at a time, a return being made to the user program between each line. If this is done in a multi-programming system and the printer is not allocated to the specific program then there is a danger of the output from the different programs becoming intermingled. The solution usually adopted in a multi-user environment is to spool the output, that is it is intercepted by the operating system and stored in a file on the disk. When the program terminates or the user signs off, the contents of the spool file associated with that program or that user are printed out.

A similar solution can be used in a multi-tasking environment providing the user task can force the printing out of the spool file for that task. This addition is needed because in a real-time multi-tasking system tasks may not terminate. Although spooling provides user tasks with the ability to control the interleaving of output, there is still the problem of what action to take if the device is in use when a user task makes a request to use it. There are several possible solutions:

1.   Suspend the task until the device becomes available.
2.   Return immediately to the task with information that the device is busy and leave it to the task to decide what action to take – normally to call a delay and try again later.
3.   Add the request to a device request queue and return to the calling task; the

calling task must check at some later time to see if the request has been completed.

There are advantages and disadvantages to each method and a good operating system will provide the programmer with a choice of actions, although not all options will be available for every device.

> *Option 1* is referred to as a non-buffered request, in that the user task and the device have to rendezvous. In some ways it can be thought of as the equivalent of hardware handshaking – the user task asks the device 'are you ready?' and waits for a reply from the device before proceeding.
>
> *Option 2* is the equivalent of polling and is rarely used.
>
> *Option 3* is referred to as a buffered request. It is a form of message passing: the user task passes to the IOSS the equivalent of a letter – this consists of both the message and instructions about the destination of the message – and then the user task continues on the assumption that eventually the message will be delivered, that is sent to the output device. Usually some mechanism is provided which enables the user task to check if the message has been received, that is a form of recorded delivery in which the IOSS records that the message has been delivered and allows the user task to check. Buffered input is slightly different in that the user task invites an external device to send it a message – this can be considered as the equivalent of providing your address to a person or to a group of people. The IOSS will collect the message and deliver it but it is up to the user task to check its 'mail box' to see if a message has been delivered.

6.9.1 Example of an IOSS

The description which follows is of a particular IOSS of an RTOS which supports both computer peripherals – VDUs, printers, disk drives, etc. – and process-related peripherals – analog and digital input and output devices. The system commands used to access the IOSS functions are listed in Table 6.3. In addition to the commands listed in the table there are commands for analog output, for pulse output devices and for incremental output devices.

The IOSS system manager maintains a device request queue for each device and is responsible for interpreting the user task request and placing the appropriate information in the device request queue. If the request is a buffered request, then a return is made immediately to the calling task. If the request is non-buffered, then the IOSS manager changes the status of the calling task to LOCKED OUT and jumps to the dispatcher to begin the search for other work. The IOSS manager, in addition to dealing with requests, has to take action on the completion of a transfer. The driver associated with a given device signals the IOSS manager on completion

Table 6.3   IOSS system commands for RTOS

DTRC01	Disk transfer request − buffered
DTRC02	Disk transfer request − non-buffered
DTRC03	Call to check for completion of buffered request
INRC01	Input request from keyboard device − buffered
INRC02	Call to check for completion of input request
OUCC01	Call to
	(a) request system data area, i.e. spool area
	(b) request user data area
	(c) check status of device
	(d) check if user data area is free
OURC01	Request output of message to printer or terminal − buffered
FMRC01	Find and reserve area of memory external to the calling task
RMRC01	Release area of memory found using an FMRC01 call
SCRC10	Check if the previously requested buffered scans have been completed
SCRC11	Request non-buffered, non-priority analog scan
SCRC12	Request non-buffered, priority scan
SCRC13	Request buffered, non-priority scan
SCRC14	Request buffered, priority scan
DORC01	Request for a normal digital output, non-priority
DORC02	Request for a normal digital output, priority
DORC03	Request for a timed digital output, non-priority
DORC04	Request for a timed digital output, priority
	(Note: all the DORC requests can be buffered or non-buffered − the selection is made by setting a parameter for the call)

of a transfer. The IOSS manager, for non-buffered requests, sets the status of the user task which made the request to ON. For buffered requests there are two possible actions: if the calling task has checked to see if the action has been completed before it was completed it will have been placed in the LOCKED OUT state and hence the IOSS treats it as a non-buffered request. If completion occurs prior to a check for completion by the user task, then the IOSS records that the transfer has been completed. When a check is made a return to the calling task will be made with an indication that the transfer is complete. The actual detail of the actions on completion varies for the different types of device.

In addition to dealing with the above, the IOSS manager following completion of a transfer by a device has to check if further requests are waiting in the device request queue; if they are it transfers information to the DCB and initiates the start of transfer before returning to the dispatcher.

6.9.2 Output to Printing Devices

The RTOS provides the programmer with a choice of spooling mechanisms:

System data areas: a number of fixed sized areas on the disk are provided and are identified by a tag number;

User data areas: the user may define data areas of any size which can be in memory or on the disk and are again identified by a tag number.

The system data areas are made available to any user task which requests a data area; the request can be only for a system data area, not one with a specified tag number. A user data area can be assigned to a particular user task. (Note that in this system the assignment is by implication only; all user tasks have to agree that a given tag number applies to a given data area and have to agree that use will be restricted to a given task.) The advantages of a user data area are:

1. the area can be in memory or on disk;
2. the area can be of any size;
3. if use of the area is restricted to one task then it can contain a mixture of permanent and variable data and the user task only needs to transfer the information which has been changed since the last output.

The sequence of operations to be carried out in order to output a message via a user data area is as follows in Example 6.5.

EXAMPLE 6.5

Input From Keyboard

```
; request for user data area
;
   LDA label
   SPB OUCC01 ; system call
   ;return here if area in use
   ;normal return, request accepted
;
label DW parameters ; type of request
         ; tag number
         ; device check yes/no
         ; device number
```

```
;
; transfer of data to data area can take place
. . . . . . . . .
;
; request for output to device
;
  SPB OURC01 ;system call
  LDK labela
;return is made to this location
;
labela DW parameters ;device number
         ;data area type
         ;address of start of data area
;
;further processing can be done during data output
;
   . . . . . . . .
;
;check for completion of transfer
;
  LDA label
  SPB OUCC01
  ;return here if not complete
  ;return here if complete
```

For input the system expects the user task to provide a buffer area in memory to contain the input. The size is limited to a maximum of 256 words and hence an input record from the keyboard, including control characters used to edit the input line, is restricted to 256 characters. The input buffer area can be part of the user task or a separate area of memory found using the FMRC01 call. In this particular RTOS it is preferable to use a system area provided by the FMRC01 call, since this allows the user task to be swapped out of memory during the input. The steps involved in the input request are shown on the flowchart in Figure 6.26 and this also shows the different layers of operation of the operating system. In outline the steps are:

1. Use FMRC01 call to obtain input buffer area.
2. Request input using INRC01.
3. Do other processing if required.
4. Check if input completed using the INRC02 call — if input is not complete the task will be suspended until it is.
5. Transfer input from buffer area to program area.
6. Release buffer area using RMRC01 call.

Note that the operating system treats the user task differently when it checks for completion of input compared with the check for completion of output. The reason is that it assumes that a check for completion of input will be made only when there

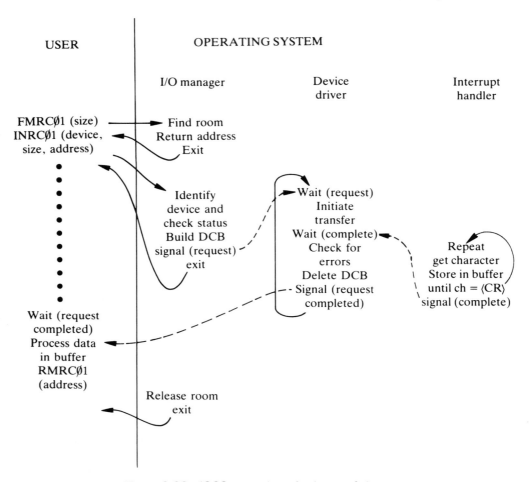

USER   OPERATING SYSTEM

I/O manager   Device   Interrupt
                            driver   handler

FMRC∅1 (size) ——— Find room
INRC∅1 (device, ◀— Return address
  size, address)   Exit

                    Identify   ▶Wait (request)
                    device and   Initiate
                    check status   transfer
                    Build DCB   Wait (complete) ◀—
                    signal (request)   Check for   Repeat
                    exit   errors   get character
                                       Delete DCB   Store in buffer
                                  — Signal (request   until ch = ⟨CR⟩
                                       completed)   signal (complete)

Wait (request
  completed)
Process data ◀— —
  in buffer
RMRC∅1
  (address)

                    Release room
                    exit

Figure 6.26   IOSS operations for input of data.

is no other work for the task to do; hence if a return to the task on non-completion was made then all the task can do is delay for a short period and then check again – an inefficient procedure. However, on output a task which finds that the previous output is not complete may be able to take some other action, for example set up another data area with a further request and continue.

6.9.3 Device Queues and Priorities

In a real-time system a simple device queue based on a first-in-first-out organisation can cause problems in that the task requesting a device effectively loses its priority.

Figures 6.27a and 6.27b illustrate this. In Figure 6.27a a number of tasks are queued waiting for the printer and one task (76) is already using the device. The higher-priority tasks including the very high-priority task 5 will not gain access to the printer until task 76 releases it. If the tasks have made a non-buffered request they will be locked out until they reach the head of the printer queue. However, if task 5 has made a buffered request it will be able to continue and, if it runs frequently, then after a short period of time the printer queue will contain several requests from task 5 as is shown in Figure 6.27b. If the system is not overloaded the printer will eventually catch up with the output from task 5. The delay between the requests

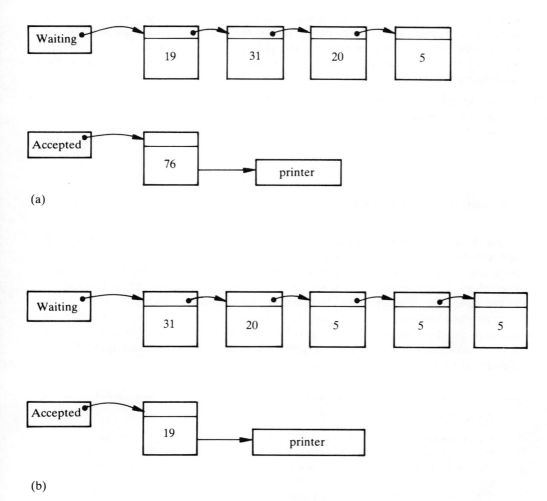

Figure 6.27   Printer queue: (a) buffered request; (b) non-buffered request.

from task 5 and the eventual output on the printer can be reduced if the printer queue is organised on a priority basis.

The position regarding priorities becomes even more complicated when the IOSS deals with many devices and hence has several device queues. Decisions on the order in which device queues are serviced are complicated and difficult.

6.10 TASK CO-OPERATION AND COMMUNICATION

In real-time systems tasks are designed to fulfil a common purpose and hence they need to communicate with each other. However, they may also be in competition for the resources of the computer system and this competition must be regulated. Some of the problems which arise have already been met in considering the input/output subsystem and they involve:

- mutual exclusion;
- synchronisation; and
- data transfer.

6.11 MUTUAL EXCLUSION

A multi-tasking operating system allows the sharing of resources between several concurrently active tasks. This does not imply that the resources can be used simultaneously. The use of some resources is restricted to only one task at a time. For others, for example a re-entrant code module, several tasks can be using them at the same time. The restriction to one task at a time has to be made for resources such as input and output devices, otherwise there is a danger that input intended for one task could get corrupted by input for another task. Similarly problems can arise if two tasks share a data area and both tasks can write to the data area. This is illustrated by Example 6.6.

EXAMPLE 6.6

Two software modules, `bottle_in_count` and `bottle_out_count`, are used to count pulses issued from detectors which observe bottles entering and leaving a processing area. The two modules run as independent tasks. The two tasks operate on the same variable `bottle_count`. Module `bottle_in_count` increments the variable and `bottle_out_count` decrements it. The modules are programmed in a high-level language and the relevant program language statements are:

```
bottle_count := bottle_count +1; (bottle_in_count)
bottle_count := bottle_count - 1; (bottle_out_count)
```

At assembler code level the high-level instructions become:

```
{bottle_in_count}        {bottle_out_count}
LD A, (bottle_count)     LD A, (bottle_count)
ADD 1                    SUB 1
LD (bottle_count), A     LD (bottle_count), A
```

Now if variable `bottle_count` contains the value 10, `bottle_in_count` is running and executes the statement `LD A, (bottle_count)` then as Figure 6.28 shows, the `A` register is loaded with the value 10. If the operating system now reschedules and `bottle_out_count` runs it will also pick up the value 10, subtract one from it and store 9 in `bottle_count`. When execution of `bottle_in_count` resumes its environment will be restored and the `A` register will contain the value 10, one will be added and the value 11 stored in `bottle_count`. Thus the final value of `bottle_count` after adding one to it and subtracting one from it will be 11 instead of the correct value 10.

A reg	bottle_in_count	Count	bottle_out_count	A reg
?	LD A, (bottle_count)	10		
10	context	10	LD A, (bottle_count)	10
10	change forced	10	SUB 1	9
10	by operating system	10	LD (bottle_count), A	9
10	ADD 1	9		9
11	LD (bottle_count), A	9		9
11		11		9

Figure 6.28 Problem of shared memory (see Example 6.6).

In abstract terms mutual exclusion can be expressed in the form

remainder 1
pre-protocol (necessary overhead)
critical section
post-protocol (necessary overhead)
remainder 2

Remainder 1 and *remainder 2* represent sequential code that does not require access to a particular resource or to a common area of memory.

Critical section is the part of the code which must be protected from interference from another task.

Pre-protocol and post-protocol called before and after the critical sections are code segments that will ensure that the critical section is executed so as to exclude all other tasks.

To benefit from concurrency both the critical section and the protocols must be much shorter than the remainders, so that the remainders represent a significant body of code that can be overlapped with other tasks. The protocols represent an overhead which has to be paid in order to obtain concurrency.

It is implicit in concurrent programming that there is 'loose connection' or 'low coupling' between tasks (see for example Pressman (1992) for a definition of 'loose connection'). Low coupling increases reliability since an error in one task which causes an abnormal termination of that task should not, if there is low coupling, cause other tasks in the system to fail. In abstract terms this can be expressed as the requirement that an abnormal termination in the code forming the remainder should not affect any other task. It would be unreasonable to demand that a failure of the protocol or the critical sections did not affect another task, since the critical section represents the code by which communication or sharing of a resource with another task is taking place.

In considering solutions to mutual exclusion problems it is normal to assume that a number of so-called primitive instructions exist. The correctness of the primitives is assumed to be guaranteed by the language or operating system supporting them. A basic assumption is that a primitive forms an indivisible instruction and hence the task invoking a primitive is guaranteed not to be pre-empted during the execution of the primitive. At the operating system level the basic primitive instruction is the machine code instruction and there is reliance on the CPU hardware to support the indivisibility of an instruction. Furthermore there is also reliance on the hardware implementation of mutual exclusion on the basic access to memory. For example, in a common memory system there will be some form of arbiter which will provide for mutual exclusion in accessing an individual memory location. The arbitration mechanism for a common bus structure was discussed in Chapter 3 in which the CPU controls access to the bus. If, for example, direct memory access is used then the hardware associated with the DMA unit has to inform the CPU when it wants to take control of the bus. Some systems have been designed with memory shared between processors, each of which has its own bus; in these cases dual ported memory devices have been used and the problem of mutual exclusion is thereby transferred to the memory itself. (Note: true dual ported memory allows concurrent access to both processors. Often the memory does not allow true concurrent access; it delays one device for a short period of time. However, to the processor the memory appears to permit concurrent access.)

6.11.1 Semaphore

The most widely used form of primitive for the purposes of mutual exclusion is the binary semaphore. The semaphore mechanism was first proposed by E. W. Dijkstra in 1968. A binary semaphore is a condition flag which records whether or not a resource is available. If, for a binary semaphore s, s=1, then the resource is available and the task may proceed; if s=0 then the resource is unavailable and the task must wait. To avoid the processor wasting time while a task is waiting for a resource to become available there has to be a mechanism for suspending the running of a task when it is waiting and for recording that the task is waiting for a particular semaphore. A typical mechanism for doing this is to associate with each semaphore a queue (often referred to as a condition queue) of tasks that are waiting for a particular semaphore. The use which the operating system makes of this queue is explained more fully in Chapter 7. For the present we will assume that such a queue is created when we declare a semaphore and tasks can be removed from and added to the queue.

There are only three permissible operations on a semaphore, Initialise, Secure, and Release, and the operating system must provide the following procedures:

Initialise	(s:ABinarySemaphore, v:INTEGER): set semaphore s to value of v (v=0 or 1).
Secure(s):	if s=1 then set s:=0 and allow the task to proceed, otherwise suspend the calling task.
Release(s):	if there is no task waiting for semaphore s then set s:=1, otherwise resume any task that is waiting on semaphore s.

The operations Secure(s) and Release(s) are system primitives which are carried out as indivisible operations and hence the testing and setting of the condition flag are performed effectively as one operation.

Example 6.7 considers a task which wishes to access a printer.

EXAMPLE 6.7

Mutual Exclusion

```
(* Mutual exclusion problem - use of binary semaphore*)
VAR
    printerAccess: SEMAPHORE;
PROCEDURE Task;
BEGIN
    (* remainder1 *)
    Secure(printerAccess)
(*
        if printer is not available task will
        be suspended at this point
*)
(*
        printer available - critical section
*)
(*
        do output
*)
    Release(printerAccess)
    (* remainder2 *)
END Task ;
```

In Figure 6.29 the underlying operations which take place as several tasks attempt to access the same resource (assumed to be a printer) are shown. The binary semaphore, printerAccess, is initialised to the value 1 in step 1. As part of this process a three-item record with the semaphore value set to 1 and the pointers to the head and tail of the semaphore queue set to null is created. In step 2, Task A performs a Secure(printerAccess) operation and the semaphore value is set to 0. Since there was no other task waiting, Task A is allowed to use the resource. Sometime later Task A suspends and Task B performs a Secure(printerAccess), step 3. Since printerAccess = 0 it cannot continue and is added to the semaphore queue by inserting pointers to the task descriptor for Task B, in the semaphore control block. If Task C now performs a Secure(printerAccess), step 4, then the pointer in the task descriptor for Task B is filled in with the address of the TD for Task C and the tail pointer entry in the semaphore control block is filled in to point to Task C. When Task A performs the Release(printerAccess) operation, step 5, Task B is removed from the semaphore queue and then obtains access to the resource. At step 6, Task B performs the Release(printerAccess) operation and hence Task C is allowed to run and at step 7 when Task C performs Release(printerAccess) the value of the semaphore is set to 1.

1. Initialise (printerAccess, 1)

printerAccess

| Value = 1 |
| Head |
| Tail |

2. Task A active Secure (printerAccess)

printerAccess

| Value = 0 |
| Head |
| Tail |

3. Task A suspends, Task B active Secure (printerAccess)

printerAccess

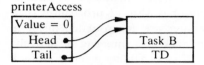

Task B is suspended and placed in printerAccess queue

4. Task C runs attempts Secure (printerAccess)

printerAccess

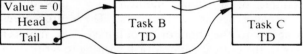

Task C is suspended and added to printerAccess queue

5. Task A runs, Release (printerAccess)

printerAccess

Task B is removed from printerAccess queue and placed in Ready queue

6. Task B runs, Release (printerAccess)

printerAccess

| Value = 0 |
| Head |
| Tail |

Task C is transferred from printerAccess queue to Ready queue

7. Task C runs, Release (printerAccess)

printerAccess

| Value = 1 |
| Head |
| Tail |

Printer is available to any task

Figure 6.29   Mutual exclusion using binary semaphore.

6.11.2 The Monitor

Although the binary and general semaphore provide a simple and effective means of enforcing mutual access they have one weakness: in use they are scattered around the code. Each task that requires access to a particular resource has to know the details of the semaphore used to protect that resource and to use it. The onus on correct use is placed on the designer and implementer of each task. In a small system this causes few problems but in larger, more complex systems where the tasks involved may be divided between several people use of the semaphore becomes more difficult and the probability of introducing errors increases.

An alternative solution which associates the control of mutual exclusion with the resource rather than with the user task is the monitor, introduced by Brinch Hansen (1973, 1975) and by Hoare (1974). A monitor is a set of procedures that provide access to data or to a device. The procedures are encapsulated inside a module that has the special property that only one task at a time can be actively executing a monitor procedure. It can be thought of as providing a fence around critical data. The operations which can be performed on the data are moved inside the fence as well as the data itself. The user task thus communicates with the monitor rather than directly with the resource.

Figure 6.30 shows an example of a simple monitor. Two procedures, WriteData and ReadData, provide access to the data. These procedures represent gates through which access to the monitor is obtained. The monitor prevents any other form of access to the critical data. A task wishing to write data calls the procedure WriteData and as long as no other task is already accessing the monitor it will be allowed to enter and write new data. If any other task was already using either the WriteData or ReadData operations then the task would be halted at the gate and suspended, since only one task at a time is allowed to be within the monitor fence.

Figure 6.31 shows a more complicated monitor with three entry points, Entry 1, Entry 2 and Entry 3, and two conditions on which tasks which have gained entry may have to wait. In the figure, one task, T15, is in the monitor and three tasks are waiting to enter, two at entry point 1 − T16 and T2 − and one at entry

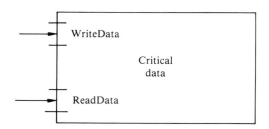

Figure 6.30   A simple monitor.

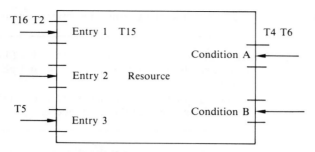

Figure 6.31   A general monitor.

point 3 — T5. Two tasks have previously entered and have been suspended waiting for condition A. There are no tasks waiting at entry point 2 or for condition B.

The advantage of a monitor over the use of semaphores or other mechanisms to enforce mutual exclusion is that the exclusion is implicit; the only action required by the programmer of the task requiring to use the resource is to invoke the entry to the monitor. If the monitor is correctly coded then an applications program cannot use a resource protected by a monitor incorrectly.

6.11.3 Intertask Communication

We can divide the issues of synchronisation and communication into three areas:

- synchronisation without data transfer;
- data transfer without synchronisation; and
- synchronisation with data transfer.

6.11.4 Task Synchronisation Without Data Transfer

Frequently one wishes to be able to inform another task that an event has occurred, or to set a task to wait for an event to occur. No data needs to be exchanged by the tasks. A mechanism that enables this to be done is the so-called *signal*:

> A signal s is defined as a binary variable such that if s = 1 then a signal has been sent but has not yet been received.

Associated with a signal is a queue and the permissible operations on a signal are:

Initialise (s:Signal: v:INTEGER) set s to the value of v (0 or 1).

Wait(s)   if s=1 then s:=0 else suspend the calling task and place it in the
              condition queue s.

Send(s)   if the condition queue s is empty then s:=1 else transfer the first
              task in the condition queue to the ready queue.

Clearly a signal is similar to a semaphore; in fact the difference between the two is not in the way in which they are implemented but in the way in which they are used. A semaphore is used to secure and release a resource and as such the calls will both be made by one task; a signal is used to synchronise the activities of two tasks and one task will issue the send and the other task the wait. (Note that a signal is sometimes implemented such that if a task is not waiting it has no effect, that is the receipt of a signal is not remembered.)

In practice two important additions to the basic *signal* mechanism are required: one is the facility to check to see if a task is waiting to send a signal, and the other is to be able to restrict the length of time which a task waits for a signal to occur. In a real-time application it is rarely correct for a task to be committed to wait indefinitely for an event to occur. The standard Wait(s) commits a task to an indefinite wait.

6.12 DATA TRANSFER (THE PRODUCER–CONSUMER PROBLEM)

6.12.1 Data Transfer Without Synchronisation

RTOSs typically support two mechanisms for the transfer or sharing of data between tasks: these are the *pool* and the *channel*.

---

pool   is used to hold data common to several tasks, for example tables of values or parameters which tasks periodically consult or update. The write operation on a *pool* is destructive and the read operation is non-destructive.

---

channel   supports communication between producers and consumers of data. It can contain one or more items of information. Writing to a *channel* adds an item without changing items already in it. The read operation is destructive in that it removes an item from the *channel*. A *channel* can become empty and also, because in practice its capacity is finite, it can become full.

It is normal to create a large number of pools so as to limit the use of global common data areas. To avoid the problem of two or more tasks accessing a pool simultaneously mutual exclusion on pools is required. The most reliable form of mutual exclusion for a pool is to embed the pool inside a monitor. Given that the read operation does not change the data in a pool there is no need to restrict read access to a pool to one task at a time.

Channels provide a direct communication link between tasks, normally on a one-to-one basis. The communication is like a pipe down which successive collections of items of data – messages – can pass. Normally they are implemented so that they can contain several messages and so they act as a buffer between the tasks. One task is seen as the *producer* of information and the other as the *consumer*. Because of the buffer function of the channel the producer and consumer tasks can run asynchronously.

There are two basic implementation mechanisms for a channel:

- queue (linked list); and
- circular buffer.

The advantage of the queue is that the number of successive messages held in the channel is not fixed. The length of the queue can grow, the only limit being the amount of available memory. The disadvantage of the queue is that as the length of the queue increases the access time, that is the time to add and remove items from the queue, increases. For this reason and because it is not good practice to have undefined limits on functions in real-time systems queues are rarely used.

The circular buffer uses a fixed amount of memory, the size being defined by the designer of the application. If the producer and consumer tasks run normally they would typically add and remove items from the buffer alternately. If for some reason one or the other is suspended for any length of time the buffer will either fill up or empty. The tasks using the buffer have to check, as appropriate, for buffer full and buffer empty conditions and suspend their operations until the empty or full condition changes.

As an example let us consider an alarm scanning task which for a period of time produces data at a rate much greater than that at which the logging task can print it out. A buffer is needed to store the data until the consuming task is ready to take it. The system is shown diagrammatically in Figure 6.32. We assume that the buffer is bounded, that is of finite size, and that the operation of storing an item of data in it is performed by the call

```
Put(x)
```

and an item of data is removed by the call

```
Get(x)
```

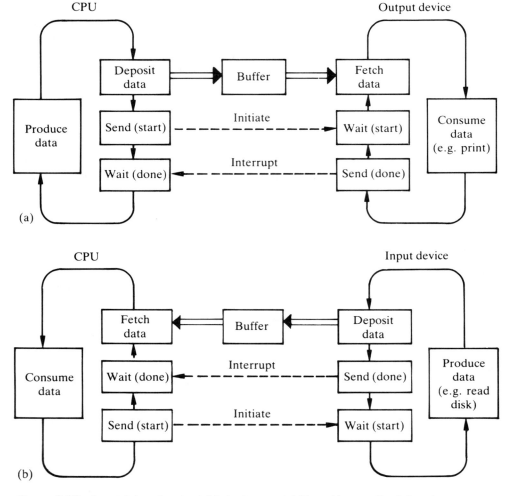

Figure 6.32   Input (a) and output (b) device model (from Young, *Real-time Languages*, Ellis Horwood (1982)).

Since the buffer is of finite size it is necessary to know when it is full and when it is empty. The following function calls are used:

> `Full` — which returns the value true if the buffer is full;
> `Empty` — which returns the value true if the buffer is empty.

Let us assume that the producer and consumer are formed by separate tasks which share a common buffer area.

EXAMPLE 6.8

Producer–Consumer Problem

```
(* Producer-consumer problem - solution 1*)
VAR commonBuffer : buffer;
TASK Producer;
VAR x:data;
BEGIN
    LOOP
        Produce(x);
        WHILE Full DO
        Wait
        END (* while *);
        Put(x);
    END (* loop *);
END Producer;
TASK Consumer;
VAR x:data;
BEGIN
    LOOP
        WHILE Empty DO
        Wait
        END (* while *);
        Get(x);
        Consume(x);
    END (* loop *);
END Consumer;
```

The producer operates in an endless cycle producing some item x and waiting until the buffer is not full to place x in the buffer; the consumer also operates in an endless cycle waiting until the buffer is not empty and removing item x from the buffer.

The solution in Example 6.8 is not satisfactory for two reasons:

1. the Put(x) and Get(x) are both operating on the same buffer and for security of the data simultaneous access to the buffer cannot be allowed — the mutual exclusion problem;
2. both the producer and the consumer use a 'busy wait' in order to deal with the buffer full and buffer empty problem.

The first problem can be solved using the semaphore, with the operations secure and release. The second problem can be solved by using the signal mechanism described above.

Because of the need to test for empty and full and to suspend the task if one or the other condition appertains then although transfer of a data item is not synchronised there is task synchronisation on the buffer full and buffer empty conditions.

This is illustrated in Example 6.9.

EXAMPLE 6.9

Producer–Consumer Problem – Solution 2

```
(*
Data transfer problem - solution 2 using semaphores and
signals
*)
VAR commonBuffer : Abuffer;
    bufferAccess : ABinarySemaphore;
    nonFull, nonEmpty : Signal;
TASK Producer;
VAR x:data;
BEGIN
    LOOP
        Produce(x);
        Secure(bufferAccess);
        IF Full THEN
           Release(bufferAccess);
           Wait(nonFull);
           Secure(bufferAccess);
        END (*if*);
        Put(x);
        Release(bufferAccess);
        Send(nonEmpty);
    END (*loop*);
END Producer;
TASK Consumer;
VAR x:data;
BEGIN
    LOOP
        Secure(bufferAccess);
        IF Empty THEN
           Release(bufferAccess);
           Wait(nonEmpty);
           Secure(bufferAccess);
        END (* if *);
        Get(x);
        Release(bufferAccess);
        Send(nonFull);
        Consume(x);
    END (*loop*);
END Consumer;
```

In this example the critical code is enclosed between secure and release operations but it is essential that the `bufferAccess` semaphore is released before executing the `Wait(nonFull)` or `Wait(nonEmpty)` primitives. If this is not done the system will deadlock. For example, if the `Producer` executes `Wait(nonFull)` while holding the access rights to the buffer then the buffer can never become non-full since the only way it can is for the `Consumer` to remove an item of data, but the `Consumer` cannot gain access to it until it is released by the `Producer`.

Both semaphores and signals can be generalised to allow a semaphore or a signal variable to have any non-negative integer value – in this form they are sometimes referred to as counting semaphores.

6.12.2 Synchronisation With Data Transfer

There are two main forms of synchronisation involving data transfer. The first involves the producer task simply signalling to say that a message has been produced and is waiting to be collected, and the second is to signal that a message is ready and to wait for the consumer task to reach a point where the two tasks can exchange the data.

The first method is simply an extension of the mechanism used in the example in the previous section to signal that a channel was empty or full. Instead of signalling these conditions a signal is sent each time a message is placed in the channel. Either a generalised semaphore or signal that counts the number of sends and waits, or a counter, has to be used.

Two examples of buffers written in Modula-2 are shown in Figures 6.33 and 6.34.

6.13 LIVENESS

An important property of a multi-tasking real-time system is *liveness*. A system (a set of tasks) is said to possess liveness if it is free from

livelock,
deadlock, and
indefinite postponement.

Livelock is the condition under which the tasks requiring mutually exclusive access to a set of resources both enter *busy wait* routines but neither can get out of the *busy wait* because they are waiting for each other. The CPU appears to be doing useful work and hence the term livelock.

Deadlock is the condition in which a set of tasks are in a state such that it is

```
IMPLEMENTATION MODULE Buffer;
(*
Title   : Implementation of a buffer using a monitor
File    : buffert1.mod
*)
FROM Monitor IMPORT
  monitorPriority;
 FROM Semaphores IMPORT
  Claim, InitSemaphore, Release, Semaphore;
 (* following is required for display of contents *)
  FROM Ansi IMPORT
    WriteCh;
 (* end of display *)
CONST moduleName='bufferT1';

MODULE BufferM [monitorPriority];
IMPORT
 Claim, InitSemaphore, Release, Semaphore, WriteCh;
EXPORT Put, Get;
 CONST
  nMax=10;
 VAR
  nFree, nTaken : Semaphore;
  in, out : [1..nMax];
  b: ARRAY [1..nMax] OF CHAR;
 (* following variables are required only for demonstration purposes *)
  row, col : CARDINAL;
  PROCEDURE Put(ch : CHAR);
  BEGIN
   Claim(nFree);
   b[in]:=ch;
   in:=in MOD nMax+1;
    WriteCh(ch, row, col+in) (* display purposes only *);
   Release(nTaken)
  END Put;
  PROCEDURE Get(VAR ch:CHAR);
  BEGIN
   Claim(nTaken);
   ch:=b[out];
   out:=out MOD nMax+1;
    WriteCh(' ', row, col+out) (* display purposes *);
   Release(nFree)
  END Get;
BEGIN
 row:=15; col:=20 (* initialise display part *);
 in:=1; out:=1;
 InitSemaphore(nFree, nMax);
 InitSemaphore(nTaken, 0);
END BufferM;
END Buffer.
```

Figure 6.33   IMPLEMENTATION MODULE of a buffer using a monitor.

```
IMPLEMENTATION MODULE Buffer;
(*
Title  : Implementation of a buffer using a semaphore
File   : buffert2.mod
*)
 FROM Semaphores IMPORT
   Claim, InitSemaphore, Release, Semaphore;
 (* following is required for display of contents *)
   FROM Ansi IMPORT
     WriteCh;
 (* end of display *)
CONST moduleName='bufferT2';
   CONST
   nMax=10;
 VAR
   nFree, nTaken, inPut, inGet : Semaphore;
   in, out : [1..nMax];
   b: ARRAY [1..nMax] OF CHAR;
   (* following variables are required only for demonstration purposes *)
   row, col : CARDINAL;
   PROCEDURE Put(ch : CHAR);
   BEGIN
    Claim(inPut);
    Claim(nFree);
    b[in]:=ch;
    in:=in MOD nMax+1;
      WriteCh(ch, row, col+in) (* display purposes only *);
    Release(nTaken);
    Release(inPut);
   END Put;
   PROCEDURE Get(VAR ch:CHAR);
   BEGIN
    Claim(inGet);
    Claim(nTaken);
    ch:=b[out];
    out:=out MOD nMax+1;
      WriteCh(' ', row, col+out) (* display purposes *);
    Release(nFree);
    Release(inGet);
   END Get;
BEGIN
 row:=15; col:=20 (* initialise display part *);
 in:=1; out:=1;
 InitSemaphore(nFree, nMax);
 InitSemaphore(nTaken, 0);
 InitSemaphore(inPut, 1);
 InitSemaphore(inGet, 1);
END Buffer.
```

Figure 6.34   IMPLEMENTATION MODULE of a buffer using a semaphore.

impossible for any of them to proceed. The CPU is free but there are no tasks that are ready to run. As an example of how deadlock can occur consider the following.

Suppose task A has acquired exclusive use of resource X and now requests resource Y, but between A acquiring X and requesting Y, task B has obtained exclusive use of Y and has requested use of X. Neither task can proceed, since A is holding X and waiting for Y and B is holding Y and waiting for X.

The detection of deadlock or the provision of resource sharing commands in such a way as to avoid deadlock is the responsibility of the operating system (see Lister (1979, pp. 94–7) for a discussion of deadlock avoidance and detection mechanisms).

Indefinite postponement is the condition that occurs when a task is unable to gain access to a resource because some other task always gains access ahead of it.

6.14 MINIMUM OPERATING SYSTEM KERNEL

As mentioned in the introduction there has been considerable interest in recent years in the idea of providing a minimum kernel of RTOS support mechanisms and constructing the required additional mechanisms for a particular application or group of applications. One possible set of functions and primitives for RTOS is:

Functions:
1. A clock interrupt procedure that decrements a time count for relevant tasks.
2. A basic task handling and context switching mechanism that will support the moving of tasks between queues and the formation of task queues.
3. Primitive device routines (including real-time clock support).

Primitives:
WAIT for some condition (including release of exclusive access rights).
SIGNAL condition and thus release one (or all) tasks waiting on the condition.
ACQUIRE exclusive rights to a resource (option – specify a time-out condition).
RELEASE exclusive rights to a resource.
DELAY task for a specified time.
CYCLE task, that is suspend until the end of its specified cyclic period.

6.15 EXAMPLE OF CREATING AN RTOS BASED ON A MODULA-2 KERNEL

The standard module Processes suggested by Wirth (1986) and supplied by most systems is not all that versatile. Many alternative versions offering a wider range of facilities have been developed. An example of one such system is described below. It was developed by Roger Henry, of Nottingham University.

The lowest-level module is `Processes` (a replacement for the Wirth module `Processes`) which provides the procedures and functions

`Cp`	— return current process identity
`Disable`	— disable a given process
`Enable`	— make a given task runnable
`MinWksp`	— return the minimum workspace size for task
`NewProcess`	— create a new task
`PriorityOf`	— return the priority of a task
`SuspendMe`	— suspend the calling task
`SuspendUntilInterrupt`	— suspend the calling task until a specific interrupt occurs.

The relationship between the procedures and the state of the tasks is shown in Figure 6.35 — `NewProcess` is used to inform `Processes` of the existence of a task, but before it can be run it must be made runnable by a call to `Enable`. In the call `NewProcess` a task is allocated a priority (an integer value — the range depending on the implementation) and the runnable task with the highest priority becomes the running task. A running task can call `Disable` which will cause a named runnable task to be changed to existent (non-runnable). If the named task is not runnable the call will be ignored; to make itself non-runnable the running task uses the call `SuspendMe`. A running task may also suspend itself to wait for a hardware

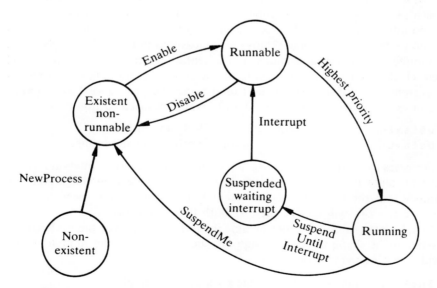

Figure 6.35   Task states and transitions for a Modula-2 kernel.

interrupt by using the call `SuspendUntilInterrupt` in which case it will not become runnable until the specified hardware interrupt occurs and is accepted.

For simple applications a high-level module, `Scheduler`, is provided which has two procedures:

> `StartProcess` — equivalent of `NewProcess` and `Enable`, it makes a
>   task known to `Processes` and makes it runnable;
> `StopMe` — stops the current task.

The separate modules of `Signals`, `Semaphores` and `Timing` allow the tasks started by `Scheduler` to synchronise and to run at specified time intervals.

The module `Signals` provides the standard operations on signals (`Initialise`, `Wait` and `Send` — the names used are `InitSignal`, `SendSignal` and `AwaitSignal`) but in addition two further operations are supported:

> `Awaited` — returns true value if at least one task is waiting for the
>   `SendSignal` operation;
> `SentWithin` — the caller will only wait for a specified length of time; the
>   function returns a true value if the signal was sent within the specified
>   time.

The module `Semaphore` supports the standard semaphore operations — the names used are `InitSemaphore`, `Claim` and `Release`.

The module `Timing` enables users to operate in absolute and relative time intervals. Absolute time begins when the `Timing` module is initialised and one value of absolute time can be said to be earlier or later than another. The difference between two values of absolute time — a time interval — is said to be relative time and one interval can be said to be longer or shorter than another.

Time — both absolute and relative — is measured in units of seconds and ticks. The number of ticks in a second is implementation defined (it depends upon the system clock used) and its value is returned by the function `TicksPerSecond`. The current time (absolute) can be found using the procedure `TellTime`. Two procedures are provided to enable tasks to wait for specified times:

> `DelayFor` — the task waits for a specified time interval
> `DelayUntil` — the task waits until a specified absolute time.

In both cases the calling task is suspended until either the time interval has elapsed or the absolute time is reached; the task is then made runnable — there is no guarantee that the task will run immediately on attaining the specified condition since `Processes` will choose the task with the highest priority.

In calculating absolute times or time intervals the value of time in seconds and ticks has to be manipulated. To support such operations the module `TimeOps` provides the following procedures:

> `IncTime` — increase a time value by a given interval
> `IncInterval` — increase an interval value by a given interval

DecTime — decrease a time value by a given interval
DecInterval — decrease an interval value by a given interval
DiffTimes — subtract second time from the first time
DiffIntervals — subtract second interval from first interval
CompareTimes — compare first time with second time
CompareIntervals — compare first interval with second interval.

A simple example of the use of the RTOS kernel facilities is given in Figure 6.36. Two tasks, the task forming the main program and the task formed by procedure

```
MODULE TwoTasks;
(*
Title   : Example of two tasks synchronising using signals
File    : sb1 twotasks.mod
*)
FROM Scheduler IMPORT
 ProcessId, Priority, StartProcess, StopMe;

FROM Signals IMPORT
 AwaitSignal, SendSignal, InitSignal, Signal;

FROM InOut IMPORT
 WriteString, WriteLn;

 CONST
  priority=2;
  worksp=600;
 VAR
  messageSent : Signal;
  count : CARDINAL;
  taskTwoId : ProcessId;

  PROCEDURE TaskTwo;
  BEGIN
   LOOP
    AwaitSignal(messageSent);
    WriteString(' message received by task two');
    WriteLn
   END (* loop *);

  END TaskTwo;

BEGIN (* body of program forms task 1 *)
 InitSignal(messageSent);
 StartProcess(TaskTwo, priority, worksp, taskTwoId);
 FOR count:=1 TO 10 DO
  WriteString('Sending message');
  SendSignal(messageSent)
 END (* for *);

END TwoTasks.
□
```

Figure 6.36   Two tasks synchronised by using signals.

TaskTwo, are run alternately, synchronised by the use of the signal messageSent. The output is ten lines of 'Sending message' − output by the main task − and 'message received by task two' − output by TaskTwo. TaskTwo is started with a priority of level 2 by the use of the procedure Scheduler.StartProcess. The main body of the program is automatically run at priority level 0.

Some of the features of the RTOS kernel can be explored by using the example program shown in Figure 6.37. Two tasks, TaskA and TaskB, are created. TaskA prints out on the screen a number of rows of the letter A; the number is specified in the constant numberOfLines. TaskB prints out a number of rows of the

```
MODULE RTS3;
(*
Title   : Demonstration of resource sharing
File    : sb1: RTS3.mod
*)
FROM InOut IMPORT
 Write, WriteLn, WriteString;

FROM Semaphores IMPORT
 InitSemaphore, Semaphore, Claim, Release;

FROM Scheduler IMPORT
 ProcessId, Priority, StartProcess, StopMe;

FROM Timing IMPORT
 DelayFor, DelayUntil, Interval, TellTime, Time, TicksPerSecond;

CONST moduleName='RTS3';
 numberOfLines=5;
VAR
 screen : Semaphore;
 endA, endB : BOOLEAN;

PROCEDURE TaskA;
CONST ch='A';
VAR
 i,j : CARDINAL;
 delayA : Interval;
BEGIN
 i:=0; j:=0; delayA.secs:=0; delayA.ticks:=1;
 LOOP
  Claim(screen); (* omit in versions 1 and 2 *)
  FOR i:=1 TO 79 DO
   Write(ch);
   DelayFor(delayA) (* omit in version 1 *)
  END (* for *);
  WriteLn;
  Release(screen); (* omit in versions 1 and 2 *)
  INC(j);
  IF j>numberOfLines THEN
   EXIT
  END (* if *);
```

```
   END (* loop *);
   endA:=TRUE;
   StopMe;
 END TaskA;
 PROCEDURE TaskB;
 CONST ch='B';
 VAR
   i,j : CARDINAL;
   delayB : Interval;
 BEGIN
   i:=0; j:=0; delayB.secs:=0; delayB.ticks:=1;
   LOOP
     Claim(screen); (* omit in versions 1 and 2 *)
     FOR i:=1 TO 79 DO
       Write(ch);
       DelayFor(delayB); (* omit in version 1*)
     END (* for *);
     WriteLn;
     Release(screen); (* omit in versions 1 and 2*)
     INC(j);
     IF j>numberOfLines THEN
       EXIT
     END (* if *);

   END (* loop *);
   endB:=TRUE;
   StopMe;
 END TaskB;
CONST
 priorityA=1;
 priorityB=1;
 wkspSizeA=1000;
 wkspSizeB=1000;

VAR
 taskAId, taskBId : ProcessId;
BEGIN
 WriteString(moduleName);
 WriteLn;
 endA:=FALSE; endB:=FALSE;
 InitSemaphore(screen, 1);
 StartProcess(TaskA, priorityA, wkspSizeA, taskAId);
 StartProcess(TaskB, priorityB, wkspSizeB, taskBId);
 LOOP
   (* Idle process *)
   IF endA AND endB THEN
     EXIT
   END (* if *);
 END (* loop *);
 WriteLn;
 WriteString('Program end');
END RTS3.
□
```

Figure 6.37 Example showing resource sharing.

letter B. The display which is obtained on the screen depends on the way the tasks are scheduled and whether the tasks are given exclusive access to the screen. By compiling the module in version 1 form (not using the semaphore and leaving out the DelayFor calls), the scheduler treats the two tasks as coroutines. Hence TaskA, which is started first, gains control and runs to completion; only then does TaskB run. The result of this is that first several rows of As are displayed on the screen followed by several rows of Bs. The reason for this behaviour is that the scheduler continues to run a task until either a higher-priority task wishes to run or until the running task suspends or ends. Introducing the DelayFor statements to form version 2 of the program causes the two tasks to run alternately giving an alternating sequence of As and Bs. You are invited to work out what the output will be if the Claim(screen) and Release(screen) statements are included.

The creation of modules for specific purposes is in keeping with the Modula-2 philosophy. The aim is that the core language should remain fixed and standard and that any extensions required for special purposes should be provided in the form of library modules. It should be noted that all the input/output, file handling and other operations are not handled as part of the language, but by standard procedures imported from modules which are assumed to be provided as part of the system.

6.16 SUMMARY

In this chapter we have concentrated on describing the features to be found in traditional operating systems. Such operating systems are usually specific to a particular computer, or range of computers. Examples are the Digital Equipment Corporation's RT/11 and RSX/11 operating systems for the PDP-11 series; the Data General RTOS and RDOS for the Nova range; and more recently RMX-80 for the Intel 8080 range and OS-9 for the Motorola 68XXXX series.

The advantage of many of the traditional operating systems is their wide user base and the fact that they have seen extensive use in control applications. There has, however, been a tendency for the size of the operating systems to increase with each successive upgrade and it is often difficult to create small subsets for a particular application. Another disadvantage with many is that access to the system from high-level languages is very restricted and the addition of new devices normally requires hardware drivers to be written in assembler.

The development of the MASCOT (see Chapter 9) environment represents one way in which some of the problems of lack of standardisation and difficulty of accessing operating system functions have been addressed. A similar approach has been taken by Baker and Scallon (1986) with the Rex architecture. As with MASCOT the Rex system presents the user with a virtual machine which hides the details of the operating system and the hardware. The detailed procedures required for carrying out the various functions of the application program are written in a conventional language and compiled using a standard compiler. A separate language is used to describe how the components of the system should be connected together to form a multi-tasking system and to describe how the data sets can be shared. At this stage decisions on the number of processors to be used are made.

The system has been designed for use in the aerospace industry and the problem of overheads involved in context switching has been carefully considered. Individual processes are short procedures which once started are not interrupted; they are considered to be the equivalent of a single assembler instruction. The allocation of storage for data and code for processes is static.

EXERCISES

6.1   Draw up a list of functions that you would expect to find in a real-time operating system. Identify the functions which are essential for a real-time system.

6.2   Why is it advantageous to treat a computer system as a virtual machine?

6.3   Discuss the advantages and disadvantages of using
(a) fixed table
(b) linked list
methods for holding task descriptors in a multi-tasking real-time operating system.

6.4   A range of real-time operating systems are available with different memory allocation strategies. The strategies range from permanently memory-resident tasks with no task swapping to fully dynamic memory allocation. Discuss the advantages and disadvantages of each type of strategy and give examples of applications for which each is most suited.

6.5   What are the major differences in requirements between a multi-user operating system and a multi-tasking operating system?

6.6   What is meant by context switching and why is it required?

6.7   What is the difference between static and dynamic priorities? Under what circumstances can the use of dynamic priorities be justified?

6.8   Choosing the basic clock interval (tick) is an important decision in setting up an RTOS. Why is this decision difficult and what factors need to be considered when choosing the clock interval?

6.9   List the minimum set of operations that you think a real-time operating system kernel needs to support.

7

Design of Real-time Systems – General Introduction

As we said at the end of Chapter 4, there is much more to designing and implementing computer control systems than simply programming the control algorithm. In this chapter we first give an outline of a general approach to the design of computer-based systems (it actually applies to all engineering systems). We will then consider, as an example, the hot-air blower system described in Chapter 1. In designing the software structure we illustrate three approaches:

- single task;
- foreground/background; and
- multi-tasking.

We end the chapter by considering in detail some of the problems that arise when using a multi-tasking approach. We deal with both multi-tasking on a single computer and the case in which the tasks are distributed across several computers.

The objectives are:

- To show how to approach the planning and design of a computer-based system.
- To illustrate the basic approaches for the top level design of real-time software.
- To illustrate some of the problems associated with real-time, multi-tasking software.

7.1 INTRODUCTION

The approach to the design of real-time computer systems is no different in outline from that required for any computer-based system or indeed most engineering systems. The work can be divided into two main sections:

- the planning phase; and
- the development phase.

The planning phase is illustrated in Figure 7.1. It is concerned with interpreting user

264

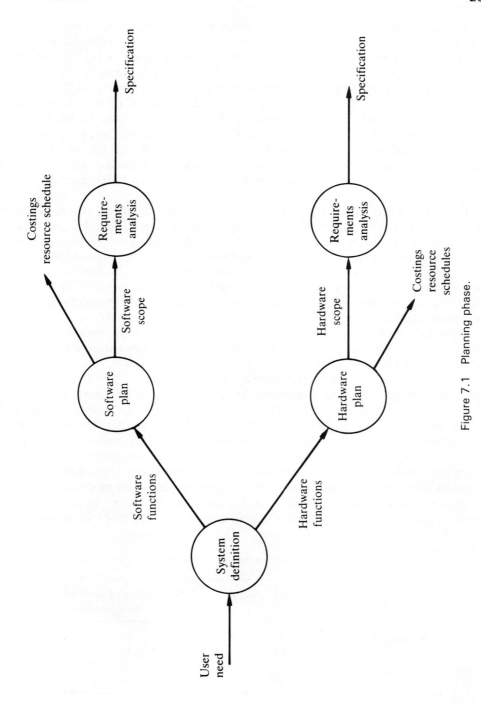

Figure 7.1 Planning phase.

requirements to produce a detailed specification of the system to be developed and an outline plan of the resources – people, time, equipment, costs – required to carry out the development. At this stage preliminary decisions regarding the division of functions between hardware and software will be made. A preliminary assessment of the type of computer structure – a single central computer, a hierarchical system, or a distributed system – will also be made. The outcome of this stage is a *specification* or *requirements* document. (The terminology used in books on software engineering can be confusing; some refer to a specification requirement document as well as to specification document and requirements document. It is clearer and simpler to consider that documents produced by the user or customer describe requirements, and documents produced by the supplier or designer give the specifications.)

It cannot be emphasised too strongly that the specification document for both the hardware and software which results from this phase must be complete, detailed and unambiguous. General experience has shown that a large proportion of *errors* which appear in the final system can be traced back to unclear, ambiguous or faulty specification documents. There is always a strong temptation to say 'It can be decided later'; deciding it later can result in the need to change parts of the system which have already been designed. Such changes are costly and frequently lead to the introduction of errors.

Some indication of the importance of this stage can be seen by examining Table 7.1 which shows the distribution of errors and cost of rectifying them (the figures are taken from DeMarco, 1978).

The stages of the development phase are shown in Figure 7.2. The aim of the preliminary design stage is to decompose the system into a set of specific sub-tasks which can be considered separately. The preliminary design stage is also referred to as the high-level design stage. The inputs to this stage are the high-level specifications; the outputs are the global data structures and the high-level software architecture. During this stage extensive liaison between the hardware and software designers is needed, particularly since, in the case of real-time systems, there will be a need to revise the decisions on the type of computer structure proposed and if, for example, a distributed system is to be used, to decide on the number of processors, communication systems (bandwidth, type), etc. The control strategy will

Table 7.1 Distribution of errors and of costs of correcting errors

Stage	*Distribution of errors %*	*Distribution of costs of rectifying errors %*
Requirements	56	82
Design	27	13
Code	7	1
Other	10	4

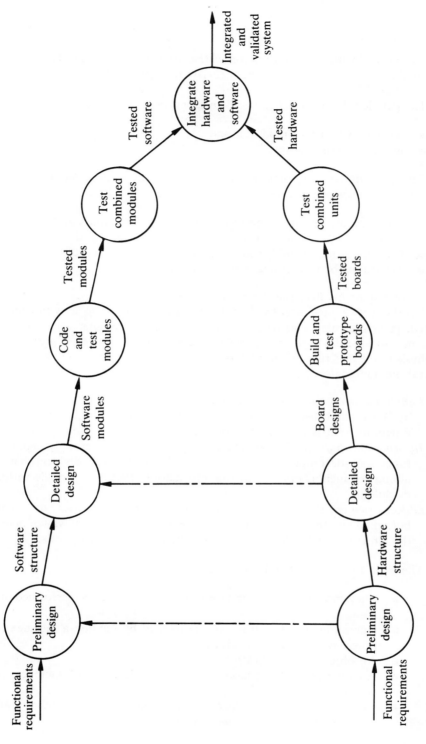

Figure 7.2   Development phase.

need reviewing including consideration of the control algorithms to be used. At the end of the preliminary design stage a review of both the hardware and software designs should be carried out.

The detailed design is usually broken down into two stages:

- decomposition into modules; and
- module internal design.

For hardware design, the first of these stages involves questions on the board structure of the system such as:

- Are separate boards going to be used for analog inputs and digital inputs or are all inputs going to be concentrated on one board?
- Can the processor and memory be located on one board?
- What type of bus structure should be used?

The second stage involves the design of the boards.

For the software engineer the first stage involves identifying activities which are related. Heuristic rules have been developed to aid the designer with decisions on division into modules (Pressman, 1992). The various heuristics are given differing emphasis in the different design methodologies. A brief description of some of the general methodologies is given below.

Functional decomposition: the top-down approach which has been advocated by Wirth and others leads to module subdivision based on a separation into functions, that is each module performs a specific function.

Information hiding: Parnas (1972a, b) has argued strongly against the functional decomposition approach and for module division based on hiding as much as possible of the information used by a module within the module.

Object oriented: this is the division of the system into entities which contain both data and the functions that operate on the data. It is a way of achieving the aims set out by Parnas in his information hiding proposals.

The main properties of objects are as follows:

- They encapsulate data and the operations that can be performed on that data.
- Operations on the object are performed by sending a message to the object requesting the operation.
- An object is an instance of a *class*.
- A class defines a common structure that describes all members of the class.
- Classes are organised hierarchically and sub-classes inherit all the features of the superordinate class.

As an example of the differences between the functional and information hiding (object-oriented) approaches consider a program which has to read a block of text from a device. The text has to be sorted into words and the words sorted into alphabetical order with duplicate words eliminated. The sorted list has to be printed. The simple *functional decomposition* approach would be to divide the system initially into three modules: input, sort and print. All the modules would need access to a shared data structure in which the text was held. Each module would thus know what form of data structure (array, linked list, record, or file) was being used to hold the data.

The Parnas approach would be to subdivide so that one module, say `StoreManager`, deals with the storage of the data. The other modules can access the data only through functions provided by the `StoreManager`. For example, it may provide two functions:

```
Put (word);
```

and

```
Get (word);
```

to enable information to be put into store and retrieved from store. The other modules need not know how the storage is organised. The advantage claimed for this approach is that design changes to one module do not affect another module.

The above approaches do not give any guidance on how to decide on boundaries between function modules or object modules. The following are commonly used heuristics for guiding module subdivision:

Coupling and cohesion: the maximising of module cohesion and the minimising of coupling between modules are the heuristics underlying data-flow design methodologies. The heuristic and methodologies have been developed by Constantine and Yourdon (1979), Myers (1978) and Stevens *et al.* (1974).

Partition to minimise interfaces: this heuristic was proposed by DeMarco (1978) and can be combined with data-flow methods. It suggests that the transformations indicated on the data-flow diagram should be grouped so as to minimise the number of interconnections between the modules.

For real-time systems additional heuristics are required, one of which is to divide modules into the following categories:

- real-time, hard constraint;
- real-time, soft constraint; and
- interactive.

The arguments given in Chapter 1 regarding the verification and validation of different types of program suggest a rule that aims to minimise the amount of software that falls into the hard constraint category since this type is the most difficult to design and test.

The major differences in software design between real-time and standard systems occur in preliminary design and decomposition into modules, and in this chapter we concentrate on these areas. Module internal design, coding, and testing are similar in both types of system.

From the description given above the whole process of specification, design and construction appears sequential; this is a simplistic view. In practice top-level design decisions frequently cannot be made until lower-level design decisions have been made. For example, we cannot decide what type of computer is required (processing power, memory requirements) until either we have coded the software or we have made detailed estimates of the amount of code and the type of computation required. A control algorithm which requires extensive arithmetic operations on real numbers results in a different computational load from one involving simple logic operations.

Modern software development methodologies address this problem and we shall examine some of them in detail in the later chapters. For the purposes of this chapter we will assume that we can proceed step-by-step through specification, preliminary design and detailed design.

7.2 SPECIFICATION DOCUMENT

To provide an example for the design procedures being described we shall consider a system comprising several of the hot-air blowers described in Chapter 1. It is assumed that the planning phase has been completed and a specification document has been prepared. A shortened version of such a document is given in Example 7.1.

EXAMPLE 7.1

Hot-Air Blower Specification

Version 3.1
Date 10 January 1992

1.0 *Introduction*
The system comprises a set of hot-air blowers arranged along a conveyor belt. Several different configurations may be used with a minimum of 6 blowers and a maximum of 12.

2.0 *Plant interface*

2.1 *Input from plant*
Outlet temperature: analog signal, range 0–10 V, corresponding to 20°C to 64°C, linear relationship.

2.2 *Output to plant*
Heater control: analog signal 0 V to −10 V, corresponding to full heat (0 V) to no heat (−10 V), linear relationship.

3.0 *Control*

A PID controller with a sampling interval of 40 ms is to be used. The sampling interval may be changed, but will not be less than 40 ms. The controller parameters are to be expressed to the user in standard analog form, that is proportional gain, integral action time and derivative action time. The set point is to be entered from the keyboard. The controller parameters are to be variable and are to be entered from the keyboard.

4.0 *Operator communication*

4.1 *Display*

The operator display is as shown below:

Set temperature	:nn.n$^\circ$C	Date	:dd/mm/yyyy
Actual temperature	:nn.n$^\circ$C	Time	:hh.mm
Error	:nn.n$^\circ$C		
Heater output	:nn% FS	Sampling Interval	:nn ms
Controller settings			
Proportional gain	:nn.n		
Integral action	:nn.nn s		
Derivative action	:nn.nn s		

The values on the display will be updated every 5 seconds.

4.2 *Operator input*

The operator can at any time enter a new set point or new values for the control parameters. This is done by pressing the 'ESC' key. In response to 'ESC' a menu is shown on the bottom of the display screen:

1.	Set temperature = nn.n	2.	Proportional gain = nn.n
3.	Integral action = nn.nn	4.	Derivative action = nn.nn
5.	Sampling interval = nn	6.	Management information
7.	Accept entries		

Select menu number to change

In response to the number entered, the present value of the item selected will be deleted from the display and the cursor positioned ready for the input of a new value. The process will be repeated until item 7 – *Accept entries* – is selected at which time the bottom part of the display will be cleared and the new values shown in the top part of the display.

5.0 *Management information*

On selection of item 6 of the operator menu a management summary of the performance of the plant over the previous 24 hours will be given. The summary provides the following information:

    (a)   Average error in °C in 24 hour period.
    (b)   Average heat demand %FS in 24 hour period.
    (c)   For each 15 minute period:
        (i)   average demanded temperature;
        (ii) average error; and
        (iii) average heat demand.
    (d)   Date and time of output.

6.0 *General information*
There will be a requirement for a maximum of 12 control units. A single display and entry keyboard which can be switched between the units is adequate.

7.3 PRELIMINARY DESIGN

7.3.1 Hardware Design

There are many different possibilities for the hardware structure. Obvious arrangements are:

1.   Single computer with multi-channel ADC and DAC boards.
2.   Separate general purpose computers on each unit.
3.   Separate computer-based microcontrollers on each unit linked to a single general purpose computer.

Each of these configurations needs to be analysed and evaluated. Some points to consider are:

Option 1: given that the specification calls for the system to be able to run with a sample interval for the control loop of 40 ms, can this be met with 12 units sharing a single processor?

Option 2: is putting a processor that includes a display and keyboard on each unit an expensive solution? Will communication between processors be required? (Almost certainly the answer to this is yes; operators and managers will not want to have to use separate displays and keyboards.)

Option 3: what sort of communication linkage should be used? A shared high-speed bus? A local-area network? Where should the microcontrollers be located? At each blower unit or together in a central location?

Each option needs careful analysis and evaluation in terms of cost and performance. The analysis must include consideration of development costs, performance operating and maintenance costs. It should also include consideration of reliability and safety.

To provide a basis for consideration of the widest range of approaches to software design we will assume that option 1 above is chosen.

7.3.2 Software Design

Examining the specification shows that the software has to perform several different functions:

- DDC for temperature control;
- operator display;
- operator input;
- provision of management information;
- system start-up and shut-down; and
- clock/calendar function.

The various functions and type of time constraint are shown in Figure 7.3. The control module has a hard constraint in that it must run every 40 ms. In practice this constraint may be relaxed a little to, say, 40 ms ± 1 ms with an average value over 1 minute of, say, 40 ms ± 0.5 ms. In general the sampling time can be specified

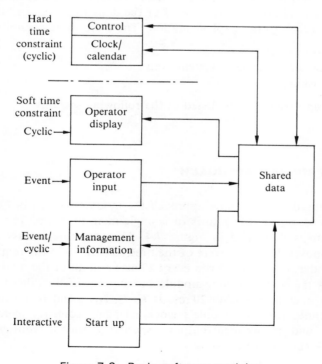

Figure 7.3 Basic software modules.

as $T_s \pm e_s$ with an average value, over time T, of $T_s \pm e_a$. The requirement may also be relaxed to allow, for example, one sample in 100 to be missed. These constraints will form part of the test specification.

The clock/calendar module must run every 20 ms in order not to miss a clock pulse. This constraint can be changed into a soft constraint if some additional hardware is provided in the form of a counter which can be read and reset by the clock/calendar module. The constraint could now be, say, an average response time of 1 second with a maximum interval between reading the counter of 5 seconds. (For these values what size of counter would be required?)

The operator display, as specified, has a hard constraint in that an update interval of 5 seconds is given. Common sense suggests that this is unnecessary and an average time of 5 seconds should be adequate; however, a maximum time would also have to be specified, say 10 seconds.

Similarly soft constraints are adequate for operator input and for the management information logs. These would have to be decided upon and agreed with the customer. They should form part of the specification in the requirements document. The start-up module does not have to operate in real time and hence can be considered as a standard interactive module.

There are obviously several different activities which can be divided into sub-problems. The sub-problems will have to share a certain amount of information and how this is done and how the next stages of the design proceed will depend upon the general approach to the implementation. There are three possibilities:

- single program;
- foreground/background system; and
- multi-tasking.

Each of these approaches is discussed in the following sections.

7.4 SINGLE-PROGRAM APPROACH

Using the standard programming approach the modules shown in Figure 7.3 are treated as procedures or subroutines of a single main program. The flowchart of such a program is illustrated in Figure 7.4. This structure is easy to program; however, it imposes the most severe of the time constraints − the requirement that the clock/calendar module must run every 20 ms − on all of the modules. For the system to work the clock/calendar module and any one of the other modules must complete their operations within 20 ms. If t_1, t_2, t_3, t_4 and t_5 are the *maximum* computation times for the module's clock/calendar, control, operator display, operator input and management output respectively, then a requirement for the system to work can be expressed as

$$t_1 + \max(t_2, t_3, t_4, t_5) < 20 \text{ ms}$$

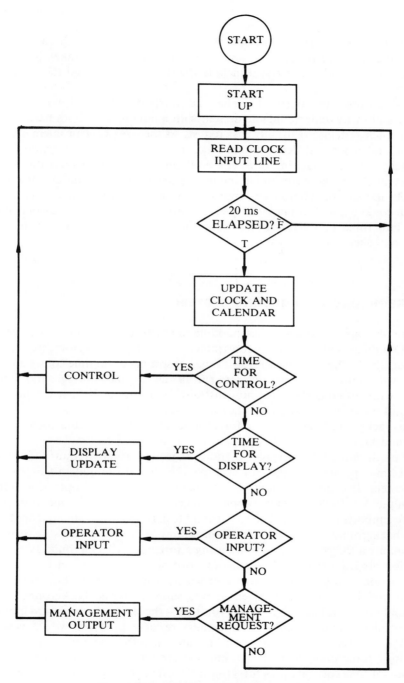

Figure 7.4 Single-program approach.

(Note: (a) The *control* module provides the control computations for each of the 12 units; (b) the values of t_1, t_2, t_3, t_4, and t_5 must include the time taken to carry out the tests required and t_1 must also include the time taken to read the clock input line.)

The single-program approach can be used for simple, small systems and it leads to a clear and easily understandable design, with a minimum of both hardware and software. Such systems are usually easy to test. As the size of the problem increases, there is a tendency at the detail design stage to split modules not because they are functionally different but simply to enable them to complete within the required time interval. In the above example the management output requirement makes it unsuitable for the single-program approach; if that requirement is removed the approach could be used. It may, however, require the division of the display update module into three modules: display date and time; display process values; and display controller parameters.

7.5 FOREGROUND/BACKGROUND SYSTEM

There are obvious advantages — less module interaction, less tight time constraints — if the modules with hard time constraints can be separated from, and handled independently of, the modules with soft time constraints or no time constraints. The modules with hard time constraints are run in the so-called 'foreground' and the modules with soft constraints (or no constraints) are run in the 'background'. The foreground modules, or 'tasks' as they are usually termed, have a higher priority than the background tasks and a foreground task must be able to interrupt a background task.

The partitioning into foreground and background usually requires the support of a real-time operating system, for example the Digital Equipment Corporation's RT/11 system. It is possible, however, to adapt many standard operating systems, for example MS-DOS, to give simple foreground/background operation if the hardware supports interrupts. The foreground task is written as an interrupt routine and the background task as a standard program.

If you use a PC you are in practice using a foreground/background system. The application program that you are using (a word processor, a spreadsheet, graphics package or some program which you have written yourself in a high-level language) is, if we use the terminology given above, running in the background. In the foreground are several interrupt-driven routines — the clock, the keyboard input, the disk controller — and possibly some memory-resident programs which you have installed — a disk cacheing program or an extended memory manager. The terminology foreground and background can be confusing; literature concerned with non-real-time software uses foreground to refer to the application software and background to refer to interrupt routines that are hidden from the user.

Using the foreground/background approach the structure shown in Figure 7.4

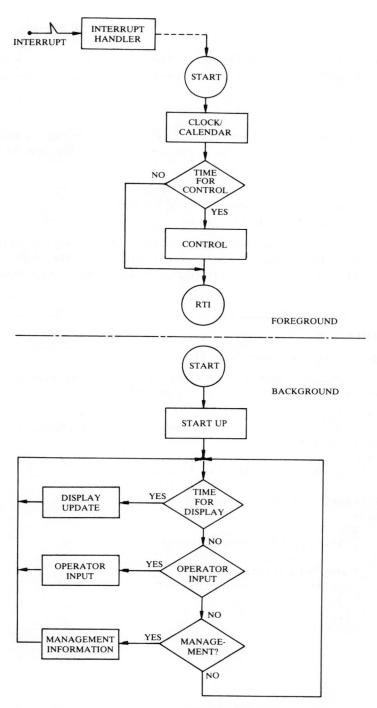

Figure 7.5   Foreground/background approach.

can be modified to that shown in Figure 7.5. There is now a very clear separation between the two parts of the system. A requirement for the foreground part to work is that

$$t_1 + t_2 < 20 \text{ ms}$$

where t_1 = maximum execution time for clock/calendar module and t_2 = maximum execution time for the control module. A requirement for the background part to work is that:

1. max $(t_3, t_4, t_5) < 10$ s;
2. display module runs on average every 5 s; and
3. operator input responds in < 10 s.

Although the time constraints have been relaxed the measurements to be made in order to check the performance are more complicated than in the single-program case and hence the evaluation of the performance of the system has been made more difficult.

EXAMPLE 7.2

Foreground/Background System Using Modula-2

Using the facilities provided by SYSTEM a simple foreground/background structure can easily be created to handle real-time control applications.

```
MODULE Main;
FROM SYSTEM IMPORT ADR, SIZE, WORD, PROCESS, NEWPROCESS,
TRANSFER, IOTRANSFER;
VAR
    main, operator, control : PROCESS;

PROCEDURE Control;
BEGIN
    LOOP
        IOTRANSFER(control, operator, clockVector);
            ...
    (* control actions go here
       routine should keep track of time as well *)
        ...
    END;
END Control;
```

```
PROCEDURE Display;
*** ...
(* insert the display update code here *)
*** ...
END Display;
PROCEDURE Keyboard;
*** ...
(* insert keyboard code here *)
*** ..
END Keyboard;

PROCEDURE Operator;
BEGIN
    LOOP
        IF time = displayTime THEN Display;
        Keyboard;
    END (* LOOP *);
END Operator;

BEGIN
    NEWPROCESS (Control, ADR(controlWksp),
    SIZE(controlWksp),
            control);
    NEWPROCESS (Operator, ADR(operatorWksp),
    SIZE(operatorWksp),
            operator);
    TRANSFER (main, control);
END Main.
```

Note that because we have used the low-level facilities of the language directly and simply none of the problems of data sharing and mutual exclusion discussed in the previous sections have been solved. All the variables required by the controller are assumed to be stored in common storage and hence are accessible at any time to either the operator task or the control task. Also in the above example the control task, which is entered on an interrupt, can return only to a specific named task and hence there can be only one background task. However, as the next section shows the low-level facilities provided can be used to create a much more powerful set of real-time multi-tasking support routines.

Although the foreground/background approach separates the *control* structure of the foreground and background modules, the modules are still linked through the data structure as is shown in Figure 7.6. The linkage occurs because they share data variables; for example, in the hot-air blower system, the control task, the display task and the operator input task all require access to the controller parameters. In the single program (sometimes called single tasking) there was no difficulty in controlling access to the shared variable since only one module (task) was active at

Foreground

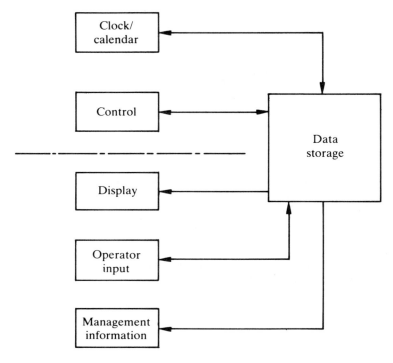

Background

Figure 7.6 Software modules for foreground/background system showing data storage.

any one time, whereas in the foreground/background system tasks may operate in parallel, that is one foreground module and one background module may be active at the same time. (Note: active does not mean 'running' since if one CPU is being used only one task can be using it at any instant; however, both the foreground and background tasks may have the potential to run.)

In this particular example the variables can be shared between the control, display and operator input modules without any difficulty since only one module writes to any given variable. The operator input module writes the controller parameters and set point variables, the clock/calendar module writes to the date and time variables and the control module writes to the plant data variables (error and output temperature). However, the input from the operator must be buffered and only transferred to the shared storage when it has been verified. Example 7.3 shows a method of doing this.

EXAMPLE 7.3

Buffering of Parameter Input Data

```
MODULE HotAirBlower;
VAR
    p1, p2, p3 : REAL; (* Controller parameters declared as
global variables *)
PROCEDURE GetParameters (VAR x, y, z : REAL);
BEGIN
    ...
    (* get new parameters from terminal and store in x, y, z *)
    ...
END GetParameters;

PROCEDURE OperatorInput;
VAR
    x, y, z : REAL;
BEGIN
    GetParameters (x, y, z);
    (* insert code to verify here *)
    p1: = x; (* transfer parameters to global variables *)
    p2 : = y;
    p3 : = z;
END OperatorInput;
BEGIN
    (* main program *)
END HotAirBlower.
```

To understand the reasons for buffering, let us consider what would happen if, when a new value was entered, it was stored directly in the shared data areas. Suppose the controller was operating with $p_1 = 10$, $p_2 = 5$ and $p_3 = 6$ and it was decided that the new values of the control parameters should be $p_1 = 20$, $p_2 = 3$ and $p_3 = 0.5$. As soon as the new value of p_1 is entered the controller begins to operate with $p_1 = 20$, $p_2 = 5$, $p_3 = 6$, that is neither the old nor the new values. This may not matter if the operator enters the values quickly. But what happens if, after entering p_1, the telephone rings or the operator is interrupted in some other way and consequently forgets to complete the entry? The plant could be left running with a completely incorrect (and possibly unstable) controller.

The method used in Example 7.3 is not strictly correct and safe since an interrupt could occur between transferring x to p_1 and y to p_2, in which case an incorrect controller would be used. For a simple feedback controller this would have little effect since it would be corrected on the next sample. It may be more serious if the change were to a sequence of operations. The potential for serious and possibly dangerous consequences is not great in small, simple systems (a good reason for keeping systems small and simple whenever possible); it is much greater in large systems.

The transfer of data between the foreground and background tasks, that is the statements

```
p1:=x;
p2:=y;
p3:=z;
```

form what is known as a *critical section* of the program and should be an indivisible action. The simple way of ensuring this is to inhibit all interrupts during the transfer:

```
InhibitInterrupts;
p1:=x;
p2:=y;
p3:=z;
EnableInterrupts;
```

However, it is undesirable for several separate modules each to have access to the basic hardware of the machine and each to be able to change the status of the interrupts. From experience we know that modules concerned with the details of the computer hardware are difficult to design, code and test, and have a higher error rate than the average module. It is good practice to limit the number of such modules. Ideally transfers should take place at a time suitable for the controller module, which implies that the operator module and the controller module should be synchronised or should rendezvous.

7.3 MULTI-TASKING APPROACH

The design and programming of large real-time systems is eased if the foreground/background partitioning can be extended into multiple partitions to allow the concept of many active tasks. At the preliminary design stage each activity is considered to be a separate task. (Computer scientists use the word *process* rather than task but this usage has not been adopted because of the possible confusion which could arise between internal computer processes and the external processes on the plant.) The implications of this approach are that each task may be carried out in parallel and there is no assumption made at the preliminary design stage as to how many processors will be used in the system.

The implementation of a multi-tasking system requires the ability to:

● create separate tasks;
● schedule running of the tasks, usually on a priority basis;
● share data between tasks;

- synchronise tasks with each other and with external events;
- prevent tasks corrupting each other; and
- control the starting and stopping of tasks.

The facilities to perform the above actions are typically provided by a real-time operating system (RTOS) or a combination of RTOS and a real-time programming language. We dealt with these in detail in the previous two chapters.

We now examine some examples to illustrate the problems that arise with multitasking and why real-time systems require special language and operating system facilities. For simplicity we will assume that we are using only one CPU and that the use of this CPU is time shared between the tasks. We also assume that a number of so-called primitive instructions exist. These are instructions which are part of a programming language or the operating system and their implementation and correctness is guaranteed by the system. All that is of concern to the user is that an accurate description of the syntax and semantics is made available. In practice, with some understanding of the computer system, it should not be difficult to implement the primitive instructions. Underlying the implementation of primitive instructions will be an eventual reliance on the system hardware. For example, in a common memory system some form of arbiter will exist to provide for mutual exclusion in accessing an individual memory location.

7.7 MUTUAL EXCLUSION

EXAMPLE 7.4

Mutual Exclusion

Consider the transfer of information from an input task to a control task as shown in Figure 7.7. The input task gets the values for the proportional gain, the integral action time and the derivative action time. From these it computes the controller parameters KP, KI and KD and these are transferred to the CONTROL task. A simple method is to hold the parameter values in an area of memory which has been declared as being COMMON and hence is accessible to both tasks. Unless the input task is given exclusive rights to this COMMON data area while it writes the parameter values there is a danger that the control task will read one new value, say KP, and two old values, KD and KI. Giving exclusive rights to the input task is not a satisfactory solution in this case as will be seen later.

As another example of the need for mutual exclusion consider the problem in Example 7.5.

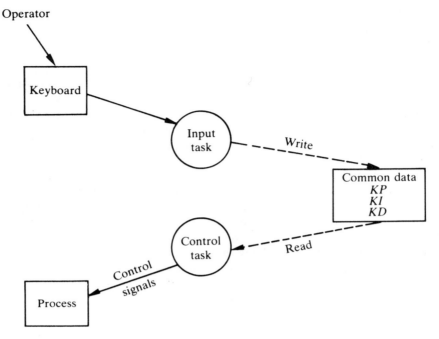

Figure 7.7   Data sharing using common memory.

EXAMPLE 7.5

As part of the maintenance procedures a record is kept of accesses to a particular device (after a specified number of accesses some preventative maintenance has to be carried out). The system designer arranges that each task in the system which uses the device will increment a common variable, deviceUse, by using the code

```
deviceUse := deviceUse+1
```

The hardware will resolve the problem of simultaneous access to the memory location in which deviceUse is stored, but this is not sufficient to guarantee the correct functioning of the counter.

Consider the following scenario. The compiler generates machine code in the following form:

```
load deviceUse
add 1
store deviceUse
```

Suppose the value of deviceUse is 38 and task A executes the load deviceUse; then the tasks reschedule and task B executes load deviceUse: both tasks now have in their own environments a register containing the current value of deviceUse, that is 38. Task B now executes the add 1 instruction and the store deviceUse instruction giving a value of 39 in deviceUse. Control is now returned to task A which executes the add 1 and the store deviceUse instructions which again give a value of 39. The final value of deviceUse is thus 39 even though it started as 38 and has been incremented twice.

In abstract terms, as we saw in the previous chapter, mutual exclusion can be expressed in the form

remainder 1
pre-protocol
critical section
post-protocol
remainder 2

where *remainder 1* and *remainder 2* represent sequential code that does not require access to a particular resource or to a common area of memory. The critical section is the part of the code which must be protected from interference from another task. The protocols called before and after the critical sections are code that will ensure that the critical section is executed so as to exclude all other tasks. To benefit from concurrency both the critical section and the protocols must be short such that the remainders represent a significant body of code that can be overlapped with other tasks. The protocols represent an overhead which has to be paid in order to obtain concurrency.

7.7.1 Condition Flags

A simple method of indicating if a resource is being used or not is to have associated with that resource a flag variable which can be set to TRUE or FALSE (or to 0 or 1, or SET or RESET). A task wishing to access the resource has to test the flag before using the resource. If the flag is FALSE (0 or RESET) then the resource is available and the task sets the flag TRUE (1 or SET) and uses the resource. The procedure is illustrated in Example 7.6.

EXAMPLE 7.6

Mutual Exclusion Using a Condition Flag

```
MODULE MutualExclusion1;
(* Mutual exclusion problem Condition Flag solution 1*)
VAR
    deviceInUse: BOOLEAN;
```

```
PROCEDURE Task; (* task assumed to be running in
parallel with other tasks *)
BEGIN
    (*      remainder1  *)
    WHILE deviceInUse DO
        (* test and wait until available *)
    END (* while *)
    deviceInUse:= TRUE;  (*claim resource*)
    (*......
    use the resource - critical section
    ......*)
    deviceInUse := FALSE;
    (*      remainder2 *)
END Task;
    (* main program *)
END MutualExclusion1.
```

In this solution there are two problems:

1. The WHILE statement forms a *busy wait* operation which relies on a pre-emptive interrupt to escape from the loop. If the task which has already claimed the resource cannot interrupt the *busy wait* then the task will continue to use the CPU and will exclude all other tasks.
2. The testing and setting of the flag are separate operations and hence the task could be suspended and replaced by another task between checking the flag and setting the resource unavailable. A consequence could be that, as is shown in Figure 7.8, two tasks could both claim the same resource.

The two tasks A and B shown in Figure 7.8 both share a printer. It is assumed that the flag variable printerInUse, which is set to 1 when the printer is in use and to 0 when it is available, controls access to the printer. Task A checks the printerInUse flag and finds that the printer is available, but before it can execute the next instruction, which would be to set the printerInUse flag to 1 and hence claim the printer, the dispatcher forces a task status change and task B runs. Task B also wishes to use the printer and checks the flag: it finds that the printer is available, sets the printerInUse flag to 1 and begins to use the printer. At some time later it requires some other resource and the dispatcher suspends it and makes task A the active task. Task A now claims the printer and begins to use it. Thus both tasks think that they have the exclusive use of the printer, whereas they are both using it and the output from the two tasks will be mixed up. After some time task A is again suspended and task B continues; it now finishes with the printer and releases it by setting printerInUse to 0, making the printer available to any other task even though task A still thinks that it has exclusive use of the printer. At task change 4, task A again uses the printer and eventually releases it although it has in fact already been released by task B.

For a condition flag to work securely it is therefore vital that the operations of test condition/set condition are indivisible. If a primitive instruction at the machine

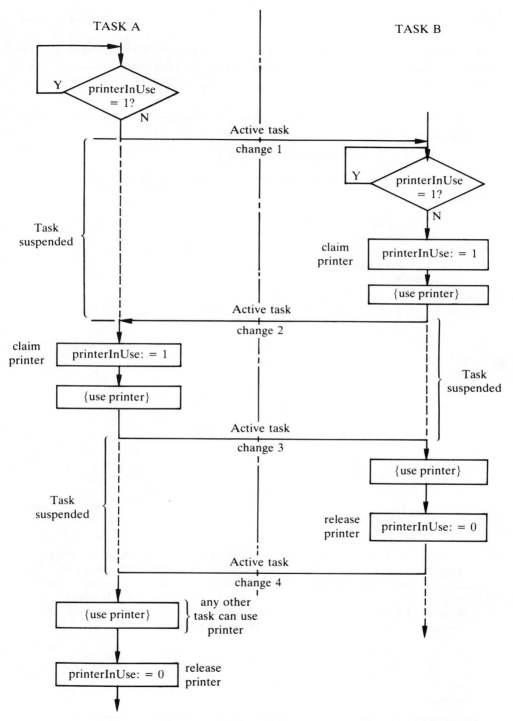

Figure 7.8 Attempt at mutual exclusion using condition flags.

code level which provides a combined test and set operation does not exist then the test/set operation must be made indivisible by the use of the enable/disable interrupt instructions.

EXAMPLE 7.7

Mutual Exclusion Condition Flags – Solution 2

```
(* Mutual exclusion problem Condition flag solution 2*)
VAR
    deviceInUse:BOOLEAN;
    deviceClaimed: BOOLEAN;
PROCEDURE Task;
BEGIN
    (* remainder1 *)
    REPEAT
        DisableInterrupts; (* procedure *)
        IF deviceInUse THEN
           deviceClaimed:= FALSE
        ELSE
           deviceInUse:= TRUE;
           deviceClaimed:= TRUE
        END (* IF *)
        EnableInterrupts;
    UNTIL deviceClaimed;
    (*......
    use resource (*critical section*)
    .....*)
    DisableInterrupts;
    deviceInUse:=FALSE;
    deviceClaimed:= FALSE;
    Enable interrupts;
    (* remainder2 *)
END Task;
```

Solution 2 is an improvement in that it will prevent two tasks gaining access to the same resource. There is still the problem of being in an endless loop waiting for the resource to become available and thus relying on some form of pre-emption to allow other tasks to run. If this approach is used in practice it would be sensible to incorporate a request for a short delay between each testing of the condition flag. The insertion of a call delay statement between lines 13 and 14, that is

```
enable interrupts;
delay(delayTime);
UNTIL deviceClaimed;
```

would be appropriate.

Because errors in the interrupt enable/disable status are potentially dangerous (a failure to enable interrupts at the end of a critical section will cause the whole system to fail), manipulation of the interrupt status flag should be restricted as much as possible and preferably should not occur in application level programs; therefore the solution in Example 7.7 is not recommended.

7.7.2 Semaphores

It is possible to devise a safe and reliable flag-based mutual exclusion system. One such system is the *turn flag* technique in which the flag, instead of showing if the resource is free or in use, indicates which task can *next* use the resource. The problem with this technique is that the tasks must run in strict sequence. The most reliable solution is to use Dekker's algorithm but the method becomes unwieldy as the number of tasks increases (see Cooling (1991, pp. 306–9) for a discussion of the methods). The most commonly used approach is the use of some form of semaphore (Example 7.8).

EXAMPLE 7.8

Use of Semaphores to Solve Transfer of Controller Parameters Problem

```
MODULE Controller;
(*Mutual exclusion-transfer of controller parameters
solution 1*)
TYPE
    AParameterRecord = RECORD
        kp : REAL;
        kd : REAL;
        ki : REAL
        END;
VAR
    mutex : SEMAPHORE;
    controlParameters : AParameterRecord;
    inputBlock : AParameterRecord;

TASK DataTransfer;
(* transfers input data to the controller *)

BEGIN
    Secure (mutex);
    controlParameters := inputBlock;
    Release (mutex);
END DataTransfer;
```

```
TASK Control;

BEGIN
    Secure (mutex);
    DoControl (*actual routines to perform control would be
    placed here*)
    Release (mutex);
END Control;

BEGIN (*main body of program *)
    Initialise (mutex, 1);
    StartTask (DataTransfer,5);
    StartTask (Control,1);
(*
DataTransfer is allocated priority level 5 which is a lower
priority than Control at priority level 1
*)
END Main.
```

This is an acceptable solution for co-operating tasks in a non-real-time environment, but for real-time work there are several problems. The first is in TASK DataTransfer. The use of a semaphore does not prevent the task being suspended and another task being run during the critical section: it prevents any other task accessing the ControlParameters record (providing that the task checks the semaphore mutex before proceeding). If, for example, it is time to run the Control task, which has a higher priority than the DataTransfer, the Control task will check the mutex semaphore and will then be suspended awaiting completion of the data transfer by the DataTransfer task. The consequences of the delay could be unpredictable: if the DataTransfer task is the only other task waiting to run, or has the highest priority of any of the waiting tasks, then the solution could be acceptable in that the transfer will be completed and the control task will run immediately the operation Release (mutex) is performed. However, if we introduce a third task, DisplayUpdate, with a priority level of, say, 3 then the sequence of events could be as illustrated in Figure 7.9.

In this figure we assume that the DataTransfer task has just secured the semaphore mutex, at which time the clock interrupt forces a rescheduling of the tasks and the Control task runs. The Control task will be suspended when it attempts to execute Secure(mutex) and there will again be a rescheduling of the tasks. Assume now that the DisplayUpdate task is ready to run; because it is of higher priority than DataTransfer, it will be run. The task DataTransfer cannot run until DisplayUpdate suspends or finishes running, and the Control task cannot run until DisplayUpdate has run and released mutex. The consequence of this delay may be that the Control task is not run within the specified sampling period.

At first sight it would seem that the use of a semaphore in this manner for mutual exclusion is not appropriate for real-time control system applications.

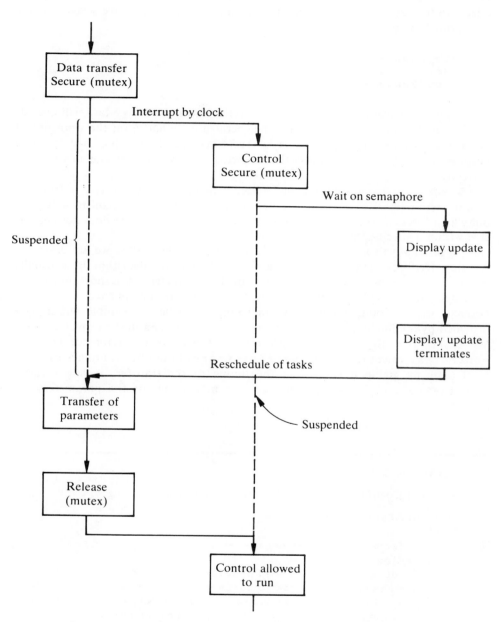

Figure 7.9 Transfer of controller parameters – use of semaphore.

However, you should be able to see that by changing the priorities of the task an adequate solution can be obtained. Try running through the above scenario if the task priorities are

```
DataTransfer = 1;
Control = 2;
DisplayUpdate =5.
```

Although the allocation of a high priority to the DataTransfer task will provide a solution it is not a safe solution for general use. Reliance on the priorities of tasks is unsafe as once the number of tasks becomes larger than five to ten it becomes very difficult to construct the scenarios to prove that the system will work correctly.

A semaphore-based solution can be used if there is a means of escape from the commitment to wait for a resource to become free. A suitable mechanism is the time out which forces a return from the task suspension if the resource does not become free within a specified period of time.

An alternative solution is to exploit the fact that the system we are designing is using feedback control and for a feedback control algorithm it is usually preferable to continue using the old values of the parameters for a short time rather than delaying the calculation of the next value of the manipulated variable. A possible way of doing this is shown in Example 7.9 below. Double buffering is provided by holding a local copy of controlParameters in the Control task. We ensure that the Control task does not wait if the reference copy of controlParameters held by DataTransfer is in use but continues using its local copy by giving it control of the timing of the transfer. A flag variable messagePresent is used to indicate that new values of the parameters are available.

EXAMPLE 7.9

Solution of Controller Parameters Problem Using Double Buffering

```
MODULE Controller;
(*
Transfer of controller parameter using double
buffering.
Delay(delayTime)is assumed provided by another
module and enables a task to suspend itself for a
given period of time. The variables kp, kd, ki, and
r are assumed to be obtained from the operator by
another task and held in some data area - a pool -
which is accessible to DataTransfer.
*)
```

```
TYPE
    AParameterRecord = RECORD
        kp : REAL;
        kd : REAL;
        ki : REAL
        r : REAL (* set point *)
        END;

VAR
    messagePresent : BOOLEAN;
    controlMessage : AParameterRecord;
    inputBlock : AParameterRecord;

TASK DataTransfer;

CONST
    delayTime = 20;
BEGIN
    (* wait if previous message has not been taken *)
    WHILE messagePresent DO
        Delay(delayTime)
    END (* while *);
    controlMessage.kp := kp;
    controlMessage.ki := ki;
    controlMessage.r := r;
    messagePresent := TRUE;
END DataTransfer;

PROCEDURE Control;

VAR
    controlParameter : AParameterRecord;
BEGIN
    DoControl; (*actual control statements would go
here *)
    IF messagePresent THEN
        controlParameters := controlMessage;
        messagePresent := FALSE
    END (* if *);
END Control;
BEGIN (*main body*)
    messagePresent := FALSE;
    StartTask(DataTransfer,5);
    StartTask(DisplayUpdate,2);
    StartTask(Control,1);
END Main.
```

The flag variable `messagePresent` is used to signal that a data transfer should take place. The timing of the transfer is left to the task Control and hence there should be no danger of the Control task being interrupted while it is transferring the data since it is running at the highest priority in the system. The penalties involved in using this solution are (a) that the task Control has to execute an additional instruction

```
IF messagePresent THEN
```

every time it runs (and because tasks such as Control are typically the most frequently run tasks in the system it is usually desirable to keep their execution time as short as possible); and (b) the use of a small amount of extra memory.

In this particular example – passing of the controller parameters – you may be wondering why we are not just updating kp, kd, ki and r without any mutual exclusion provision since the probability of the task Control interrupting the transfer is low. If we assume that the transfer of the parameter takes 20 microseconds, that Control runs every 100 ms with an execution time of 1 ms and if a transfer of parameters takes place during every run of Control then assuming that they are not synchronised in any way the chance of interference is approximately 1 in 5000. Given that the controller parameters are not going to be changed at this rate the chance of task Control interrupting a transfer is very low and since for this type of feedback control the disturbance to the controller from using a wrong set of parameters for one sample is likely to be small we can conclude that mutual exclusion for the examples shown above is unnecessary.

However, even in this simple system we must ensure that the operator does not directly change the values of the variables kp, ki, kd and r used by the Control task. The reason for this restriction should be obvious. The variables form a related data set, that is the individual values are not independent (see Chapter 4 if you want to know why this is so), and so we must protect against the possibility that an inconsistent set is being used. For example, consider the scenario where the operator enters kp and ki and is then interrupted by a colleague or by a telephone call; it could be several minutes before the entry of the new values is completed. We must therefore always ensure that there is buffering between the operator input and the transfer of input data to the set of variables available to the Control task and for security the variables used by the Control task should be hidden from the input task. Example 7.10 shows in outline one method of doing this by using the features of Modula-2.

EXAMPLE 7.10

Buffering of Operator Input

```
MODULE Operator;
IMPORT FROM Controller
    PutData (* Procedure *)
    . . .
```

```
TASK Operator
(* get, assemble and check parameters *)
   PutData(controlParameters)
   ...
END (* TASK Operator *)
BEGIN
   ...
END Operator.

DEFINITION MODULE Controller;
EXPORT
TYPE
   AParameterRecord = RECORD
      kp : REAL;
      kd : REAL;
      ki : REAL
      r : REAL (* set point *)
   END (* RECORD*);

PROCEDURE PutData(VAR ControlParameters:
AParameterRecord);
END Controller.
```

A final word of warning: real-time control systems can rapidly become complex and it then becomes difficult to work out all possible what-if scenarios. For example, assume that the designer/programmer of the `Controller` module decided not to use any form of mutual exclusion on the parameter transfer on the grounds that the chance of an interruption to the transfer was very small, but the system also includes several high-priority alarm condition tasks. There is now the possibility that the alarm occurs at the very instant when a partial transfer of values has occurred. The control loop continues running but completion of the transfer is held up because the alarm tasks take up all the available processor time. We are thus attempting to correct some problem on the plant and may well be making the problem worse because our control loop is running with an inconsistent set of controller parameters. A remote possibility perhaps, but if we are to produce safe systems we must expect that such remote possibilities will occur and plan for them. In general it is safest to assume that if something can possibly occur it will.

7.7.3 Notes on Using Semaphores

A further word of caution: in using the semaphore construct for mutual exclusion it is perhaps natural to assume that suspended tasks gain access to the resource in the order in which they performed the `secure(s)` operation, that is the task which has been waiting longest is served first. However, the order in which waiting tasks are selected when a resource becomes available is a matter for the designer of the operating system. Possible schemes are:

First in, *first out*: the task that has been waiting longest is chosen.
Priority order: the highest-priority waiting task is chosen.
Non-deterministic: any waiting task may be chosen arbitrarily.

The software designer must know what selection mechanism is being used. This is an example of just one of the many practical difficulties of separating out design from implementation details. Ideally the designer should be able to choose which mechanism she/he wants to use but often the choice will be restricted because of a decision to use a particular operating system or family of operating systems.

The semaphore provides an elegant mechanism for mutual exclusion but it is a low-level primitive and like the simple use of enable and disable interrupts it is error prone. One omitted or one misplaced semaphore instruction will cause the whole mutual exclusion protection to fail and the collapse of the whole system. Semaphores are historically important but for real-time systems a safer approach, for example the use of monitors as discussed below, is required.

7.8 MONITORS

The basic idea of a monitor was explained in Chapter 6. In Example 7.11 the implementation of a monitor in Modula-2 to protect access to a buffer area is shown. Monitors themselves do not provide a mechanism for synchronising tasks and hence for this purpose the monitor construct has to be supplemented by allowing, for example, signals to be used within it.

EXAMPLE 7.11

Monitor Using Signals

```
MODULE Buffer[monitorPriority];
(*solution to producer-consumer problem, also
implementation of a simple CHANNEL *)
    FROM Signals IMPORT
        Signal, InitSignal, AwaitSignal, SendSignal;
    EXPORT
        Put,Get;
    CONST
        nMax = 32;
    VAR
        nFree, nTaken: [0..nMax];
        in, out: [1..nMax];
        b: ARRAY [1..nMax] OF INTEGER;
        notFull, notEmpty: Signal;
```

```
        PROCEDURE Put(i: INTEGER);
        BEGIN
            IF nFree = 0 THEN
            AwaitSignal(nonFull)
            (* another task can call Put during the wait - it will
            also find nFree=0 and will wait *)
            ELSE
                DEC(nFree)
            END;
            b[in] := i;
            in := in MOD nMax + 1;
            IF Awaited(nonEmpty) THEN
                SendSignal(nonEmpty)
                (* a higher priority task waiting for this signal
                will run now and may lead to another call of Put *)
            ELSE
                INC(nTaken)
            END
        END Put;

    PROCEDURE Get(VAR i: INTEGER);
    BEGIN
        IF nTaken = 0 THEN
            AwaitSignal(nonEmpty)
            (* another task can call Get during the wait - it will
            also find nTaken = 0 *)
        ELSE
            DEC(nTaken)
        END;
        i := b[out];
        out := out MOD nMax + 1;
        IF Awaited(nonFull) THEN
            SendSignal(nonFull);
            (* a higher priority task waiting for this signal will
            run now and may lead to another call of Get *)
        ELSE
            INC(nFree)
        END
    END Get;
    BEGIN
        nFree := nMax; nTaken := 0;
        in := 1; out := 1;
        InitSignal(nonFull); InitSignal(nonEmpty);
    END Buffer;
```

To prevent deadlock when using signals within a monitor for task synchronisation, a task that gains access to a monitor procedure but then executes a wait(signal) operation must be suspended and placed outside the monitor. This procedure is necessary to allow another task to enter. Referring to the producer–consumer problem above, suppose the producer task enters the monitor with a call to Put but is forced to wait because the buffer is full; then the buffer can become non-full only if another task, the consumer, is able to enter the monitor and remove an item from the buffer using the Get procedure. The consumer will then issue a send(signal) to awaken the producer task and unless the send(signal) operation is the last executable statement in the Get procedure then two tasks could be active within the monitor thus breaching the mutual exclusivity rule. Hence the use of signals within a monitor requires the rule that the Send operation must be the last executable statement of a monitor procedure.

The standard monitor construction outlined above, like the semaphore, does not reflect the priority of the task trying to use a resource; the first task to gain entry can lock out other tasks until it completes. Hence a lower-priority task could hold up a higher-priority task in the manner described in Example 7.8. The lack of priority causes difficulties for real-time systems. Traditional operating systems built as monolithic monitors avoided the problem by ensuring that once an operating system call was made (in other words, when a monitor function was invoked) then the call would be completed without interruption from other tasks. The monitor function is treated as a critical section. This does not mean that the whole operation requested was necessarily completed without interruption. For example, a request for access to a printer for output would be accepted and the request queued; once this had been done another task could enter the monitor to request output and either be queued, or receive information from the monitor as to the status of the resource. The return of information is particularly important as it allows the application program to make a decision as to whether to wait for the resource or take some other action.

Preventing lower-priority tasks locking out higher-priority tasks through the monitor access mechanism can be tackled in a number of ways. One solution adopted in some implementations of Modula-2 is to run a monitor with all interrupts locked out; hence a monitor function once invoked runs to completion. In many applications, however, this is too restrictive and some implementations allow the programmer to set a priority level on a monitor such that all lower-priority tasks are locked out – note that this is an *interrupt* priority level, not a task priority.

The monitor has proved to be a popular idea and in practice it provides a good solution to many of the problems of concurrent programming. The benefits and popularity of the monitor constructs stem from its modularity which means that it can be built and tested separately from other parts of the system, in particular from the tasks which will use it. Once a fully tested monitor is introduced into the system the integrity of the data or resource which it protects is guaranteed and a fault in a task using the monitor cannot corrupt the monitor or the resource which it protects. Although it does rely on the use of signals for intertask synchronisation it does have the benefit that the signal operations are hidden within the monitor.

The monitor is an ideal vehicle for creating abstract mechanisms and thus fits in well with the idea of top-down design. However, the nested monitor call problem – calling procedures in one monitor from within another monitor – can lead to deadlock. Providing that nested monitor calls are prohibited the use of the monitor concept provides a satisfactory solution to many of the problems for a single-processor machine or for a multi-processor machine with shared memory. It can also be used on distributed systems.

The monitor's usefulness in some real-time applications is restricted because a task leaving a monitor can only signal and awaken one other task – to do otherwise would breach the requirement that only one task be active within a monitor. This means that a single controlling synchroniser task, for example a clock level scheduler, cannot be built as a monitor. The problem can be avoided by allowing signals to be used outside a monitor but then all the problems associated with signals and semaphores re-emerge.

7.9 RENDEZVOUS

The rendezvous, developed by Hoare (1978) and Brinch Hansen (1973), provides an alternative to the use of monitors and signals to ensure mutual exclusion and synchronisation in intertask communication. In the rendezvous the actions of synchronisation and data transmission are seen as inseparable activities. The fundamental idea is that if two tasks A and B wish to exchange data, for example if A wishes to transmit data to B, then A must issue a transmit request and B a receive request. If task A issues the transmit request before B has issued the receive, then A must wait until B issues its request and vice versa. When both tasks have synchronised the data is transferred and the tasks can then proceed independently.

A problem with the original formulation is that both tasks must name each other and hence general library tasks cannot be created. The solution adopted in the language Ada is to use an asymmetric rendezvous in which only one task, known as the caller, names the other task, known as the server. In the descriptions that follow it is assumed that the language which supports the rendezvous concept does so by means of a construct of the form

```
ACCEPT name(parameter list)
    statements
END
```

The statements within the `ACCEPT...END` are assumed to be a critical section and are executed in a mutually exclusive manner. They would normally be executed by the server task. The `ACCEPT` statement represents an entry point and the calling task specifies the name of the entry point when it wishes to synchronise with the server task.

EXAMPLE 7.12

Simple Rendezvous

```
TASK A;
    VAR x:ADataItem;
    BEGIN
        ...
        B.Transfer(x);
        ...
    END;

TASK B;
    VAR y:ADataItem;
    BEGIN
        ...
        ACCEPT Transfer(IN item:ADataItem);
        y:=item;
    END;
END;
```

TASK A wishes to pass information held in variable x to a variable y in TASK B. The actual data transfer takes place using the normal parameter passing mechanisms — the actual parameters supplied in the call (in this case the variable x) are bound to the formal parameters of the ACCEPT statement (in this case item). The synchronisation of the two tasks is obtained by the requirement that the entry procedure call — B.Transfer(x) — cannot be completed until the corresponding ACCEPT statement — ACCEPT Transfer — is executed and conversely the execution of the ACCEPT statement cannot be completed until the entry call is executed. The actual transfer is completed within the body of the ACCEPT statement; in this case the data supplied by the entry call is transferred to a variable which is local to TASK B.

Note that in the ACCEPT statement the direction of the transfer is specified, in this case IN. Variables can be declared as being for input (IN) or output (OUT) or as bidirectional (IN OUT).

When using the rendezvous the two tasks have to synchronise in order to transfer information. The task which is producing the information cannot leave it in a buffer and continue but must wait for the consumer task to arrive before the transfer can take place. The position is equivalent to that which the motorist would face if there were no filling stations but only roving petrol tankers. The motorist and the petrol tanker driver would have to arrange for a rendezvous; when both arrived at the designated place the motorist would fill up with petrol from the tanker, and then both would continue on their respective ways. The requirement of the rendezvous that tasks synchronise in order to exchange data is too severe a constraint for many applications.

One solution to strict synchronisation is to introduce a buffer task between the two tasks which wish to exchange data. A simple buffer requires that the tasks have to call the buffer task in strict rotation. Continuing the filling station analogy this is the equivalent of demanding that the tanker and motorist alternate visits to the same filling station – clearly impractical as the tanker will deliver a much larger quantity of fuel than a single motorist will receive. The problem can be solved by introducing indeterminacy into the task by means of a SELECT statement (Example 7.13).

EXAMPLE 7.13

Illustrating Use of a SELECT Statement

```
TASK AFillingStation;
    VAR y:AFuel;
    BEGIN
        DO
            SELECT
                ACCEPT deliver(IN fuel:AFuel);
                y:=fuel;
                END;
            OR
                ACCEPT receive(OUT fuel:AFuel);
                fuel:=y;
                END;
                END SELECT;
            END DO;
        END;
```

The key to the operation of the above task is the action of the SELECT statement: each time the SELECT statement is executed there are four possible states in which the entry points to the task can be:

1. a call to deliver is pending;
2. a call to receive is pending;
3. calls to deliver and to receive are pending; and
4. no calls are pending.

For cases 1 and 2 then the appropriate ACCEPT statement is immediately executed. In case 3 one of the ACCEPT statements is selected at random and executed. In case 4 the task is suspended until a call is made to either of the ACCEPT statements at which time the task is resumed and the appropriate statement is executed. The SELECT statement ensures that only one ACCEPT statement will be executed at any one time but the order is not predetermined and there can be successive calls to the same ACCEPT.

As was seen in Example 7.8, involving the transfer of control parameters from an input task to the control task itself, there is a need to be able to test if another task is waiting, or if another task has left data to be collected in order to avoid committing a high-priority repetitive task to wait for an event. There is also frequently the requirement in real-time systems to have some form of time out such that a task only commits itself to wait for a predetermined length of time. Two extensions to the rendezvous primitive provide facilities to support these actions.

The time-out facility is provided in a simple and natural way by extending the SELECT statement to allow a delay option in the possible choices within the SELECT construct. This is illustrated in Example 7.14 for the control parameter problem. It is assumed that when an input task has gathered the new parameters it makes a call to a Put entry point in the control task. The control task includes the following code.

EXAMPLE 7.14

Use of Time Out

```
TASK Control;
    . . .
BEGIN
    . . .
    (*control action*)
    . . .
(*start of section to check if update of parameters is
required*)
    SELECT
        ACCEPT Put(IN parameters:AControlParRecord);
            kp := parameter.kp;
            kd := parameter.kd;
            ki := parameter.ki;
        END
    OR
        DELAY 1 (*delay in milliseconds*)
            . . .
        END
    END
END.
```

In the above code fragment if the control task reaches the SELECT statement when a call to the entry point Put is pending, then the ACCEPT part of the SELECT statement is executed and the parameter values are transferred to the control task. However, if no call is pending then the DELAY part of the SELECT statement is executed. The action of the delay is to cause the control task to wait for the length of time specified in the delay; during this period of suspension any call to the ACCEPT statement will be recognised and the ACCEPT statement executed. If no calls are received, then at the end of the delay period the statements following the DELAY statement are executed.

An alternative to the delay part within a SELECT statement is an else part (this can be thought of as a delay 0). The above problem could be coded using the ELSE statement as in Example 7.15.

EXAMPLE 7.15

Use of ELSE with SELECT

```
TASK Control;
   . . .
BEGIN
   . . .
   (*control action*)
   . . .
   (*start of section to check if update of parameters is
   required*)
   SELECT
      ACCEPT Put(IN parameters:AControlParRecord);
         kp := parameter.kp;
         kd := parameter.kd;
         ki := parameter.ki;
      END
   OR
      ELSE
         . . .
      END
   END
END.
```

In this example, if there is no call pending for the ACCEPT statement when the SELECT statement is reached, then the ELSE part of the SELECT statement is executed immediately. The use of the SELECT...OR...ELSE construct is the most appropriate for the control parameters problem.

The DELAY statement is useful in many applications; for example, on detection of an alarm condition the operator may be alerted and expected to acknowledge the alarm and take appropriate action within a predetermined time. If the operator does not respond, then the computer system has to take further action, possibly by sounding an audible alarm or by beginning to close down the plant. The SELECT...DELAY construct provides a natural and simple way of expressing the requirement:

```
SELECT
    ACCEPT OperatorAcknowledge;
OR
    DELAY 30 (*delay 30 seconds*)
    ...
    AlternativeAction;
    ...
END
```

Another example of the use of the DELAY statement is to provide time out in communications with peripherals or other computers.

The rendezvous concept provides the most flexible and easily understood mechanism for handling multi-tasking problems and in the SELECT mechanism provides facilities which none of the other concepts have. It has been implemented as part of the Ada language.

7.10 SUMMARY

In this chapter we have dealt informally with the basic approaches to the design of real-time systems. We have emphasised the division of the system into subsystems – modules – and briefly considered the heuristics commonly used to guide this process. An important aspect of subdivision is that the modules should be used to hide information.

There are three models on which the implementation of real-time software can be based. These are:

- single task;
- foreground/background; and
- multi-tasking.

For small, simple systems the first two models should be used. They result in a simple implementation that can be easily understood and tested. However, only the single-task model, without interrupts, can be formally proved correct: once interrupts are permitted the system immediately becomes non-deterministic and its correctness cannot be formally proved.

As systems become larger and more complex they can be most easily implemented if a multi-tasking model is adopted. Multi-tasking introduces problems of mutual exclusion, intertask communication and intertask synchronisation. These problems are now well understood. Modern real-time languages and operating systems provide primitive instructions and various mechanisms that support multi-tasking. Some of the standard problems and their solution were described.

Detailed knowledge of the application and judicious use of the application characteristics can simplify some multi-tasking problems as we illustrated when considering the transfer of the controller parameters. Simplifications of this sort are part of the art of engineering; however, they must be used with care and must be *documented* – in particular the conditions for which the simplification is valid must be clearly stated.

EXERCISES

7.1 The standard input routines in languages such as FORTRAN, Pascal and BASIC cannot be used within a timed loop to obtain information from the keyboard. This is also true of Modula-2. Why can't we use the standard Modula-2 routines?

7.2 A plant operating in a remote location is controlled by an embedded computer control system. The plant operates in two modes referred to as Amode and Bmode. The control algorithm for Amode is of the form

$$m(n) = Ae(n) + Be(n - 1) + Ce(n - 2) + Dm(n - 1) + Em(n - 2)$$

and for Bmode

$$m(n) = K_1 e(n) + K_2 e(n - 1) + K_3 e(n - 2)$$

where $e(n) = R - c(n)$
R = set point
$c(n)$ = the measured output of the plant at interval n.

The change-over from Amode to Bmode is to be made when $c(n) > ChangeA$ for five successive readings. The change-over from Bmode to Amode is to be made when $c(n) < ChangeB$ for five successive readings. The parameters *ChangeA* and *ChangeB* and the set point R can all be changed from a central station.

A change to the value of R requires a change in the values of *ChangeA* and *ChangeB*. The controller parameters A, B, C, D, E and K_1, K_2 and K_3 also need changing. They must be changed as the set $\{A, B, C, D, E, K_1, K_2, K_3\}$ and not as individual elements. The data transmission link to the remote station has a slow transmission speed and is subject to frequent bursts of interference. You can assume that the data transmission system support software contains error checking software and organises retransmission of erroneous data.

Discuss the problems of designing the software for the embedded computer system and discuss possible ways of dealing with the slow and unreliable data transmission system.

7.3 What is the principal difference between a pool and a channel? Explain why you would use (a) a pool and (b) a channel.

8

Real-time System Development Methodologies − 1

This chapter begins with an overview of the general approach now being adopted in the specification, design and construction of complex real-time systems, followed by a brief description of some of the standard methodologies. The Yourdon methodologies are then described in detail. The aims of the chapter are to:

- Show how specification, design and implementation can be considered as a process of *modelling*.
- Describe the major methodologies.
- Provide a more detailed understanding of one methodology, the Yourdon methodology.

8.1 INTRODUCTION

The production of robust, reliable software of high quality for real-time computer control applications is a difficult task which requires the application of engineering methods. During the last ten years increasing emphasis has been placed on formalising the specification, design and construction of such software, and several methodologies are now extant. The major ones are shown in Table 8.1. All of the methodologies address the problem in three distinct phases. The production of a *logical* or *abstract* model − the process of *specification*; the development of an *implementation* model for a *virtual machine* from the logical model − the process of *design*; and the construction of software for the virtual machine together with the implementation of the virtual machine on a physical system − the process of *implementation*. These phases, although differently named, correspond to the phases of development generally recognised in software engineering texts. Their relationship to each other is shown in Figure 8.1.

> *Abstract model*: the equivalent of a requirements specification, it is the result of the requirements capture and analysis phase.
>
> *Implementation model*: this is the equivalent of the system design; it is the product of the design stages − architectural design and the detail design.

Table 8.1 Summary of design methodologies

MASCOT	Design, construction, operation and test tools	Jackson and Simpson (1975)
CORE	Controlled Requirements Expression	British Aerospace, Systems Designers (1979)
PAISLey	Specification and simulation tools	Zave (1982)
DARTS	Design and Analysis of Real-Time Systems	Gomaa (1984)
JSD	Jackson System Development	Jackson (1983)
Yourdon (a)	Structured design and development of real-time systems	Ward and Mellor (1986)
Yourdon (b)	Strategies for real-time system specification	Hatley and Pirbhai (1988)
HOOD	Hierarchical Object-Oriented Design	CISI Ingenierie, CRA A/S Matra Aerospace (1989)

Implementation: the process of mapping the implementation model onto the physical hardware, identifying the software modules and coding them.

Although there is a logical progression from abstract model to implementation model to implemented software, and although three separate and distinct artifacts – abstract model, implementation model, and deliverable system – are produced, the phases overlap in time. The phases overlap because complex systems are best handled by a hierarchical approach: determination of the detail of the lower levels in the hierarchy of the logical model must be based on knowledge of higher-level design decisions, and similarly the lower-level design decisions must be based on the higher-level implementation decisions. Another way of expressing this is to say that the higher-level design decisions determine the requirements specification for the lower levels in the system.

All the methodologies require the support of CASE (Computer-Aided Software Engineering) tools for their effective use. Without such tools the methods become too laborious for use on large systems and many of their benefits in terms of enforcing consistency are lost. The number and range of CASE tools available is growing rapidly and some simple ones are now available for PCs (McClure, 1989).

We will consider MASCOT and the two Yourdon methodologies in some detail in later sections and also discuss the use of PAISLey. An introduction to CORE, JSD and HOOD can be found in Cooling (1991, Chapter 10).

CORE (Mullery, 1979) is specifically designed for the requirements capture and analysis phase of the development, that is the construction of the requirements specification. Subsequent design and implementation has to use other methodologies. It is an attempt to find an approach that will reduce the amount of information that the customer 'forgot to tell one about' – of course this will never be reduced to zero.

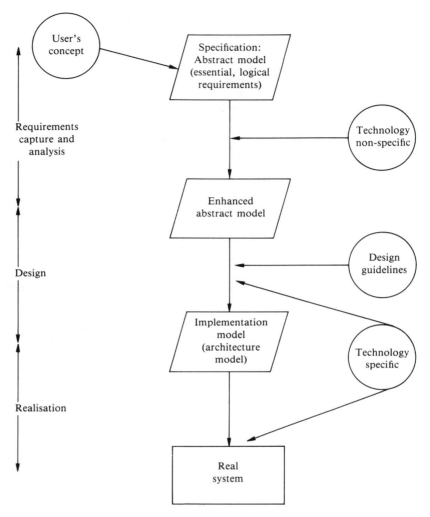

Figure 8.1   Software modelling.

The Jackson System Development methodology has been widely used in non-real-time systems, as was the Jackson Structured Programming methodology, but has only recently been applied to real-time systems (Cooling, 1991, p. 358). It is a data-driven method and is now supported by a number of CASE tools (Program Design Facility, Speedbuilder and Network Builder) which are supplied by the Michael Jackson Company.

The abstract model in JSD is a network of processes. There are three types of process:

1. *Input*: these detect actions taking place in the environment (events) and pass these events to the internal system.
2. *Output*: these pass the system response back to the environment.
3. *Internal*: these deal with events reported by the input processes and actions resulting from other internal processes; they pass events and actions to output and other internal processes.

The processes are connected in two ways to form the network:

1. by buffered asynchronous data streams;
2. by state vector connections.

By using the state vector one process can inspect the internal state of another process.

In common with other methodologies originally developed for non-real-time applications there is no formal method of incorporating timing constraints. Also, as we will see when dealing with the Yourdon methodologies, JSD provides a notation for representing a specification and a design. It does not perform the design. JSD is a data-flow method and the basic design technique is to use functional decomposition and to preserve *natural* data flows.

Once the design has been produced and an implementation model obtained, then the realisation of this model proceeds in a systematic manner. With the appropriate CASE tools much of the code for JSD designs can be automatically generated, particularly if the implementation language is Ada or occam 2.

HOOD is a new addition to the real-time methodologies and is based on an object-oriented approach. It is targeted at implementations based on the use of Ada but can be adapted for use with other programming languages. Like MASCOT it is meant to be a design method which takes a requirements specification obtained by other means as its starting point. Also like MASCOT the diagrams used for the design have a direct textual equivalent which formalises the design and which can be used in the implementation stage of the development.

In recent years there has been extensive consideration of formal (mathematical) techniques for the specification of systems. The most widely used formal method is VDM and the language Z is also gaining support. These techniques are aimed at producing a *formal specification* of a system. The benefit of a formal specification is that it is possible to prove that it is consistent (it is not possible to prove that it is complete – there is still the 'we forgot to mention' factor – but formal analysis of consistency may well reveal incompleteness). The second advantage of formal techniques is the possibility (theoretically) of transforming a specification into a realised implementation and proving that each step in the transformation is correct. In practice the proofs are very difficult. A major limitation of the formal specification language approach is that at present none of the languages make any provision for the incorporation of timing constraints. In the next chapter we describe one methodology (PAISLey) that has attempted to combine formality and timing constraints.

8.2 YOURDON METHODOLOGY

The Yourdon methodology has been developed over many years. It is a structured methodology based on using *data-flow modelling* techniques and *functional decomposition*. It supports development from the initial analysis stage through to implementation. Both Ward and Mellor (1986) and Hatley and Pirbhai (1988) have introduced extensions to support the use of the Yourdon approach for the development of real-time systems and the key ideas of their methodologies are:

- subdivision of system into activities;
- hierarchical structure;
- separation of data and control flows;
- no early commitment to a particular technology; and
- traceability between specification, design and implementation.

Although the original method was developed as a pencil and paper technique most benefit can be obtained if there is CASE tool support and many software engineering CASE tools now support both the Ward and Mellor and the Hatley and Pirbhai versions of Yourdon. Examples are Software Through Pictures and EasyCase Plus.

There are many similarities between Ward and Mellor and Hatley and Pirbhai but to avoid confusion we will deal with them separately. As an example we shall use the system which is described in the next section.

8.3 REQUIREMENTS DEFINITION FOR DRYING OVEN

1.   Introduction

1.1   Components are dried by being passed through an oven. The components are placed on a conveyor belt which conveys them slowly through the drying oven. The oven is heated by three gas-fired burners placed at intervals along the oven. The temperature in each of the areas heated by the burners is monitored and controlled. An operator console unit enables the operator to monitor and control the operation of the unit. The system is presently controlled by a hardwired control system. The requirement is to replace this hardwired control system with a computer-based system. The new computer-based system is also to provide links with the management computer over a communication link.

1.2   The general arrangement of the system is shown in figure 1.1 [Figure 8.2].

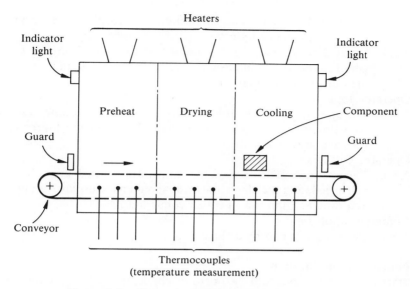

Figure 8.2   General arrangement of drying oven.

2.   Input/Output

2.1   The inputs come from a plant interface cubicle and from the operator. There will need to be inputs obtained from the communication interface.

2.2   Plant Inputs

    2.2.1   A thermocouple is provided in each heater area – the heater areas are *pre-heat*, *drying*, and *cooling*. The inputs are available as voltages in the range 0 to 10 volts at pins 1 to 9 on socket j22 in the interface cubicle.

    2.2.2   The conveyor speed is measured by a pulse counting system and is available on pin 3 at socket j23 in the interface cubicle. It is referred to as *con-speed*.

    2.2.3   There are three interlocked safety guards on the conveyor system and these are *in-guard*, *out-guard*, and *drop-guard*. Signals from these guards are provided on pins 4, 5, 6 of socket j23. These signals are set at logic HIGH to indicate that the guards are locked in place.

    2.2.4   A *conveyor-halted* signal is provided on pin 1 of socket j23. This signal is logic HIGH when the conveyor is running.

2.3   Plant Outputs

    2.3.1   Heater Control: each of the three heaters has a control unit. The input to the control unit is a voltage in the range 0 to 10 volts which corresponds to no heat output to maximum heat output.

    2.3.2   Conveyor Start-up: a signal *convey-start* is output to the conveyor motor control unit. A second signal *convey-stop* is also output to the motor control unit. The connections are to pins 1, 2, 3 on j10 for convey-start, and to pin 5 on j10 for convey-stop.

2.3.3   Guard Locks: asserting the *guard-lock* line, pin 8 on j10, causes the guards to be locked in position and turns on the red indicator lights on the outside of the unit.

2.4   Operator Inputs

2.4.1   The operator sends the following command inputs: *Start, Stop, Reset, Re-start,* and *Pause.* The operator can also adjust the desired set points for each area of the dryer.

2.5   Operator Outputs

2.5.1   The operator VDU displays the temperature in each area, the conveyor belt speed, and the alarm status. It should also display the current date and time and the last operator command issued.

2.6   Communication Inputs

2.6.1   These have yet to be defined.

3.   Functional Specification

3.1   Start-up: starting from cold the operator checks that all the guards are closed and issues the Start command. The guards are locked and if locking is correctly achieved the heaters are switched to on. Under this condition maximum heat is supplied. The temperature is monitored and when each area reaches a temperature within 20% of its set point control of the heaters is switched to normal.

3.2   Conveyor Start-up: when all areas have switched to normal control the conveyor start-up sequence is initiated. The conveyor is stepped through the start-up procedure. Motor position 1 is selected and held until the speed reaches 0.5 ft/min; then position 2 is selected. This is held until the speed reaches 1.5 ft/min; at this point the normal running position 3 is selected. If the conveyor fails to reach the desired speed within 30 seconds the conveyor is stopped and the *conveyor-fault* signal is asserted. The normal conveyor speed is 8 to 10 ft/min. If at any time the speed drops below this for more than 30 seconds the *conveyor-alarm* signal should be asserted.

3.3   Temperature Monitoring: the temperature measurement for each area is read at 2 second intervals. If the temperature for an area varies by more than 5% from the set point for that area then an alarm should be asserted.

3.4   Each area is controlled using a PID control algorithm.

3.5   Conveyor Failure: if the conveyor fails to start the operator can issue a Reset command which closes the whole system down. When it has been closed down and the system checked for obstructions the operator can issue the Start command again. If the conveyor stops or slows during normal running the operator can issue either a Re-start command which causes the conveyor first to be stopped and then to enter the full conveyor start-up cycle, or a reset signal that causes the whole system to be closed down.

3.6   Conveyor Pause: during normal running the operator may issue a Conveyor

Pause command. This halts the conveyor. It may for example be used to permit clearance of a blockage. The conveyor can be re-started by the operator issuing the Re-start command.

3.7 At any time during normal running the operator may issue the Stop command. The response is to turn off the heaters and the conveyor. When the conveyor stopped signal is asserted the guards are unlocked and the display lights are turned to green.

8.4 WARD AND MELLOR METHOD

The outline of the Ward and Mellor method is shown in Figure 8.3. The starting point is to build, from the analysis of the requirements, a software model representing the requirements in terms of the abstract entities. This model is called the *essential model*. It is in two parts: an *environmental* model which describes the relationship of the system being modelled with its environment; and the *behavioural*

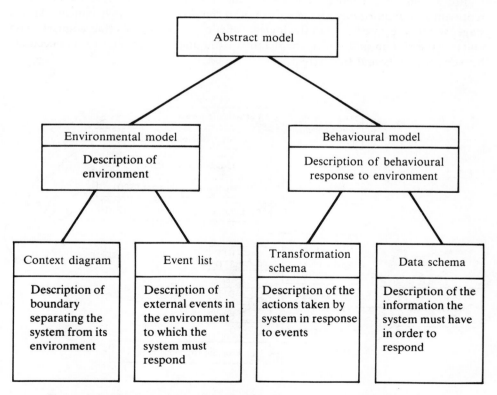

Figure 8.3 Outline of abstract modelling approach of Ward and Mellor.

model which describes the internal structure of the system. The second stage – the design stage – is to derive from the essential model an *implementation model* which defines how the system is implemented on a particular technology and shows the allocation of parts of the system to processors, the subdivision of activities allocated to each processor into tasks, and the structure of the code for each task.

The essential model represents *what* the system is required to do; the implementation model shows *how* the system will do what has to be done. The implementation model provides the design from which the implementors of the physical system can work. Correct use of the method results in documentation that provides *traceability* from the physical system to the abstract specification model. The type of documentation produced is shown in Figure 8.4.

8.4.1 Building the Essential Model – the Environmental Model

For most real-time systems the environmental model will comprise a *context* diagram and an *event* list and the entity relationship diagram will not be used. Figure 8.5 shows a context diagram for the drying oven. The rectangular boxes represent *terminator blocks* which are entities that exist in the environment. At this stage we are interested only in the logical function of the signals that connect these units to the system and not in the details of the units or the physical connections between the units and the system.

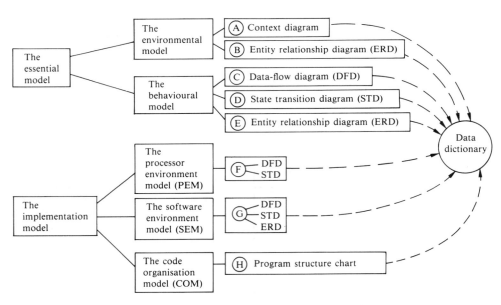

Figure 8.4   Relationship between models and diagrams.

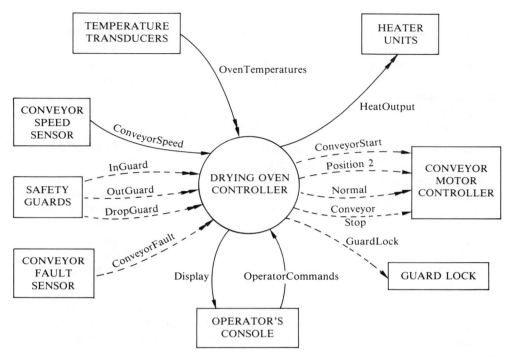

Figure 8.5 Drying oven — context diagram.

The system that we are building is represented as a single *bubble*. The directed lines represent *data* and *control* information that is passing between the system and the environment. The flows are classified as shown in Table 8.2.

Continuous flows are used to represent data or control signals for which there is always a value available to the system. *Discrete* flow is used for data or control which is generated as separate items or transactions and once the system has received or transmitted the item it is no longer available. A loose analogy is that discrete flows are the equivalent of a connection based on a channel and continuous flows with a connection based on a pool. As the context diagram is drawn, all flows should be named and each should be entered into a *data dictionary*. A data dictionary (also

Table 8.2 Flow notations — Ward and Mellor

Data	Continuous	Solid line, double arrow head
	Discrete	Solid line, single arrow head
Control	Continuous	Dotted line, double arrow head
	Discrete	Dotted line, single arrow head

known as a *requirements dictionary*) contains entries describing, in a rigorous manner, all the data elements of the system such that the user and system analyst will have a common understanding of them. Data elements in this sense include control flows as well as data flows and it is also common practice to include in the dictionary descriptions of the activities (see below) that the system performs (hence the use of the name requirements dictionary).

A data dictionary always forms part of a CASE tool. The exact structure can take many forms but must include the following information:

- *Name* — the primary name of the data or control item, or of an external entity, or of a process (activity).
- *Alias* — other names used instead of the primary name.
- *Usage* — where and how it is used, that is a listing of the processes/activities that use it and whether it is an input or output to a process or a store or external entity.
- *Content description* — a description using a standard notation of the content.
- *Other information* — data types, preset values, range of values, and other restrictions.

When using a CASE tool once an item has been entered into the dictionary consistency of naming can be enforced. For example, if a name already exists in the dictionary any attempt to use the name for another flow will be detected and the user will be warned.

An example of a data dictionary is given in Table 8.3. The symbols CD, DC are used to indicate continuous data and discrete control flows respectively and T is used to indicate that the entity is a terminator block. We would need to use symbols to indicate discrete data (DD) and continuous control (CC) as well as data and control stores (DS and CS) and processes (P). The first entry in Table 8.3 shows

Table 8.3   Example of a data dictionary

Name		*Description*
OvenTemperatures	CD	PreHeatTemp + DryingTemp + CoolingTemp *output from TEMPERATURE TRANSDUCERS, input to DRYING OVEN CONTROLLER*
PreHeatTemp	CD	range 0–100°C * measurement of temperature in PreHeat area of oven *
ConveyorStart	DC	range [on/off] * output from DRYING OVEN CONTROLLER, input to CONVEYOR MOTOR CONTROLLER *
GUARD LOCK	T	* external unit controls operation of guard locks *

a compound or group data flow made up of three elements representing the temperature measurement flows from each area of the oven. Group flows are widely used since they provide a concise means of describing the information that is being handled. For example, we could have used a group flow **GuardStatus** to describe the three control flows **InGuard**, **OutGuard** and **DropGuard** shown in the context diagram (Figure 8.5).

Associated with the context diagram is an event list. This is a table which lists all the events that can cause a change in the system and result in a change in an output. An event list for the drying oven is shown in Table 8.4. Note that the event list shown in Table 8.4 also shows the time response required for the response to the various events including the cycle time. This is not part of the Ward and Mellor requirement but is important and should be added. If possible the type of time constraint — hard or soft — and any tolerances should also be indicated at this stage.

8.4.2 Building the Essential Model — the Behavioural Model

The behavioural model shows how the system should respond to events taking place in the environment. A hierarchical approach to building the model is used. The system is divided into the various functions or activities that it has to perform. These functions are referred to as *transformations* in the Ward and Mellor method and are shown on a *transformation diagram*. Figure 8.6 shows the first-level transformation diagram for the **Drying Oven**.

The bubbles drawn with solid lines represent *data transformations* and those drawn with dotted lines represent *control transformations*. The transformation **ControlAreaTemp** has a double line round part of the bubble which indicates that

Table 8.4   Event list for drying oven

Event	Action	Response	Time
Start	Lock guards	Guardlock	< 0.5 s
InGuard	Start heat up	Set maximum	
OutGuard	cycle	heat output	< 0.5 s
DropGuard	when heat normal	ConveyorStart	?
Pause	Stop conveyor	ConveyorStop	< 0.5 s
ConveyorFault	Raise alarm	ConveyorAlarm	< 0.1 s
Stop	Close system down	ConveyorStop	
		Heaters off	< 0.5 s
OvenTemperature	Do control	HeatOutput	Cyclic 1.0 s
ConveyorSpeed	Check for normal	ConveyorAlarm	Cyclic 5.0 s

there are multiple instances of this transformation. It is shown in this way as the temperature control has to be replicated for each of the three areas of the oven. The dotted lines labelled **ENABLE** and **DISABLE** that enter **ControlAreaTemp** are known as *prompts* and indicate whether a particular transformation is *active* (that is, running) or not. Transformations without a prompt attached are assumed to be running all the time the system is running. The two entities placed between parallel

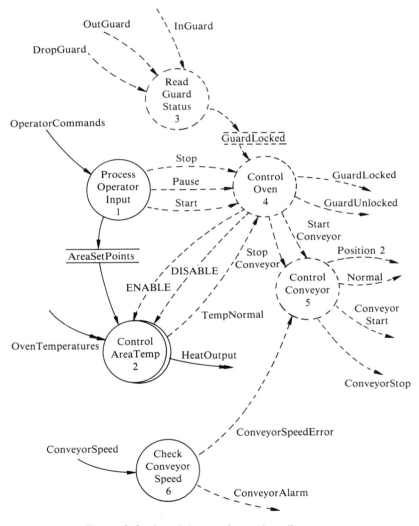

Figure 8.6   Level 1 transformation diagram.

lines and labelled **AreaSetPoints** and **GuardLocked** are respectively a *data store* and a *control store*. Flows entering and leaving stores are not named as they are assumed to take the name of the store. Flows that enter and leave the diagram must appear on the context diagram. Each transformation is given a number, for example **ControlConveyor** is numbered as 5 in the diagram. The single transformation in the context diagram is assigned the number 0 but this is usually not shown. All level 1 transformations have single-digit numbers (this indicates that they are level 1).

Building the model proceeds by taking each transformation in the level 1 diagram and breaking it down into smaller units. For example, Figure 8.7 shows the expansion of **ControlAreaTemp**. This diagram contains one new name, **TRIGGER**. This is a prompt which is used to indicate that the transformation is run once each time the **TRIGGER** becomes true. It is thus a way of indicating that a transformation runs in response to either a periodic signal or an event. You should

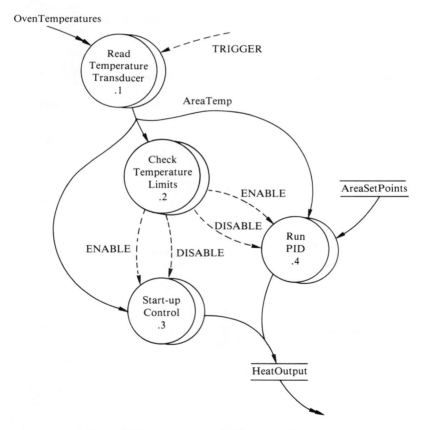

Figure 8.7 Level 2 transformation diagram.

notice that the input to **ReadTemperatureTransducer** is a continuous data flow and the output is a discrete data flow; this is a consequence of the transformation being triggered. Using the data store **HeatOutput** provides the reverse process of converting a discrete data flow into a continuous data flow.

Two other conventions are illustrated in the diagram: a common data flow passing to several transformations – **AreaTemp** – and a data flow being supplied from either of two transformations – **HeatOutput**. (The flow notation is summarised in Figure 8.8.) The transformations on this level 2 diagram (Figure 8.7) are numbered with a full stop in front of the number; this indicates that for full identification the number of the transformation diagram should be added. Thus the transformation **CheckTemperatureLimits** is 2.2 (the presence of two digits indicates that it is a level 2 transformation).

The process of subdivision continues until the analyst decides that no useful purpose is served by splitting up a transformation into smaller units. At this point a transformation specification is drawn up. For data transformations a *process specification* (PSPEC) is produced and for control transformations a *control specification* (CSPEC) is produced. These are described in sections 8.4.4 and 8.4.5 below.

8.4.3 Behavioural Model – Rules and Conventions

There are a number of rules and conventions associated with transformation

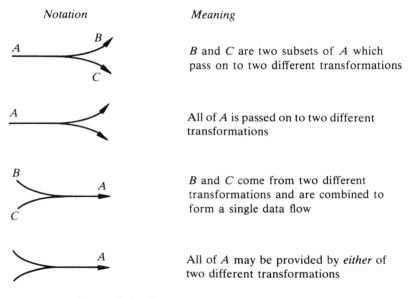

Notation	Meaning
	B and *C* are two subsets of *A* which pass on to two different transformations
	All of *A* is passed on to two different transformations
	B and *C* come from two different transformations and are combined to form a single data flow
	All of *A* may be provided by *either* of two different transformations

Figure 8.8   Summary of data-flow notation.

diagrams and in order to construct and interpret the diagrams these rules must be understood.

1. *Data transformations* with discrete input flows are assumed to be data triggered and hence there can be only one input from another transformation (inputs from data stores and prompts do not count). The reason

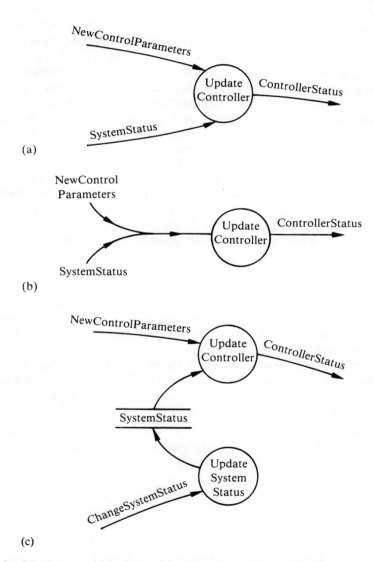

Figure 8.9 Synchronous data flows: (a) ambiguous diagram; (b) merged flows; (c) use of data store.

for this restriction is to avoid ambiguity. Consider the transformation shown in Figure 8.9a: should this be interpreted to mean that the transformation executes only if both inputs are present simultaneously? Or is it to be interpreted that one input will be stored until the other arrives and the transformation executed?

To avoid this ambiguity the transformation must be represented as shown in either Figure 8.9b or 8.9c. Figure 8.9b shows the input as a composite data flow and the convention is that the transformation can only be triggered when all elements of a composite data flow are present. In Figure 8.9c a data store is used to hold one of the data flows. The interpretation of this diagram is that **UpdateSystemStatus** runs when a transaction **ChangeSystemStatus** appears and **SystemStatus** is held in the data store. **UpdateController** runs when a transaction **NewControlParameters** appears and it obtains the current **SystemStatus** from the data store. This implies that the data store will have to have the characteristics of a *pool* since it may be read many times.

There is no need to restrict the number of continuous data flows entering a transformation since by definition a data value is always available; however, it is normal to show only one such input from another transformation and not to mix discrete and continuous data flows to the same transformation.

2. *Control transformations* – there is no restriction on the number of inputs to a control transformation.

3. *Input and output types*: the permitted mixture of input and output flows from the transformations is summarised in Table 8.5.

4. *Balancing* – since transformation diagrams at a given level represent the same information as the diagram at the next higher level (they simply provide more detail) the inputs and outputs must match the inputs and outputs of the higher-level diagram. The process of checking that the inputs and outputs correspond is referred to as balancing. Referring to Figures 8.6 and 8.7 above, the input to TFD 2 is **OvenTemperatures** and the output is **HeatOutput**; thus they balance with the input and output to **ControlAreaTemp** in TFD 0.

Table 8.5   Summary of inputs and outputs for
transformations

Transform	Inputs	Outputs
Data	Data flow	Data flow
	Prompt	Control flow
Control	Control flow	Control flow
	Prompt	Prompt

8.4.4 Process Specifications

The process specification (PSPEC) is a description of the actions that the data transformation has to carry out. The description can be given in any form that the user wishes. Ward and Mellor (1986) discuss a number of methods of specifying data transformations including the standard procedural techniques of program design languages, pseudo-code and structured English. They also discuss and give examples of a non-procedural method based on the use of precondition–postcondition statements (Heniger, 1980). If the goal of producing an essential model devoid of implementation constraints is to be achieved it would seem important that specifications at this stage are expressed non-procedurally. The choice of methodology is open: any of the rapidly developing formal techniques can be used.

At some point in the design the transformation specification will have to be expressed procedurally and often a transformation will be specified using a procedural notation. The most common method is to use some form of *pseudo-code*. Pseudo-code can be thought of as an informal programming language. For example, the PSPEC for **ReadTemperatureTransducer** in Figure 8.7 could be written as follows:

```
PSPEC 2.1 ReadTemperatureTransducer

INPUTS: OvenTemperatures
OUTPUTS: AreaTemp

Every 1.0 seconds DO
    read OvenTemperatures
    convert to internal data representation
    output converted value as AreaTemp
END.
```

Most CASE tools provide a means by which PSPECs can be stored in text files which can be manipulated using a text editor. Many automatically insert in the file the names of the inputs and outputs (obtained from the transformation diagram) of the transformation for which the PSPEC is being created. As development proceeds the PSPEC can gradually be elaborated to form a code segment in the appropriate programming language.

8.4.5 Control Specifications

Control transformations are described using CSPECs. The most usual form for a CSPEC is a *state transition diagram* (STD) and/or a *state transition matrix* (STM). The general form of the STD is shown in Figure 8.10 and of the associated STM in Figure 8.11. The action resulting from an event may be the generation of an event signal (control flow) or the generation of a prompt, either an ENABLE/DISABLE

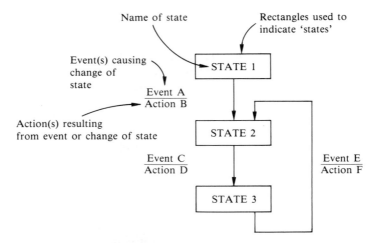

Figure 8.10   State transition diagram.

Present state	Event A	Event C	Event E	
STATE 1	STATE 2 – – – – Action B			Next State – – – – Action
STATE 2		STATE 3 – – – – Action D		Next State – – – – Action
STATE 3			STATE 2 – – – – Action F	Next State – – – – Action

Figure 8.11   State transition table.

or a TRIGGER. An example of a CSPEC is given in Figure 8.12. The associated STM is shown in Figure 8.13.

The blank entries in the STM represent non-operational states or undefined states. The value of using an STD to represent the CSPEC is that it leads to a concentration on the normal behaviour of the system. However, the associated STM should always be drawn up as the STM reveals the undefined states which frequently

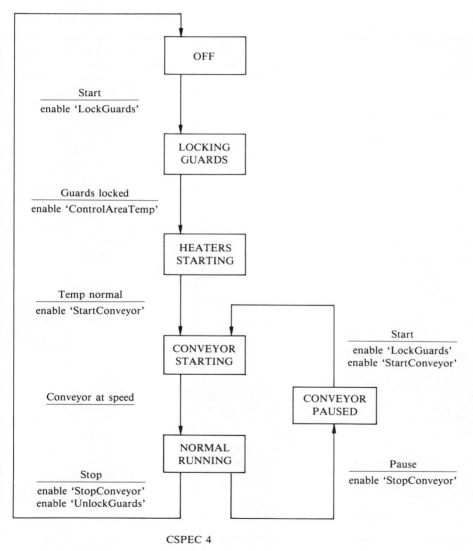

CSPEC 4

Figure 8.12   Example of a CSPEC in state transition diagram form.

represent exceptions which the designer must take into account. For example, referring to Figure 8.13 we find that there is only one entry in the column for the event **Stop**. It should be clear that we need to know what to do if the **Stop** occurs when the system is in all other states. What action should be taken if the event **TempNormal** occurs when in state **LockingGuards**? It should be clear that this should not occur and hence it must represent a fault in the system. We will examine

	Start	Stop	Pause	Guards locked	Temp normal	Conveyor at speed
OFF	LOCKING GUARDS -- - - - - lock guards					
LOCKING GUARDS				HEATERS STARTING -- - - - - control area temp		
HEATERS STARTING					CONVEYOR STARTING -- - - - - start conveyor	
CONVEYOR STARTING						NORMAL RUNNING -- - - - -
NORMAL RUNNING		OFF -- - - - - stop conveyor unlock guards	CONVEYOR PAUSED -- - - - - stop conveyor			
CONVEYOR PAUSED	CONVEYOR STARTING -- - - - - lock guards start conveyor					

Figure 8.13 Example of state transition matrix.

an alternative approach to the use of state transition diagrams and state transition matrices in Chapter 10.

8.4.6 Checking the Essential Model

Ward and Mellor recommend checking the transformation schema of the behavioural model in two ways. The first is to use the rules for data flow to check for consistency. This is the equivalent of checking the syntax of a program and can be done by hand or, given the advances in graphics processing capabilities in recent years, it is now feasible to construct a graphics compiler to perform the necessary checks. The second level of checking is to determine whether the model can be executed – can it in some sense generate outputs from a given set of inputs? The approach suggested by Ward and Mellor is based on ideas derived from work on Petri nets. We shall deal with this in Chapter 10.

At this stage the abstract modelling is complete and we are ready to move to the implementation stage.

8.4.7 Building the Implementation Model

The construction of the implementation model divides into four phases:

● enhancing (or elaborating) the environmental model;
● allocation of processors;
● allocation of activities (transformations) to tasks for each processor; and
● definition of the structure of each task.

The latter three can be considered as being concerned with the allocation of resources and will be dealt with together.

8.4.8 Enhancing the Model

Enhancing the model is concerned with:

● clarifying the boundaries between the system and the environment and determining what activities the system – as defined by the behavioural model – will carry out and what will be done as part of the environment;
● elaborating data descriptions; and
● adding timing and process activation information.

Design begins at this stage and we have to begin to take into account the technology involved in the system.

In the context diagram we treated the terminal units as virtual devices which we assumed provided a clean input signal to the behavioural model. We now have to examine each detail to find out what sort of signal they provide and how much processing of that signal is required before it can be passed to the behavioural model. For example, consider the terminal unit TEMPERATURE TRANSDUCERS and take just one element for the **PreHeat** area of the oven. From the data dictionary we find that **PreHeat** is assumed to be provided as a temperature measured in degrees Centigrade in the range 0 to 100 and that it is a continuous data value. From the requirements document we find that the transducer is a thermocouple, the signal of which is amplified and is available as a voltage in the range 0 to 10 volts. We thus have to sample and digitise the thermocouple output, using an analog-to-digital converter, and then convert it to degrees Centigrade. The actions are shown in Figure 8.14, which represents part of the terminator block TEMPERATURE TRANSDUCERS; the rest of the block will consist of the conversion units for the other temperature measurements. In Figure 8.14 we have assumed that a precision of 12 bits will be adequate for the ADC. The data dictionary entry for **PreHeatTemp** will be updated to include information on the resolution of the temperature measurement (1 in 4096).

If we examine Figure 8.7 we find that the TRIGGER prompt does not have any timing information attached to it; as part of the enhancement we would add the comment (* 1 second, cyclic *) to it to indicate that the **ReadTemperatureTransducer** transformation has to be executed every 1 second.

Enhancing the model is a process of adding detail and beginning to take into account the possible technologies that might be used in implementing the system. Ward and Mellor regard the process as largely being concerned with providing an interface shell round the internal software system as shown in Figure 8.15. It separates the functional operations of the interfaces from their electrical and physical manifestations and also serves to hide many of the details of how the

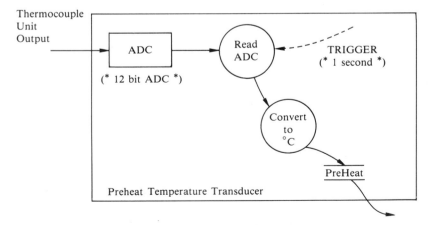

Figure 8.14 Example of a virtual transducer.

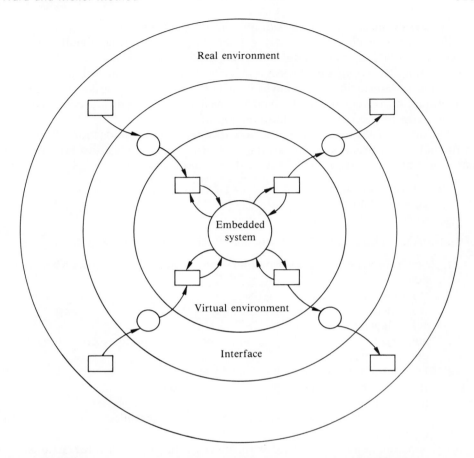

Figure 8.15 Relationship of real environment and virtual environment.

functions are implemented. The implementation may involve both hardware and software. The internal system is seen as communicating with virtual devices; the details of the actual devices are hidden within the interface modules. This approach encourages good design in that non-essential details are hidden and also technology-dependent operations are confined to specific parts of the system and not distributed throughout the software.

8.4.9 Allocation of Resources

The first stage of resource allocation is to decide on how many and what type of processing units are required and how the various functions to be performed are to be allocated to each. Processing units may be digital computers, logic circuits, analog devices, mechanical devices or human beings. Typically the transformations specified in the behavioural model will be carried out using digital computer elements as the processing units, although some functions may be allocated to human beings. Use of analog devices and hardwired logic systems will usually occur in relation to interfacing to the environment. For example, in order to obtain the temperature measurements for the various oven areas we need to use an ADC and we may need to precede this with an analog filter.

The next stage is to decide on the task structure and the allocation of the tasks to the individual processors. In carrying out this process we need to keep in mind both the standard software engineering design heuristics of:

1. information hiding,
2. coupling and cohesion, and
3. interface minimisation,

and also some additional rules of guidance needed for real-time systems:

1. Separate actions (transformations) into groups according to whether the action is:
    (a) time dependent;
    (b) synchronised;
    (c) independent;
    and try to minimise the size and number of modules containing time-dependent actions.
2. Divide the time-dependent actions into:
    (a) hard time constraint
    (b) soft time constraint
    and try to minimise the size and number of modules with a hard time constraint.
3. Separate actions concerned with the environment from other actions.

The recommended design strategy can be expressed simply as: minimise the part of the system which falls into the category of having a *hard time constraint*.

The simplest processor allocation is to allocate one processor for the whole system. The choice of processor is then based on its ability to perform all the activities, and factors such as processor power, memory size, ability to handle interface devices, and reliability predominate. The more usual case is when it is necessary to distribute units of the essential model across a set of processors. The choice of appropriate processor may be dependent on, for example, the need for an extended instruction set, special memory requirements, or the ability to interface to special devices. The number and type of processors may also depend on the environment, the need or appropriateness of distributed processing, the need for low power consumption, etc. Also at this stage units of the essential model that might be best performed by special purpose hardware or by a human operator will be identified.

The second stage of the design process is to group units of the essential model

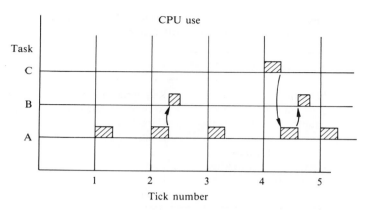

Figure 8.16   Example of a task activation diagram.

that have been allocated to a specific processor into *tasks*. Tasks are concurrent activities (even if run on a single processor); however, units of the essential model when grouped into a single task lose their potential concurrency and hence the grouping of essential model units to form a task introduces a distortion between the essential and implementation models. Also the essential model supports continuous data flows and data transforms; allocation to a task introduces *sampling* and the implementation constraint that a task takes a finite computation time.

Task allocation thus requires analysis of the system to estimate its performance in relation to some of the time constraints given in the specification. A simple way of analysing the timing requirements is to use a task activation diagram. An example of such a diagram is shown in Figure 8.16. The activation diagram shows the use of the CPU by each task that has to run at a fixed cycle time during each clock cycle (tick) of the real-time clock. Using the diagram the effects of task priority and pre-emption strategies are clearly seen and assessed.

8.5 HATLEY AND PIRBHAI METHOD

As might be expected the general approach of the Hatley and Pirbhai methodology is very close to that of Ward and Mellor. There are some differences in terminology which are summarised in Table 8.6.

8.5.1 Requirements Model

The basic structure of the requirements model is shown in Figure 8.17. The major

Table 8.6   Differences between the Ward and Mellor
and the Hatley and Pirbhai methodologies

Ward and Mellor	*Hatley and Pirbhai*
Essential model	Requirements model
Implementation model	Architecture model
Transformation schema	Data-flow diagram
	Control flow diagram
Data transformations	Process model
Control transformation	Control model
Data dictionary	Requirements dictionary
	Architecture dictionary

differences between this and the essential model of Ward and Mellor are:

- separate diagrams are used for data and control;
- only one CSPEC can appear at any given CFD level; and
- all data flows and control flows are shown with single arrow heads; the distinction between continuous and discrete flows is determined by the way in which a process is activated. The normal assumption is that a flow is

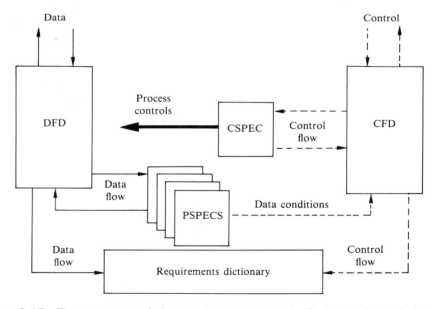

Figure 8.17   The structure of the requirements model. (Redrawn from Hatley and Pirbhai, *Strategies for Real-time System Specification*, Dorset House (1988).)

continuous (it is implicitly assumed that if the activity is implemented on a digital processor it will be carried out frequently enough to appear continuous).

Figure 8.18 shows the Drying Oven Controller in the Hatley and Pirbhai notation. There are several points to note:

1. The process bubbles (transformations) appear on both the DFD and the CFD. This is because, as the CFD shows, a process can produce a control

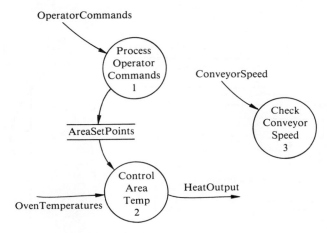

DFD 0 Drying Oven Controller

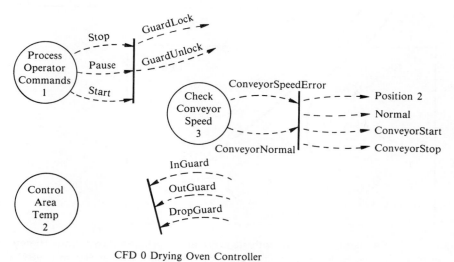

CFD 0 Drying Oven Controller

Figure 8.18   Hatley and Pirbhai notation.

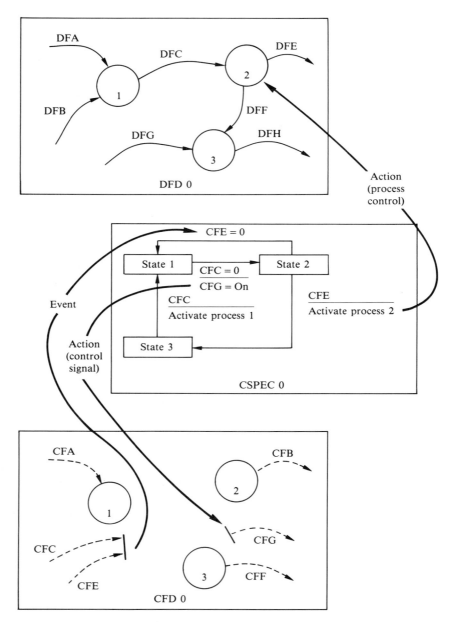

Figure 8.19 Sequential CSPEC, with its DFD and CFD. (Redrawn from Hatley and Pirbhai, *Strategies for Real-time System Specification*, Dorset House (1988).)

flow as an output. It usually arises as a result of some form of comparison which generates an event.

2. The CSPEC is represented by a bar. Although three bars are shown they form one CSPEC and, because there is only one, it does not need to be named on the diagram – it takes the number and name of the diagram. In this case it is numbered CSPEC 0.

3. There is no process activation information shown in the diagram. The process activation information is held in the CSPEC. The relationship between CSPECs and the DFD and CFD diagrams is shown in Figure 8.19.

8.5.2 Architecture Model

The general structure of the architecture model is shown in Figure 8.20 and as with the requirements model it is a hierarchical layered structure. In developing the architecture model a procedure based on using an *architecture template* is suggested. Figure 8.21 shows the form of the template. It is akin to the Ward and Mellor method for enhancing the model by developing virtual terminators but in this case it is suggested that the template be applied at each level in the requirements model hierarchy. The architecture model also includes diagrams showing the

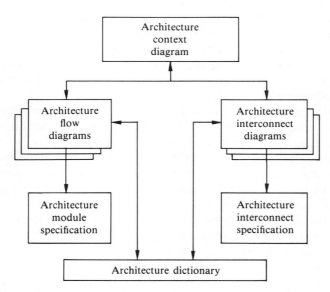

Figure 8.20 Architecture model components. (Redrawn from Hatley and Pirbhai, *Strategies for Real-time System Specification*, Dorset House (1988).)

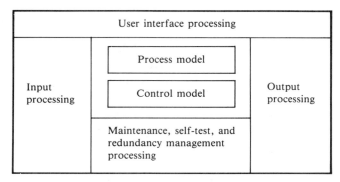

Figure 8.21   Architecture template. (Redrawn from Hatley and Pirbhai, *Strategies for Real-time System Specification*, Dorset House (1988).)

interconnection technology between various elements of the system. Figure 8.22 shows in more detail the style of the architecture model elements.

8.6 COMMENTS ON THE YOURDON METHODOLOGIES

Both methodologies – Ward and Mellor and Hatley and Pirbhai – are simple to learn and have been widely used. They are founded on the well-established structured methods developed by the Yourdon organisation and hence over the years a lot of experience in using the techniques has been gained. For serious use on large-scale systems they both require the support of CASE tools. The labour involved in checking the models by hand is such that short cuts are likely to be taken and mistakes are bound to occur.

It can be argued that the methods are really only a set of procedures for documenting a specification and a design and to some extent this is true. The analysis procedures are minimal and adequate checking for consistency can be performed only with the support of a CASE tool. However, the methodologies are still useful in that the procedures they recommend provide a sensible way of preparing both a specification and a design in that they encourage the development of hierarchical, modular structures.

Of the two, the Hatley and Pirbhai method is the more structured and formalised in its approach. Its diagrams are less cluttered than those of the Ward and Mellor method and, once the separation is understood, are easier to follow. Many CASE tools provide alternative displays which allow a choice of either separate diagrams or a combined diagram with switching between the two forms.

The weakness of both methods lies in the allocation of processors and tasks. The suggestion that one allocates activities to processors and then subdivides the

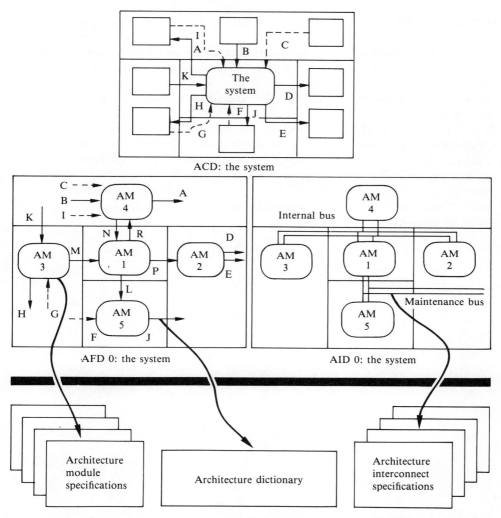

Figure 8.22 The structure of the architectural model. (Redrawn from Hatley and Pirbhai, *Strategies for Real-time System Specification*, Dorset House (1988).)

activities into tasks allocated to each processor appears at first sight a sensible way to proceed. However, when it is tried one soon realises that the information required to do this is not available. How can one determine the processor requirements until at least some detailed coding has been done? How can tasks be structured until some estimate of the feasibility of finding a suitable task schedule has been carried out? Both Ward and Mellor and Hatley and Pirbhai remain silent about these issues.

8.7 SUMMARY

The structured methodologies on which both Ward and Mellor and Hatley and Pirbhai are based are a widely used method of producing a requirements model. For serious use, however, the support of a CASE tool is essential. It is only through the use of a CASE tool that consistency, correctness and full adherence to the standards can be maintained.

The extension of the method to attempt to support design through the development of an implementation or architecture model is less successful. Of the two, the Hatley and Pirbhai approach is most well developed and useful. The lack of analysis tools is a weakness but this will eventually be remedied through the development of such tools within CASE environments. For a full understanding of the methods the books by Ward and Mellor (1986) and by Hatley and Pirbhai (1988) must be consulted.

9

Real-time System Development Methodologies − 2

In this chapter we will consider two further methodologies − MASCOT and PAISLey. MASCOT is one of the older methodologies but it has recently been extensively revised. It is interesting in that it assumes the system is being designed for a particular virtual machine and that the implementation of this virtual machine on a specific computer or set of computers is a separate problem. Hence at the design stage there is no need for consideration of specific technologies.

PAISLey is primarily a specification technique which is based on formal mathematics. Its most interesting features are: a means of specifying the timing constraints of the system; and execution of the specification.

9.1 MASCOT

MASCOT was the first formal real-time software development methodology. The first version of MASCOT was developed by Jackson and Simpson during the period 1971−5 (Jackson and Simpson, 1975). The official definition of MASCOT 1 was published in 1978 and a revised version − MASCOT 2 − was issued in 1983. Between 1983 and 1987 extensive changes to the technique were made and the official standard for MASCOT 3 was published in 1987.

The official handbook states that MASCOT is a Modular Approach to Software Construction Operation and Test which incorporates:

- a means of design representation;
- a method of deriving the design;
- a way of constructing software so that it is consistent with the design;
- a means of executing the constructed software so that the design structure remains visible at run-time; and
- facilities for testing the software in terms of the design structure.

9.2 BASIC FEATURES OF MASCOT

The application software is designed for a specific virtual machine and the problem of mapping the MASCOT machine onto a real computer is treated as a separate problem. In MASCOT software is represented as:

- a set of concurrent operations; and
- the flow of data between such operations.

The operations are referred to as *components*. The system consists of a set of interconnected but independent *components* that make no direct reference to each other. Each *component* has specific, user-defined, characteristics that determine how it can be connected to other *components*. *Components* are created from *templates*, that is patterns used to define the structure of the *component*. Two *classes* of *templates* are fundamental to MASCOT: (a) *activity* and (b) *intercommunication data area* – *IDA*.

An *activity template* is used to create one or more *activity components* each of which is a single sequential program thread that can be independently scheduled. It is assumed that at the implementation stage each *activity* will be mapped onto a software task. Such a task may run on its own processor or be scheduled by a run-time system (usually referred to as the MASCOT kernel) to run on a processor shared with other *activities*. The *activities* communicate through *IDAs*. The *IDA* provides the necessary synchronisation and mutual exclusion facilities.

An *IDA* is a passive element with the sole purpose of servicing the data communication needs of *activity components*. It can contain its own private data areas. It provides procedures which *activities* use for the transfer of data. Within an *IDA*, and only within an *IDA*, the designer has access to low-level synchronisation procedures and thus is not limited to using high-level operations such as monitors, message passing, or rendezvous, provided by the implementation language, but is able to use any technique appropriate to the problem. A structure containing *activity components* connected by means of one or more *IDAs* is referred to as a *network*.

MASCOT supports three forms of *IDA*: a generalised *IDA*; a *channel*; and a *pool* – their behaviour is defined as follows:

channel: supports communication between producers and consumers. It can contain one or more items of information. Writing to a *channel* adds an item without changing items already in it. The read operation is destructive – it removes an item from the *channel*. A *channel* can become empty and also, because its capacity is finite, it can become full.

pool: is typically used to represent a table or dictionary which *activities* periodically consult or update. The write operation on a *pool* is destructive and the read operation is non-destructive.

9.2.1 Simple Example

MASCOT can be used at a simple level to provide a virtual machine supporting *activities*, *pools* and *channels*. A design is constructed in the form of an *activity*, *pool* and *channel network* – an ACP diagram – as is shown in Figure 9.1.

The diagram represents part of a system for the control of a plant. The activity **Heater1Input** gets data from a plant interface. The data is held in a pool **Heater1In** from where it is read by activity **Heater1Alarm** and **Heater1Con**. The required output to the plant and the alarm status are held in a pool **Heater1Status**. An activity **Heater1Report** gets data from the pool holding status information and sends it via a channel **Heater1Ch** to some other activity (not shown). Also not shown are the activities required to pass the data to the plant control. This ACP differs from a MASCOT 2 ACP since the components now contain *ports* and *windows* (shown as solid circles and rectangles) the significance of which is explained below.

Once the ACP diagram has been produced, design of the templates for the

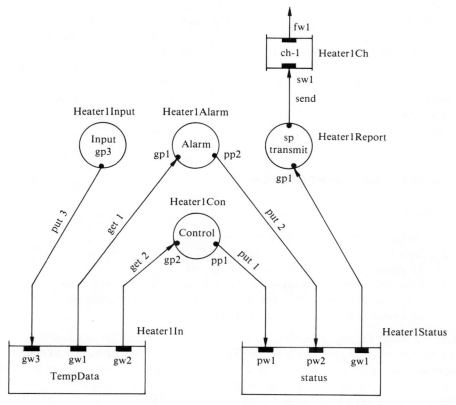

Figure 9.1 An example of a MASCOT ACP diagram.

individual components can proceed. Many component templates will be reusable and hence only application-specific ones will need to be designed. Instances of the component are created when the network is constructed by translating the ACP diagram to textual form and entering it into the MASCOT database.

At this level a design in MASCOT may be represented in either graphical (ACP diagram) or textual form. Both forms are equivalent and may be derived from each other. The textual form stored in the database can be progressively updated as the design proceeds.

9.2.2 Communication Methods

Entities in MASCOT communicate by means of *paths*. A path connection is made between a port and a window. A port is represented by a solid circle and a window by a solid rectangle. Thus in Figure 9.1 activity **Input** has a port **gp3** and is connected by a path labelled **put 3** to a window **gw3** in an IDA **TempData**. In path **put 3** the port is the source of the data and the window is the sink. However, a port can act as a sink and a window as a source as is the case with path **get 1**.

Windows are *passive* devices which provide a set of operations for use by an active device for the transmission of data. Ports are *active* devices which specify a set of operations required to transmit data. Windows are normally found in communication components whereas ports are found in both activities and communication components. (The use of ports in communication components enables IDAs to be connected together.)

9.3 GENERAL DESIGN APPROACH

The general approach to the design of the system is hierarchical:

1. Define system and external devices.
2. Decompose system into a network of concurrent subsystems, IDAs and hardware interface units – *servers*.
3. Continue decomposition of subsystems until further decomposition is not desirable. At this point – the component level – a particular subsystem will be composed of activities and IDAs.

Figure 9.2 shows the top-level network diagram for the Drying Oven which was used as an example in the previous chapter. Two subsystems – **ControlAreaTemp** and **GeneralOvenControl** – are used and they connect to the external devices through servers. The connection to the **Temperature transducer** is through the server **TT** which uses a path **fetch** to connect from window **TT.W** in the server to port **P1** in the subsystem.

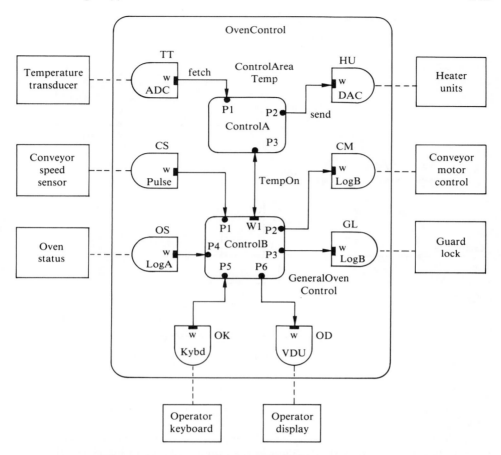

Figure 9.2   Drying oven – MASCOT network level.

Figure 9.3 shows subsystem **ControlAreaTemp** decomposed to the component level. There are two activities **Act1** and **Act2** which are named **ReadTemp** and **Control** respectively. The names inside the circles are the template names and those outside are the component names. A component is an instantiation of a template. **ReadTemp** is assumed to read the value **ControlOn** by using the access procedure **get** and if it is true it reads **Temp** and uses **send** to store the value in a pool **AreaTemp**. **Control** gets the value of the temperature from **AreaTemp** and calculates the heat output.

An activity represents a task and cannot be subdivided into smaller, separately schedulable units (they would themselves be activities) but it can be divided into smaller modules as shown in Figure 9.4. One module within the activity is defined as the *root* component, that is the main module, and it provides coding for the initial

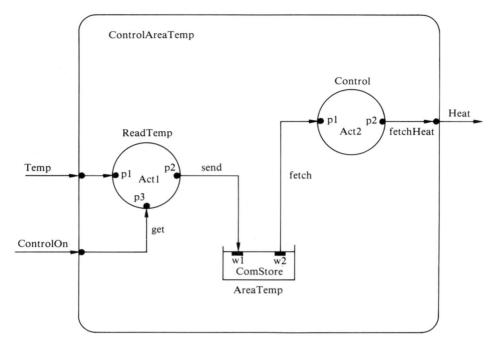

Figure 9.3   Subsystem – **ControlAreaTemp**.

entry and calls on the services provided by the other units which are known as *subroots*.

Subroots usually comprise a collection of procedures and can be designed using the standard structured methods. Thus in the example shown in Figure 9.4 the main root **M** calls **CheckTemp** which gets the **AreaTemp** from the pool (see Figure 9.3) and checks to see if it is within the normal range. If it is not within the normal range **MaxControl** (subroot **SR3**) is called, otherwise **PIDControl** (subroot **SR2**) is called. Finally subroot **SR4** which outputs the heat demand is called.

There are two general comments to make about this example. One is that the design procedures in MASCOT assume that a general set of templates will be created and used; thus all entities are created from generic types by a process of instantiation, and hence even if only one instance is required a template has to be created and then instantiated. This is in fact less cumbersome than it seems. The advantages are obvious even in this small system: subsystem **ControlAreaTemp** is a template and three instances can be created to deal with the **PreHeat**, **Drying** and **Cooling** areas of the oven. One of the purposes of this approach is to encourage the reuse of software.

The second comment is that although the subsystem **ControlAreaTemp** contains two activities that are run at periodic intervals this is not shown on the design

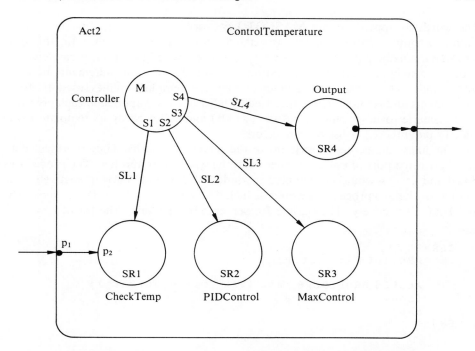

Figure 9.4 Diagram showing activities in **ControlTemperature**.

diagrams. The scheduling support is provided by the MASCOT kernel and the information is entered when the activities are made known to the kernel.

9.4 TEXTUAL REPRESENTATIONS OF MASCOT DESIGNS

In MASCOT there is a direct correspondence between the diagrams and their textual representation; in fact the text is the formal description of the system, not the diagrams. The description of the system shown in Figure 9.2 takes the form

```
SYSTEM OvenControl;
    {specification part goes here}
USES    ControlA, ControlB, ADC, DAC, Pulse,
        LogA, LogB, Kybd, VDU;
    SUBSYSTEM ControlAreaTemp:ControlA(p1=TT.w,
        p2=HU.w, p3=ControlB.w1);
    SUBSYSTEM GeneralControl:ControlB(p1=CS.w,
        p2=CM.w,p3=GL.w, p4=OS.w, p5=OK.w, p6=OD.w);
    END;
END.
```

The words in upper case are MASCOT key words. Each textual unit or module
follows a similar pattern: it has associated with it an explicit class (in this case
SYSTEM); a name part (OvenControl); a specification part (in this case empty);
and, for a template, an implementation part (in this case the statements between
USES and END). The USES statement lists the template names used to form
components and is immediately followed by a list of the components, each preceded
by its generic form (in this case SUBSYSTEM) and followed by its template name
and, in parentheses, a list of connectors.

The lines connecting subsystems to the servers are paths. They represent data
flow between a port of one component and the window of another. The procedures
read and write would be available at windows **TT.w** and **HU.w** respectively. The
coding for the two procedures would be included in the module for the servers **ADC**
and **DAC** respectively. The outline for server **ADC TT** takes the form:

```
SERVER TT;
    PROVIDES w:   fetch;

    ACCESS PROCEDURE read (VAR item:PlantData);
    {body of procedure }
    END;
END.
```

A server is the only MASCOT design entity which is permitted to communicate with
a device. It has all the features of an IDA but also is allowed to contain one or more
handlers which can be invoked by hardware interrupts. It thus provides the means
for low-level direct communication with the system hardware.

The identifier used to label a path indicates its type which is defined in a module
called an *access interface*. Consider the subsystem **ControlAreaTemp** shown
in Figure 9.3: the path connecting port **p1** in activity **Control** to the window **w2**
in IDA **AreaTemp** is of type **fetch** and the path connecting port **ReadTemp.p2**
to window **AreaTemp.w1** is of type **send**. The modules defining them take the
form

```
ACCESS INTERFACE send;
    WITH PlantData;
    PROCEDURE write(item: PlantData);
END.

ACCESS INTERFACE fetch;
    WITH PlantData;
    PROCEDURE read(VAR item:PlantData);
END.
```

The WITH clause used in the specification indicates that the type definition
PlantData is held in a common source accessible to all modules. The declaration

is made in a specification statement, for example

```
DEFINITION PlantData;
    TYPE
        PlantData = RECORD
            {put definition here}
        END {RECORD};
END.
```

The textual description of the composite activity shown in Figure 9.3 takes the form

```
ACTIVITY ControlTemp;
    {specification part}
    REQUIRES    p1:fetch;
                p2:send;
    USES M, SR1, SR2, SR3, SR4;
    ROOT Controller: M
    SUBROOT CheckTemp: SR1(p1=p1);
    SUBROOT PIDControl: SR2;
    SUBROOT MaxControl: SR3;
    SUBROOT Output: SR4(p1=p2);
END.
```

The templates for the root and subroots are defined as follows:

```
ROOT main;
    NEEDS   s1:sl1;
            s2:sl2;
            s3:sl3;
            s4:sl4;
    {code goes here}
END.
```

The NEEDS section specifies the links connecting the root module to the other components. As shown below the corresponding subroot will name the links in a GIVES statement. For example the subroot SR1 takes the form

```
SUBROOT SR1;
    REQUIRES p1:fetch;
    GIVES sl1;
    {coding goes here}
END.
```

9.5 OTHER FEATURES OF MASCOT

9.5.1 Constants

The usefulness of the template method of creating components is enhanced by the facility to create from the same template components which differ in minor ways. In the specification of a template dummy constants are declared; their actual value is supplied when a component is created. The dummy constants are known as template constants. They can be considered to be the equivalent of dummy arguments in a macro declaration. For example, they permit servers with different device addresses and different interrupt levels to be created from a single template; or they can enable components with different buffer sizes or different iteration counts to be created.

9.5.2 Direct Data Visibility

In most real-time applications certain functions cannot be satisfactorily performed if the designer is restricted to using the data hiding approach provided by the IDA construct of MASCOT. A typical example is a module providing direct feedback control subjected to a hard time constraint. If a module of this type requires access to external data, for example in order to update controller parameters, it must have guaranteed access at all times and must not be kept waiting because another module is accessing the data. There are a variety of solutions to this problem; one is to allow the module to access the data directly without using the standard access procedures. MASCOT provides the designer with a means of providing direct access through a construct called an access interface.

9.5.3 Qualifiers

Software design techniques and implementation languages impose certain general constraints on what can and cannot be done. However, in most software systems there are areas in which the designer would wish either to relax the constraints or to impose locally more stringent constraints. For these purposes MASCOT provides a set of qualifiers which can be used to modify the behaviour of parts of the system.

1. *Connectivity constraints*: in the default mode windows are open to one or more ports; qualifiers can be used to restrict access to a single port, or to allow a window to exist without a port connection (used normally for test purposes).
2. *Data access constraints*: can be used to limit variables made directly accessible via an access interface.
3. *Data flow*: permits the direction of data flow to be shown in the textual form of the design.

4. *Context qualifiers*: provide a means of restricting certain functions provided by the support environment to particular types of template, for example certain low-level functions may be restricted to servers.
5. *Code generation constraints*: allow the designer to force the compiler to generate in-line code for an access procedure to a data area and hence avoid the overheads in making a procedural call.

9.6 DEVELOPMENT FACILITIES

As has already been mentioned the MASCOT design is captured by entering data relating to the design in textual form into a database. The method of constructing textual modules and the database have been designed so as to enable a design to be built up incrementally. As information is added, either in the form of a new module, or as an addition to an existing module, checks are made on the validity of the information. For example, the first stage is to register a module and for this process to be successful the name part must be defined and legal, and no module with the same name must have been previously registered. The various stages are listed in Table 9.1.

Table 9.1   MASCOT status conditions (reproduced from *The Official Handbook of MASCOT*)

Operation	Status to be achieved	Module class	Preconditions
Register	Registered	All	Name part defined and legal No other module with same name
Introduce	Partially introduced	All	Registered preconditions satisfied Specification dependencies registered Specification part defined and legal
	Fully introduced	All	Partially introduced preconditions satisfied Specification dependencies fully introduced
Enrol	Partially enrolled	Composite templates	Partially introduced preconditions satisfied Implementation dependencies introduced Implementation part defined and legal
	Fully enrolled	Simple template	Fully introduced preconditions satisfied Implementation dependencies fully introduced Implementation part defined and legal
		Composite templates	Partially enrolled preconditions satisfied Implementation dependencies fully enrolled

The progress status of each module in the design can be listed. The database thus provides management and designers with a support environment for project development.

9.7 THE MASCOT KERNEL

The MASCOT design procedures are based on the assumption that the design will be implemented using the features provided by a piece of software known as a MASCOT kernel. The implementation of this software on a particular computer using a particular operating system is considered to be a separate issue outside the application development. It is useful for an understanding of the MASCOT methodology to be aware of the main features of the kernel, which represents a *virtual* machine on which the MASCOT application will run.

> *Scheduling*: the kernel must allocate processor time to the parallel activities that constitute a MASCOT system. It must provide a real-time clock and must support primitive synchronisation procedures. The synchronisation procedures are precisely defined and are similar to semaphores and signals.
>
> *Interrupt handling*: the kernel must support the handling of hardware interrupts.
>
> *Subsystem control*: a MASCOT design may include groups of activities (tasks) which form a subsystem. The kernel must provide a means of adding and removing such subsystems from the attention of the scheduler.
>
> *Monitoring*: the kernel must provide a comprehensive set of facilities to aid testing and optimisation of the implementation.

In MASCOT 2 the exact structure of the support required was mandatory. The specific schemes for synchronisation, device handling, interrupts, process scheduling and priorities are analysed and a comparison with alternatives is given in Sears and Middleditch (1985). If the language being used in the implementation supports concurrency then the designer should consider mapping *activities* onto the appropriate language feature. Budgen (1985) describes a Modula-2 implementation of the MASCOT kernel. MASCOT 2 imposes one restriction: *activities* should not be created dynamically; the *system network* (*activities*, *IDAs* and *servers*) must remain invariant at run-time. The designer must document how the language features have been used to support the MASCOT virtual machine.

In MASCOT 3 the specific form of the kernel is not mandatory; it is only a recommendation and the implementer can provide the support in any appropriate way. MASCOT 3, however, requires the virtual machine to support additional features. For example, an activity providing direct feedback control that has a hard time constraint may require access to external data in order to update controller parameters. To meet its time constraint it must have guaranteed access at all times and must not be kept waiting because another activity is accessing the data. A simple solution to this problem is to allow the module to access the data directly without using the standard access procedures. MASCOT provides for this through a construct called an *access interface* that must be supported by the kernel.

9.8 SUMMARY OF MASCOT

Although MASCOT is not widely used – its use has largely been for military applications within the UK – it has been dealt with at some length because it demonstrates some valuable ideas and features. Some of the more important are considered below.

> *Templates*: the template construct encourages the reuse of software components, which contributes to increased reliability. Templates are generic entities and the hierarchical structure of systems, subsystems and activities lends itself to the development of knowledge-based support tools based on using the frame paradigm for the representation of knowledge (Bennett, 1992).
>
> *Encapsulation*: entities in MASCOT have many of the features of objects in that the method forces the designer to encapsulate procedures for accessing data within communication units (IDAs) and to adopt message passing as a major means of communication between activities. So although MASCOT does not claim to be an object-oriented design method it contains many of the features of such methods.
>
> *Virtual machine*: MASCOT systems are designed for a specific virtual machine, the MASCOT kernel. This has several advantages: the designer becomes familiar with the characteristics of the machine on which the system is to run; portability of designs is enhanced as the designer cannot utilise the peculiarities of one specific operating system or type of hardware; there is a clear separation of application design from system implementation. The disadvantage of the approach is the possibility of a less efficient implementation.

MASCOT 3 provides an excellent design methodology. It is sufficiently rich in concepts to provide design flexibility but has sufficient constraints for creating safe and reliable software. The introduction of hierarchical structures and the support for the generation of networks has overcome the limitations of MASCOT 2. However, the diagrams used in MASCOT 3 are much more complex to draw than those of the previous version and it is therefore more difficult to use the method quickly to sketch out ideas by hand. CASE tool support for both manipulating the diagrams and handling the textual representation is essential.

9.9 FORMAL METHODS

One of the features of MASCOT (and also of HOOD) is the way in which diagrams used as the basic design element are formalised by the use of textual equivalents. The textual form of the diagram is then successively elaborated to form the code for the system the diagram represents. Through the use of CASE tools conformity of the diagrams and textual representation to the rules can be checked and consistency can be enforced, as can the transformations between various representations. The weakness lies in showing that the design conforms to the specification and this issue is not addressed in MASCOT.

The aim of formal software engineering methods is to be able to transform a formal specification into implementation code and to be able to prove that each transformation step is correct. A first requirement for doing this is that the specification should be expressed in a formal (mathematical) language. There are a growing number of formal specification languages. Cooling (1991, p. 203) has listed the main ones as:

Model based	*Axiom based*
VDM (Vienna Development Method)	LARCH
Z	ACT-ONE
INA-JO	OBJ
me too	CLEAR

The most widely used of the model-based methods are VDM and Z and attempts have been made to use these for the specification of real-time systems. However, as yet they contain no facilities for specifying timing or concurrency. Cooling (1991, Chapter 7) gives a simple and brief introduction to the basic ideas of formal specification languages. There is one formal specification method that was developed specifically for real-time systems, namely PAISLey.

9.10 THE *PAISLEY* SYSTEM FOR REAL-TIME SOFTWARE DEVELOPMENT METHOD

PAISLey (Process-oriented, Applicative and Interpretable (executable) Specification Language) has been developed by Pamela Zave at the University of Maryland and the Bell Laboratories of AT & T. The specification is developed in terms of an explicit model of the environment interacting with an explicit model of the proposed system. Both the environment and the proposed system are modelled as a set of 'asynchronous interacting digital processes'. Non-digital objects in both the environment and the proposed system are modelled as discrete simulations of the object.

The behaviour of the asynchronous processes (which will be referred to as processes or process in the following sections) is described in a formal language. The language statements are executable and an interpreter is provided as part of the system. It is the intention (not yet achieved) that implementation code should be generated by applying formal transformation rules to the specification. The language is based on two well-established models of computing, asynchronous processes and functional programming, which have been merged. In doing so most of the benefits of each model have been preserved.

A system is specified in PAISLey as a set of processes that continually cycle through a set of state changes. The interval between the state changes is referred to as the *process step*; computations to be performed at each step are specified using a functional notation and each can be considered as a mapping of an input set of values to an output set of values. In order to restrict side-effects mappings should not use variables or assignment statements. Processes communicate by means of precisely defined, interprocess communication protocols.

9.10.1 A Simple System

Figure 9.5 shows a simple system module which contains two *processes* – machine and monitor. Information passes between processes by means of *channels*. A process is considered to be a finite state machine and its behaviour can be specified by defining:

1. a state space – that is, declaring a set of all possible states of the process (note that this is not the 'state space' of linear control but represents the set of discrete states of the system); and
2. a 'successor' function that defines the transition from the current_state to the next_state.

Thus if x is a member of the set X defined as $X = \{x_i \ldots x_n\}$ where n is finite, then $x(t + s) = \text{succ}\,[x(t)]$, where s is the finite interval of time required to compute the function 'succ'. The time s is referred to as the process step and is related to one basic cycle of the proposed activity (process) within the system.

Stage 1 of building the specification is to define the system structure:

```
(machine-cycle [initial-machine-state], monitor-
cycle[initial-machine-image]);
```

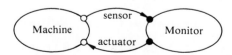

Figure 9.5   The process structure for a trivial process control system.

This states that the system is composed of two processes and that their successor mappings are `machine-cycle` and `monitor-cycle`, and their initial states are the values of `initial-machine-state` and `initial-machine-image` respectively. The set of values that can be taken by the initial states are defined by the statements

```
initial-machine-state: --> MACHINE-STATE;
    "MACHINE-STATE should be defined as the set of all
    possible states of this process."
initial-machine-image: --> MACHINE-IMAGE;
    "MACHINE-IMAGE should be defined as the set of all
    possible states of this process, capable of retaining
    whatever historical information is required."
```

The inverted commas are used to indicate a comment.

The next stage is to define each of the processes:

```
machine-cycle: MACHINE-STATE --> MACHINE-STATE;
machine-cycle: 1 --> = 0.1 s;
```

These two statements indicate that at each `machine-cycle` there is a mapping from one value in the set `MACHINE-STATE` to another, and that each `machine-cycle` takes exactly 0.1 s. The actual function evaluated to get the new value is given by

```
machine-cycle[state]
proj[(1,(simulate-,machine[(state,accept-feedback-if-
any[Null])],
offer-sensor-data[sense[state]]
))];
```

These statements indicate that each evaluation of `machine-cycle` includes the parallel evaluation of two expressions, one to compute another step of the discrete simulation algorithm and one to offer the most recent sensor data to the monitor. The value of the first expression becomes the next state of the process, while the value of the second is thrown away.

The system described in Figure 9.5 exchanges data between the processes by *channels* named `sensor` and `actuator`. Channels are specified by `Exchange Functions` which are described below. The statements describing this part of the specification for the machine process are

```
accept-feedback-if-any: FILLER --> FILLER U ACTUATOR-
SIGNAL;
accept-feedback-if-any[null] = xr-actuator[null];
simulate-machine:
MACHINE-STATE × (ACTUATOR-SIGNAL U FILLER) --> MACHINE-
STATE;
sense: MACHINE-STATE --> SENSOR-DATA;
offer-sensor-data: SENSOR-DATA --> SENSOR-DATA U FILLER;
```

The full PAISLey specification of this simple system with comments is given in Figure 9.6.

```
*------------------------------------------------------*
* SYSTEM STRUCTURE          *
*------------------------------------------------------*
(machine-cycle [initial-machine-state], monitor-cycle[initial-
machine-image1]);
"The system is composed of two processes. Their successor mappings
are machine-cycle and monitor-cycle, and their initial states are
the values of initial-machine-state and initial-machine-image
respectively."
*------------------------------------------------------*
* MACHINE PROCESS          *
*------------------------------------------------------*
initial-machine-state: --> MACHINE-STATE;
"MACHINE-STATE should be defined as the set of all possible states
of this process."
machine-cycle: MACHINE-STATE --> MACHINE-STATE;
machine-cycle: 1 --> = 0.1 s;
"Each evaluation of machine-cycle takes exactly 0.1 second."
machine-cycle[state]
proj[(1,(simulate-,machine[(state,accept-feedback-if-
any[Null])],
offer-sensor-data[sense[state]]
))];
"Each evaluation of machine-cycle includes the parallel
evaluation of two expressions, one to compute another step of the
discrete simulation algorithm and one to offer the most recent
sensor data to the monitor. The value of the first expression
becomes the next state of the process, while the value of the
second is thrown away."
accept-feedback-if-any: FILLER --> FILLER U ACTUATOR-SIGNAL;
accept-feedback-if-any[null] = xr-actuator[null];
"accept-feedback-if-any is defined as a nonwaiting interaction on
the actuator channel. If no interaction takes place the value
Null will be returned."
simulate-machine:
MACHINE-STATE x (ACTUATOR-SIGNAL U FILLER) --> MACHINE-STATE;
"This mapping should be defined as one step of the discrete
simulation algorithm."
sense: MACHINE-STATE --> SENSOR-DATA;
"This mapping should be defined to simulate the physical sensor
attached to the machine."
offer-sensor-data: SENSOR-DATA --> SENSOR-DATA U FILLER;
offer-sensor-data[data] = xr-sensor[data];
"offer-sensor-data is defined as a nonwaiting interaction on the
sensor channel."
```

Figure 9.6   The PAISLey specification listing for the trivial process control system
(continued overleaf).

```
*----------------------------------------------------------*
* MONITOR PROCESS             *
*----------------------------------------------------------*
initial-machine-image: --> MACHINE-IMAGE;
"MACHINE-IMAGE should be defined as the set of all possible states
of this process, capable of retaining whatever historical
information is required."
monitor-cycle: MACHINE-IMAGE --> MACHINE-IMAGE;
monitor-cycle: 1 --> <= 2.0 s;
"Each evaluation of monitor-cycle must take less than or equal to
two seconds."
monitor-cycle[image] = process-sensor-data[(image,get-sensor-
data[Null])];
"Each evaluation of monitor-cycle consists of getting the most
recent sensor data and then processing it."
get-sensor-data: FILLER --> SENSOR-DATA;
get-sensor-data[null] = x-sensor[null];
"get-sensor-data is defined as a waiting interaction on the sensor
channel. Its value is the most recent sensor data."
process-sensor-data: MACHINE-IMAGE x SENSOR-DATA --> MACHINE-
IMAGE;
process-sensor-data[(image,data)] =
proj[(1,(maintain-machine-image[(image,data)],
give-feedback-if-needed
[check-machine-condition[(image,data)]]
))];
"Each evaluation of process-sensor data includes the parallel
evaluation of two expressions, one to incorporate the most recent
sensor data into the historical information being saved in the
process, and one to provide feedback to the machine if it is
needed. The value of the first expression becomes the next state of
the process, while the value of the second is thrown away."
maintain-machine-image:
MACHINE-IMAGE X SENSOR-DATA --> MACHINE-IMAGE;
"This mapping should be defined to save the most recent sensor
data."
check-machine-condition:
MACHINE-IMAGE X SENSOR-DATA --> { No-Problem } U ACTUATOR-SIGNAL;
"This mapping should be defined to decide whether feedback is
needed and if so what the actuator signal should be."
give-feedback-if-needed: { No-Problem } U ACTUATOR-SIGNAL -->
FILLER;
give-feedback-if-needed[signal] =
/equal[(signal,No-Problem)]: Null,
True             : give-feedback[signal]
/;
give-feedback: ACTUATOR-SIGNAL --> FILLER;
give-feedback[signal] = x-actuator[signal];
"give-feedback is defined as a waiting interaction on the actuator
channel."
```

Figure 9.6   *continued*

9.10.2 Exchange Functions

Processes interact by sending and receiving data through channels. The behaviour of a channel is defined by an exchange function. Three primitive exchange functions referred to as x, xm and xr are defined. The syntax of the exchange function is

```
<function type> - <channelname> [argument]
```

for example x-msg[y] defines a channel of type x with the name msg and argument y. The argument provides an item to be sent and returns an item received. The function types are:

x: matches (synchronises) with a pending exchange function on its channel. If no exchange function is pending then it waits. If several requests are pending they are satisfied on a non-deterministic basis.

xm: behaves like an x-type exchange function except that two xm functions on the same channel cannot match with each other.

xr: behaves like an x-type except that it will not wait. If an xr exchange function cannot find an immediate match it will terminate and return its own argument value.

9.10.3 Timing Constraints

PAISLey supports the insertion of timing constraints into the specification. These are inserted as formal statements in the language; for example, the statement

```
machine-cycle ; ! =0.1s;
```

specifies a cyclic operation with a repetition time of exactly 0.1 s. The statement

```
monitor-cycle: ! < = 2.0s
```

specifies that monitor_cycle must compute its successor function within a time period of less than or equal to 2 seconds.

The language supports a wide range of timing constraints which include the ability to specify upper and lower bounds as well as precise hard constraints. During the execution of the specification these constraints are used to check for timing inconsistencies and conflicts and any such problems are reported.

EXAMPLE 9.1

Part of Automobile Management System

Figure 9.7 shows the state transition diagram for the main control section of an

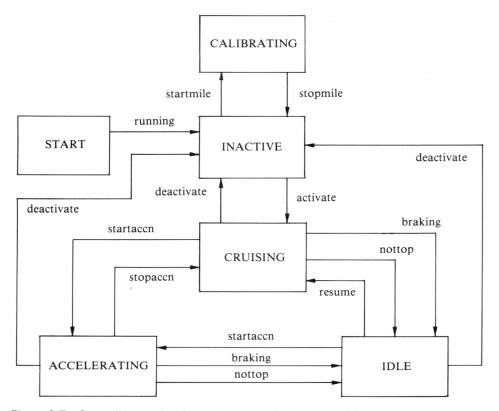

Figure 9.7   State diagram for the main states of the automobile management system.

automobile management system (this is part of a widely used case study – see Hatley and Pirbhai, 1988). The specification for this system written in PAISLey is shown in Figure 9.8.

9.11 *PAISLEY* SUMMARY

The major features of the language are:

- Support for both synchronous and asynchronous communication free from the problems of mutual exclusion.
- All computations are encapsulated; the mapping functions can be considered as black boxes.
- It is possible to execute incomplete specifications and hence rapid prototyping and incremental development are possible.

```
"Paisley file for Auto cruise state diagram"
"This file contains the Finite State Machine for the auto
cruise control. It consists of a number of states and a set
of rules for moving between them."
"A full range of driver command channels are taken by this
process and used for reading the drivers command. They are
read using the xr exchange function, so that we can pick up
one command from more than one channel. Very similar to a
multiplexor."
"The channels are;
running - Driver switches on the system
startmile - Start measured mile command
stopmile - Stop measured mile command
deactivate - deactivate the cruise control
activate - activate the cruise control
nottop - Not in top gear signal - inverted topgear signal
startaccn - start accelerating
stopaccn - stop accelerating
braking - Driver is braking
resume - Resume cruise control command
"
"The order in which these channels are read is important. If
the driver issues 'braking' and also 'startaccn' then
obviously 'braking' has to have priority."
state-machine: CRUISE-STATE --> CRUISE-STATE;
"Find out what state we're in and go off and see if there is
a change of state."
state-machine[state] =
/
equal[(state, 'START')]:          update-start,
equal[(state, 'INACTIVE')]:       update-inactive,
equal[(state, 'CALIBRATING')]:      update-calibrating,
equal[(state, 'CRUISING')]:       update-cruising,
equal[(state, 'ACCELERATING')]:  update-accelerating,
equal[(state, 'IDLE')]:           update-idle,
True:                   state     "Really an error"
/;
"Start state mapping"
update-start:    --> CRUISE-STATE;
update-start =
/
"Ordered by safety critical importance"
equal[(xr-running [Null],true)] : 'INACTIVE',
True              : 'START'
/;
"Inactive state mapping"
update-inactive:    --> CRUISE-STATE;
update-inactive =
/
"Ordered by safety critical importance"
equal[(xr-activate [Null],True)]:
proj [(1, ('CRUISING', x-motionstate['ON'],
x-select['ON']))],
equal[(xr-startmile [Null],True)]:
proj [(1, ('CALIBRATING', x-measurestate['ON']))],
True
/;
"Calibrating state mapping"
update-calibrating: --> CRUISE-STATE;
update-calibrating =
/
```

Figure 9.8   PAISLey specification for system shown in Figure 9.7 (*continued overleaf*).

```
"Ordered by safety critical importance"
equal[(xr-stopmile [Null],True)]:
proj [(1, ('INACTIVE', x-measurestate['OFF']))],
True                    : 'CALIBRATING'
/;
"Cruising state mapping"
update-cruising:    --> CRUISE-STATE;
update-cruising =
/
"Ordered by safety critical importance"
equal[(xr-braking [Null],True)]: 'IDLE',
equal[(xr-nottop    [Null],True)]:'IDLE',
equal[(xr-deactivate [Null],True)]:
proj [(1, ('INACTIVE', x-motionstate['OFF']))],
equal[(xr-startaccn [Null], True)]: 'ACCELERATING',
True
proj [(1, ('CRUISING', x-maintainv['ON']))]
/;
"Accelerating state mapping"
update-accelerating:    --> CRUISE-STATE;
update-accelerating =
/
"Ordered by safety critical importance"
equal[(xr-braking [Null],True)]: 'IDLE',
equal[(xr-nottop    [Null],True)]: 'IDLE',
equal[(xr-stopaccn  [Null],True)]:
proj[(1, ('CRUISING',
x-select['ON']))],
equal[(xr-deactivate[Null],True)]: 'INACTIVE',
True
proj [(1, ('ACCELERATING', x-maintainv['ON']))]
/;
"Idle state mapping"
update-idle:    --> CRUISE-STATE;
update-idle =
/
"Ordered by safety critical importance"
equal[(xr-braking    [Null],True)]: 'IDLE',
equal[(xr-nottop    [Null],True)]: 'IDLE',
equal[(xr-deactivate [Null],True)]: 'INACTIVE',
equal[(xr-startaccn [Null],True)]: 'ACCELERATING',
equal[(xr-resume    [Null],True)]:
proj [(1, ('CRUISING',
x-select['ON'],
x-select['ON']))],
True                    : 'IDLE'
/;
```

Figure 9.8 *continued*

- Interprocess communication is precisely defined using exchange functions which hide problems of mutual exclusion and which can be used to simulate timing constraints on communication links.
- Both hard and soft timing constraints can be specified and these constraints are automatically checked for violation when the specification is executed.
- Bounded resource usage can be guaranteed.

The major weaknesses of PAISLey are largely those that are common to many approaches to formal specification, namely that for all but the simplest systems the specification becomes long and cumbersome, making it difficult to read and follow. To deal with this problem specifications can be broken down into segments which

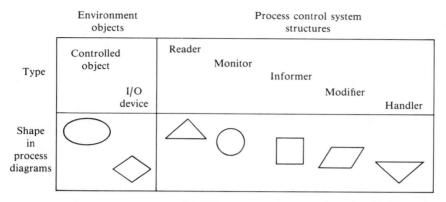

Figure 9.9   Graphical notation for PAISLey processes.

can be held in separate files. There has also been an attempt to add to PAISLey a graphical representation (see Figure 9.9) but unlike MASCOT there is no exact relationship between the graphical representations and the textual representation.

The other weakness of PAISLey is that the system is not available in a fully developed form in the sense that it lacks the good user interfaces that now characterise the majority of the CASE tools. A particular problem is that the output from the interpreter is in simple textual form which is difficult to analyse.

9.12 SUMMARY

In this chapter we have briefly examined two methods for dealing with real-time systems. MASCOT provides a well-established methodology which assumes that a limited set of basic elements are all that are required to implement the system. The additions that have been made to MASCOT 3 are such as to support an object-oriented approach to system design. A criticism of the method is that the timing requirements are not visible on the design documents and in fact the designer proceeds without direct reference to them. As MASCOT is not intended for specification and, as we have seen, it is difficult to make any use of the timing constraints until the realisation of the implementation, this is not a serious restriction on the use of MASCOT.

PAISLey is much less well developed and known than MASCOT. Its main interest lies in the attempt to support the specification of real-time systems using a formal language in which the timing constraints can be easily expressed. The other interesting feature is that the specification can be executed. It is a pity that the output from the execution is so difficult to interpret. What is required is full animation support.

10

Design Analysis

10.1 INTRODUCTION

The development methodologies considered in Chapters 8 and 9, with the exception of PAISLey, do not provide any means of analysing the design either to compare designs or to evaluate the implementation requirements. A weakness of both the Ward and Mellor and the Hatley and Pirbhai methods which we noted is that in allocating resources as part of the design process one really needs to be able to assess whether a feasible schedule, that is a schedule that meets the time constraints, exists for a particular design structure. Similarly both methods make extensive use of state transition diagrams in specifying and designing the control structure of the system. A method of analysing such diagrams to reveal unreachable states and/or undesirable states that are reachable would be useful. Ward and Mellor suggested that Petri nets might be used to analyse the essential model and hence in the next section we will look at Petri nets and their use.

10.2 PETRI NETS

Petri nets have been widely adopted as a method for *modelling* and *analysing* systems that can be described in terms of a set of states and a set of events. An event change results in a change from one state of the system to another state. Also, a change of state results in a change in the event set. The technique was originally proposed by Carl Adam Petri in 1962 as a basis for modelling computer systems with asynchronous communication between asynchronous components. It has since been used to model business systems, hardware systems and manufacturing systems. A full treatment of the technique is given in the book by Peterson (1981) which is recommended as a starting point for an in-depth study of the technique. An extensive bibliography on Petri nets can be found in the book by Reisig (1982).

362

10.2.1 Basic Ideas

A Petri net is used to model a system on the basis of two properties:

1. Condition: a Boolean description of the state of the system; a *condition* may be *true* or *false*.
2. Event: an action that depends on the state of the system.

The system model is represented by a set of conditions and a set of events. In the Petri net notation a condition is represented by a *place* and an event by a *transition*. In the graphical representation of a Petri net a place is drawn as a circle and a transition as a bar as is shown in Figure 10.1. Places and transitions are connected by arrows.

Referring to Figure 10.1, place p_1 is an *input place* for transition t_2 and an *output place* for transition t_1; whereas place p_2 is an input place for transition t_1 and an output place for transition t_2. Input places represent the necessary conditions for an event represented by a transition to occur and are referred to as *preconditions*.

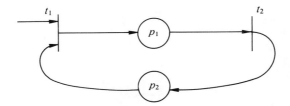

Figure 10.1 Graph representation of a Petri net.

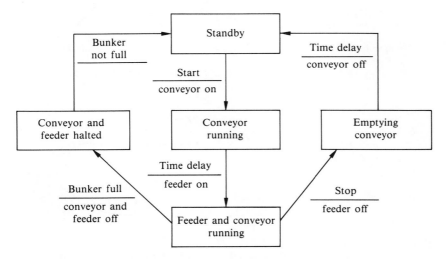

Figure 10.2 State transition diagram for a coal clearance system.

Output places represent the set of conditions that result from a transition and are called *postconditions*.

Given that Petri nets are used to model conditions and events there is obviously a similarity between state transition diagrams and Petri nets. For example, the STD shown in Figure 10.2 can be represented by the Petri net shown in Figure 10.3. A major difference arises from the idea of *executing* a Petri net. Dots are placed in

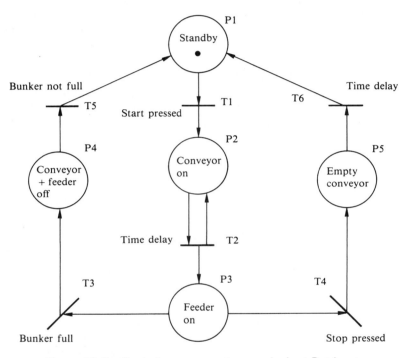

Figure 10.3   Coal clearance system equivalent Petri net.

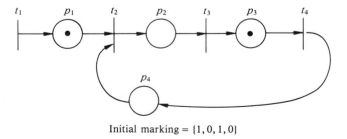

Initial marking = {1, 0, 1, 0}

Figure 10.4   A marked Petri net.

the circles representing places for which the conditions of a place are known to be true. The dots are referred to as *tokens*. A distribution of tokens is known as a *marking* of the Petri net. A marked Petri net is shown in Figure 10.4. The marking can be represented as (1, 0, 1, 0) with each number representing the number of tokens present in places p_1 to p_4 respectively. A place can contain several tokens and hence can represent a queue – for example, a buffer holding several messages or a queue of parts awaiting processing. If the number of tokens at a place is large then instead of using dots a number is written in the place. Figure 10.5 shows a Petri net with multiple tokens; the marking is (10, 1, 20, 0).

    Executing a marked Petri net causes the number and positions of the tokens to change. The rules for executing a Petri net are:

1. a transition is *enabled* if *all* its input places contain at least one token;
2. any enabled transition may *fire*;
3. firing of a transition results in one token being removed from each of its input places, and being deposited at each of its output places; and
4. execution halts when there are no enabled transitions.

Each time a transition fires the marking of the Petri net will change (usually). For example, consider the Petri net shown in Figure 10.4 where transition t_4 is enabled since it has one input place p_3 and this place contains a token. The effect of firing t_4 is to change the marking to (1, 0, 0, 1) as shown in Figure 10.6.

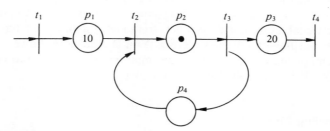

Figure 10.5 A marked Petri net with multiple tokens.

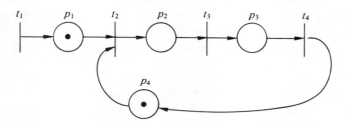

Figure 10.6 Executing a Petri net – stage 1.

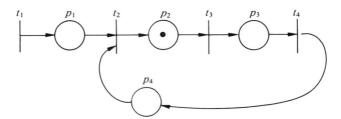

Figure 10.7   Executing a Petri net – stage 2.

Now transition t_2 can fire since places p_1 and p_4 contain tokens and as a result of t_2 firing the marking changes to (0, 1, 0, 0) and the marked net is now as shown in Figure 10.7.

Note that the firing of t_2 has changed the number of tokens in the net from two to one. This is because there are two input arrows to t_2 but only one output arrow.

In general the firing of a transition will result in a change in both the marking of a net and the number of tokens in the net. The number of tokens at an input place can never become negative since a transition can fire only if each of its input places contains at least one token. It is possible to model a system using a Petri net such that the total number of tokens in the net is kept constant. Such a model would be required if each token was being used to represent an object flowing through the system.

If several transitions are enabled then the order of firing is non-deterministic. When analysing a Petri net each possible order of firing must be considered. Use of this property enables us to use Petri nets to model systems which have concurrent events that do not have a unique ordering. You should note, however, that if the system being modelled does, in some way, guarantee precedence to certain events this must be explicitly modelled in the Petri net.

10.2.2 Modelling Mutual Exclusion

As an example of using Petri nets to model concurrent events let us consider the mutual exclusion problem. Consider two tasks **TA** and **TB** that share a resource **R**. The resource is protected by a semaphore **SR**. Assume that the tasks can be split into segments **TA1**, **TA2**, **TA3**, **TB1**, **TB2** and **TB3** where **TA2** and **TB2** represent the critical sections of each task respectively.

If we model each task segment as a place and the transition from one segment to the next as a transition (an event) we get the two independent Petri nets shown in Figure 10.8a. The presence of a token in a place will be used to indicate that a particular segment is active because there is an agreement between the task designers to insert code in each task that checks with the semaphore **SR** (for example, by executing a **SECURE (SR)** statement) before proceeding to segment **TA2** or **TB2**; the conditions for the event represented by **EA1** to occur are that **TA** should be at

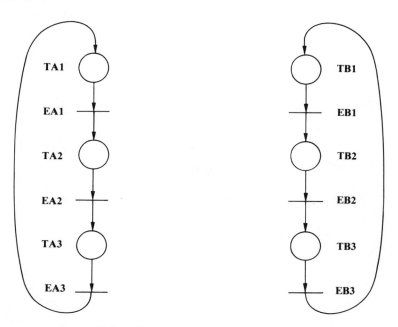

Figure 10.8a   Mutual exclusion example – stage 1.

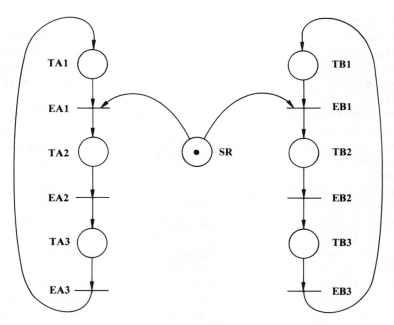

Figure 10.8b   Mutual exclusion example – stage 2.

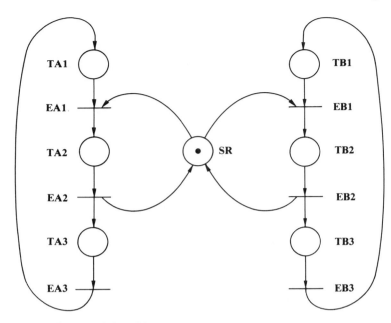

Figure 10.8c   Mutual exclusion example – stage 3.

the end of segment **TA1** and that **SR** should indicate that the resource is free. We can represent the semaphore by a place and the condition that the resource is free by the presence of a token in the place. This is shown in Figure 10.8b. The presence of a token in **TA1** or **TB1** would now cause either **EA1** or **EB1** to fire, thus removing the token from **SR** and either **TA1** or **TB1** and putting a token in either **TA2** or **TB2**.

We complete the model by considering what happens when segment **TA2** or **TB2** completes. At the end of the critical section the task executes a **RELEASE (SR)** and thus we need to return a token to **SR** and the task then continues to execute the next segment. Hence we need to add arrows from **EA2** and **EB2** to **SR** as is shown in Figure 10.8c.

To execute the Petri net model we need to make an assumption about the initial marking. Let us assume that the initial marking is (1, 0, 0, 1, 0, 0, 1) as shown in Figure 10.9a. With this marking either transition t_1 or t_4 will fire – which fires is non-deterministic. Let us assume that t_1 fires; the net marking now becomes (0, 1, 0, 1, 0, 0, 0) as shown in Figure 10.9b, since the token is removed from p_1 and from p_7. Examining Figure 10.9b we can see that task **TA** has entered the critical section **TA2** and that task **TB** is prevented from entering segment **TB2** since the conditions for transition t_4 to fire no longer hold. However, t_2 can now fire since there is a token in p_2. When t_2 has fired the net marking becomes (0, 0, 1, 1, 0, 0, 1). The conditions now exist for t_4 and t_3 to fire.

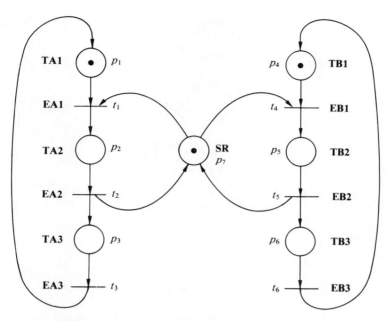

Figure 10.9a   Mutual exclusion and execution – stage 1.

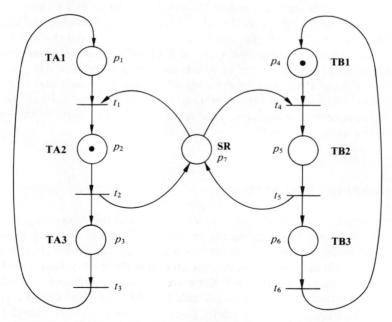

Figure 10.9b   Mutual exclusion and execution – stage 2.

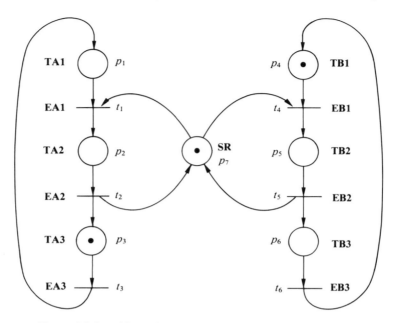

Figure 10.9c   Mutual exclusion and execution – stage 3.

In following the execution stages in Figures 10.9a, 10.9b and 10.9c you should have noticed that once a token reaches p_2 the transition t_2 is enabled and can thus fire. The implication of this is that the Petri net does not model the timing of the two tasks since execution of the task segment **TA2** will take a finite length of time. The model we really require is one in which the timing of t_1 indicates the start of execution of segment **PA2**, and the presence of a token in p_2 indicates that **TA2** is being executed. The condition for the firing of t_2 is that **TA2** has finished executing. In order to produce such a model we need to use an extended form of the Petri net notation – the timed Petri net.

10.3 ANALYSING PETRI NETS

Although it is possible to obtain some information about the behaviour of a system modelled by a Petri net by executing the net either by hand or by using a computer simulation the number of possible sequences is such that the procedure is laborious and the information obtained uncertain for all except the most simple nets. Formal methods of analysis are required and these are based on set theory formulations.

A Petri net structure C can be represented as the four-tuple $C = (P, T, I, O)$ where $P = \{P_1, P_2, \ldots, P_n\}$, the set of places; $T = \{T_1, T_2, \ldots, T_n\}$, the set of transitions; I is an input function that maps each transition to its set of input places;

and O is an output function that maps each transition to its set of output places. Thus the Petri net shown in Figure 10.9a can be represented as:

$C = (P, T, I, O)$
$P = \{p_1, p_2, p_3, p_4, p_5, p_6, p_7\}$
$T = \{t_1, t_2, t_3, t_4, t_5, t_6\}$

$I(t_1) = \{p_1, p_7\} \; O(t_1) = \{p_2\}$
$I(t_2) = \{p_2\} \quad\;\; O(t_2) = \{p_3, p_7\}$
$I(t_3) = \{p_3\} \quad\;\; O(t_3) = \{p_1\}$
$I(t_4) = \{p_4, p_7\} \; O(t_4) = \{p_5\}$
$I(t_5) = \{p_5\} \quad\;\; O(t_5) = \{p_6, p_7\}$
$I(t_6) = \{p_6\} \quad\;\; O(t_6) = \{p_4\}$

In analysing a Petri net model of a concurrent system we are concerned with obtaining answers to questions concerning:

- safeness;
- boundedness;
- conservation;
- equivalence;
- reachability;
- coverability; and
- liveness.

Safeness: A Petri net is said to be *safe* if all the places in the net are safe. A place is said to be safe if the number of tokens in the place is either 0 or 1. For example, if we were using the net shown in Figure 10.9c to find out whether we could use a binary semaphore to implement the mutual exclusion condition we would want to know if place p_7 was 'safe'.

Boundedness: A Petri net is said to be bounded if all the places are bounded. A place is bounded if the number of tokens in it can never exceed some finite integer value N. If a place is bounded then the physical element that it models can be realised using a finite storage device.

Conservation: A Petri net is said to be conservative if the number of tokens in the net remains constant, that is tokens are neither created nor destroyed when a transition fires. Strict conservation implies that for each transition the number of input places must match the number of output places. The mutual exclusion model is not conservative since the firing of transition t_1 destroys a token and the firing of t_2 creates a token. Testing strictly for conservation is important in Petri nets where tokens are used to represent objects moving around a closed system; or where tokens represent resources available to the system.

Equivalence: For two Petri nets to be said to be equivalent all possible behaviours must be equivalent. Establishing equivalence is difficult since each net has to be analysed for reachability, coverability and firing sequence. The equivalence property can be used to show that a given Petri net is a subset of another net. Its main use is in trying to optimise a system by removing redundant elements.

Reachability: Reachability is a basic property of a Petri net. It is concerned with answering the question: given an initial marking can a specified marking occur? The specified marking may be a desirable marking or it may be an undesirable marking (for example, a dangerous fault condition). In the mutual exclusion example (Figure 10.9a, b and c) we want an answer to the question: can any marking containing both p_2 and p_5 be reached from any initial marking?

Coverability: Coverability is the problem of determining if, given an initial state, there is a reachable marking that contains a particular marking subset. For example, in the mutual exclusion model we would like to know if it is possible to have simultaneously tokens in places p_2 and p_5, that is, are there any reachable markings that contain the subset $\{p_2 = 1, p_5 = 1\}$?

Liveness: A Petri net is said to be *live* if every transition can be enabled. Conversely a Petri net is *deadlocked* if one or more transitions cannot be enabled.

The two basic techniques for analysing Petri nets in order to seek answers to the above questions are:

- reachability trees; and
- matrix equations.

10.3.1 Reachability Tree

The basis of this method is: starting from an initial marking all reachable markings are found; then starting from each of these markings the reachable markings are found, etc. Figure 10.10 shows the reachability tree for the mutual exclusion net. The branches of the reachability tree are stopped *either* because with the marking of the net no further transition can be enabled *or* because the marking is equivalent to some other marking in the tree. For example, if we follow the transition firing path t_1, t_2, t_3 we reach a marking equivalent to the initial marking. Visual examination of the tree enables us to conclude that the net is safe – all the markings contain only the values 0 or 1 – and that there is no marking with tokens in both places p_2 and p_5.

The reachability tree also shows that the net possesses liveness since there is no leaf of the tree with a marking from which there is no transition. This in fact shows the net is free from deadlock: for a strict proof of liveness we have to show that the initial marking is reachable from all leaves of the tree. If a branch of the tree leads to an endless loop where one particular firing sequence repeats to the exclusion of all other firings then the system is *livelocked*.

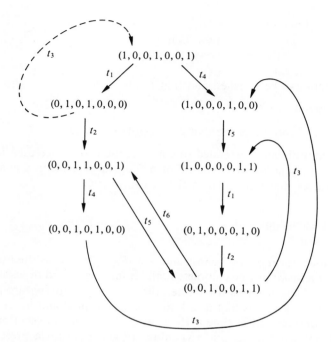

Figure 10.10 Reachability tree for mutual exclusion example.

10.4 SCHEDULING

The end point of most current real-time system development methodologies is an implementation model of the system that consists of a set of independently schedulable actions (asynchronous process) and a set of constraints (time, mutual exclusion and synchronisation). The methodologies then propose that realisation should proceed by first determining what processing elements are needed (analog circuits, general purpose digital processors, special digital processors, etc.) and by allocating groups of actions to processors. The second stage is to determine for groups of actions allocated to a general purpose digital computer which actions shall remain as independent schedulable processes and which shall be combined to form a single schedulable process.

The methodologies give no guidance on how these two stages should be carried out and on how decisions can be made on a rational basis.

Stated in general terms the problem is:

Given a set of processes
$P = \{p_1, p_2, \ldots, p_n\}$
how can a set of processors
$U = \{u_1, u_2, \ldots, u_m\}$

and process allocations
$$V = \{(p_a, u_1), (p_b, u_2), \ldots, (p_x, u_m)\}$$
be chosen so that the set of constraints
$$C = \{c_1, c_2, \ldots, c_p\}$$
on the system can be satisfied and that U and V are in some sense optimal?

An associated problem is:

Given P, U, V can we prove that constraints C are satisfied?

(Note that if the problem is extended to consider fault tolerance then U and V are not fixed. Also, in general the existence of a fault may result in a change to C.)
The constraints that have to be satisfied can involve:

- time;
- mutual exclusion; and
- precedence.

The most common form of time constraint is a *deadline*, that is the time by which the execution of a task must be completed, and it may be a hard or a soft constraint (see Chapter 1). Typical terms used when discussing task timing are as shown in Figure 10.11, which assumes that a task may be interrupted and hence executed in segments. The total execution time is the sum of the segment execution times plus any overheads involved in context switching. In general with a pre-emptive task scheduler we do not know how many segments a task will be split into and it will vary from task invocation to invocation. Some tasks may have a constraint on the *start time* as well as a deadline.

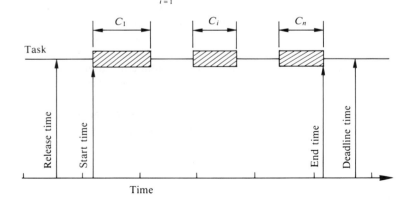

Figure 10.11   Task timing notation.

The mutual exclusion constraint arises when tasks share resources and hence task A (or a segment of task A) may not be able to run while task B (or a segment of task B) is using a particular resource. The precedence constraint arises because one task may need information generated by another task. For example, if task C (or a segment of C) requires a value produced by task D (or a segment of D) then there is no point in scheduling D to run before C (or the segment of D to run before the segment of C).

10.5 GENERAL APPROACHES TO THE SCHEDULING PROBLEM

There are two general approaches:

1. Run-time (on-line scheduling): this is the real-time scheduling (control) problem – can we design a scheduling algorithm (or algorithms) that will allocate resources (including time resources) such that the system meets its constraints?
2. Pre-run-time scheduling: this is the *design* problem – can we choose a set of resources such that a task execution schedule can be constructed that satisfies the constraints? Additional problems are: can we prove that the system does satisfy the constraints; can we choose a minimal set of resources?

The traditional approach to real-time systems has been to use an on-line scheduler. However, in recent years there has been an increased interest in the pre-run-time scheduling approach for systems in which all time constraints are hard. The difficulty with it is that any design change or, during run-time, the loss of a resource because of failure of part of the system means that a new schedule has to be found and implemented. The approach also requires systems to be designed on the basis of worst case conditions: upper bounds on execution time and communication delays, and maximum frequency of occurrence for events; hence there is overprovision of resources.

10.6 ON-LINE SCHEDULING – INDEPENDENT TASKS

The on-line scheduling approach divides into two sections:

1. assessment of schedulability; and
2. choice of scheduling algorithm.

The first of these is concerned with finding out whether there are sufficient resources allocated to the system for a feasible schedule to exist. The necessary and sufficient conditions for showing in general that a schedule that meets the constraints exists are not known. Under certain restrictions and preconditions it is possible to check if a feasible schedule might exist.

Even if it is shown that a feasible schedule exists it does not mean that in practice all the time constraints will be met; this depends on how effective the scheduling algorithm being used by the operating system is in utilising the resources, or how well it is matched to the particular requirements.

10.6.1 Schedulability

The first requirement for carrying out a schedulability analysis is to determine or estimate the execution time for each task in the system. Given that execution times are not necessarily constant we need two estimates:

1. average execution time; and
2. worst case execution time.

We can get reasonable estimates for execution times only if certain restrictions have been applied when coding the system. For example, use of the following must be avoided:

- dynamic creation of tasks;
- dynamic allocation of memory; and
- recursion.

Also the following restrictions must be applied:

- all loops must have upper bounds (periodic tasks which are written as infinite loops are permitted);
- all intertask communications must have time outs; and
- all external communications must have time outs.

Note that these restrictions apply even if the execution times are being estimated by running the software, because without them the system will be non-deterministic.

Once a set of execution times for each task has been determined we can calculate

Table 10.1   Utilisation time for cyclic processes

Task	Cycle time (s)	Execution time (s)	Utilisation (%)
ReadInputs	0.1	0.02	20
CalculateControl	0.2	0.08	40
UpdateDisplay	5.0	0.30	6
SendToActuator	0.2	0.005	2.5
Total utilisation			68.5

the processor *utilisation* time. This is easily determined for a cyclic (periodic) task. For example, if the cycle period is 0.1 s, and the execution time is 0.02 s, the processor utilisation is $(0.02/0.1) \times 100\%$, that is 20%. Thus if we have a set of tasks which run on a single processor with cycle times and execution times as shown in Table 10.1 we can easily determine how much processor time they utilise.

Liu and Layland (1973) proved that for a single-processor system running a set of n independent periodic tasks with constraints consisting only of deadlines on time of the end of execution (that is, the task can be started at any time) a feasible schedule exists if processor utilisation satisfies the condition

$$\sum_{i=1}^{n} e_i/c_i \leqslant n(2^{(1/n)} - 1)$$

and a *rate monotonic scheduling* algorithm is used. This algorithm always chooses the highest-priority task. Task priority is ordered according to the cycle time of the tasks, the highest priority being given to the task with the smallest cycle time. For a large value of n this approaches 0.693; hence we can conclude that if the processor utilisation is less than 69% all sets of tasks are schedulable.

One must **not** conclude from this that sets of tasks with process utilisation greater than 69% cannot be scheduled. For example, task set 1 shown in Table 10.2 with 100% utilisation is schedulable (under the conditions given above) whereas the task set 2 given in Table 10.3 is not (Burns and Wellings, 1990, pp. 347–9). In order

Table 10.2   Task set 1 – utilisation 100%

Task	Cycle time (ms)	Execution time (ms)	Utilisation (%)
P_1	80	40	50
P_2	40	10	25
P_3	20	5	25
Total utilisation			100

Table 10.3   Task set 2 – utilisation 82%

Task	Cycle time (ms)	Execution time (ms)	Utilisation (%)
P_1	50	12	24
P_2	40	10	25
P_3	30	10	33
Total utilisation			82

to meet the deadlines, task set 1 must be run on a processor with a pre-emptive priority scheduler and the task priorities must be ordered from highest to lowest as follows: P_3, P_2, P_1. The task activation diagram for task set 1 is shown in Figure 10.12. From this diagram it can be clearly seen that without a priority scheduler P_1 and P_2 would not be able to meet their deadlines. It should also be clear that the start of P_3 is not accurately synchronised with real time; it will always run with a 15 ms offset, and similarly P_2 will always run with a 5 ms offset from its nominal start time.

Determining schedulability becomes more difficult when there are event-driven tasks in the system. Events are aperiodic and, because there is always a non-zero probability that an event will occur within a given time interval of a previous event regardless of how small that interval is, it is not possible to carry out a worst case analysis. However, by considering the particular application, we can usually set a minimum time interval between two occurrences of a given event. Events constrained in such a way are referred to as being *sporadic*. A worst case analysis can now be carried out by converting *sporadic* events into periodic events. We do this by taking the minimum time interval between successive occurrences of the event as the cycle time for the task that responds to the event. For example, consider the addition of a task **CheckAlarms**, which has to respond within 0.15 seconds of an event, to the set given in Table 10.1. We will assume that the response interval is the minimum interval between successive occurrences of the event and hence the cycle time for **CheckAlarms** is 0.15 seconds, and if we assume the execution time

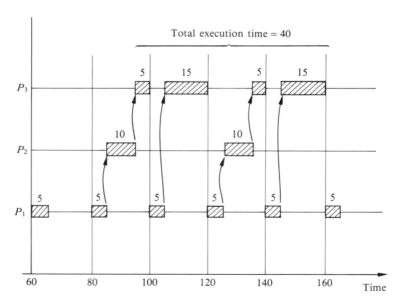

Figure 10.12 Task activation diagram.

is 0.01 seconds then the processor utilisation is 6.7%, taking the total utilisation to 75.2%.

Using the worst case analysis for sporadic tasks can lead to considerable overestimates of the processor utilisation and hence lead to low actual utilisation. The reason is easy to see – the average rate of event occurrences is likely to be much lower than the potential maximum rates.

Burns and Wellings (1990, p. 348) suggest as a guideline for assessing schedulability that the following conditions should be satisfied:

1. All tasks should be schedulable using average execution times.
2. All tasks with hard time constraints should be schedulable using worst case execution times.

A consequence of condition 1 is that there may be occasions when all tasks cannot meet their deadlines – this is referred to as *transient overload* – **but** if condition 2 is satisfied even under these conditions the tasks with hard time constraints will meet those constraints.

An analysis of processor utilisation can be useful when trying to determine the allocation of tasks in a system. In terms of simplicity of understanding, and of implementation, there are advantages in having only a small number of tasks. This may mean that some actions are performed more frequently than absolutely necessary, thus increasing total processor utilisation. However, if the set of tasks is still schedulable then such a task allocation can be adopted.

Examining Table 10.1 we can easily see that by combining tasks **CalculateControl** and **SendToActuator** we do not significantly change the processor utilisation since both tasks have the same cycle time (there will be a small saving in context switching time). But suppose we also try to combine them with **ReadInputs** and run all three at the cycle time of **ReadInputs**. The combined task has an execution time of 0.105 seconds (minus some allowance for reduced context switching) and a cycle time of 0.1 seconds and hence it is immediately clear that we cannot use this combination (the processor utilisation is 105%).

Let us consider the position when we add in the task **CheckAlarms** and let us assume that we decide to run it in combination with **ReadInputs,** that is at 0.1 second intervals rather than 0.15 second intervals. The utilisation calculation is shown in Table 10.4.

10.6.2 Scheduling Algorithms – Pre-emptive, Priority Based

We discussed some scheduling algorithms in Chapter 6 when we dealt with operating systems. In particular we assumed that for real-time applications the scheduler would use a priority-based pre-emptive algorithm. This is the simplest and most commonly used real-time scheduler. The scheduler always selects the task with the highest priority from the set of tasks that are ready to run.

Table 10.4   Effect of combining tasks on utilisation

Task	Cycle time (s)	Execution time (s)	Utilisation (%)
ReadInputs	0.1	0.03	30
CheckAlarms			
CalculateControl	0.2	0.085	42.5
SendToActuator			
UpdateDisplay	5.0	0.30	6
Total utilisation			78.5

By examining Figure 10.12 which was drawn on the assumption that a pre-emptive priority scheduler was being used, we can see that the algorithm guarantees that the highest-priority task is always run on time. This is the only assertion that we can make about this algorithm. The behaviour of a system which uses this scheduling technique is dependent on the particular choice of priority structure. Even then it is non-deterministic since the behaviour will also change according to the pattern of occurrence of events. A question then is: how should task priorities be assigned?

There are two basic approaches:

1. assign priorities according to the importance of the task; and
2. assign priorities according to the cycle time of the task with the task with the shortest cycle time being given the highest priority.

In practice designers use a mixture of the two approaches. Under normal operating conditions for a control system the tasks with the shortest cycle time will also be the ones that are the most important and hence in practice the two approaches coincide. However, under abnormal conditions the importance of some control loops may be downgraded compared to others. Hence a designer may choose a fixed order of priorities that also takes into account the requirements of abnormal running and hence departs from strict compliance with approach number 2 (alternatively the designer may choose to use dynamic priority reallocation and change to a different set of priorities during abnormal running conditions).

10.6.3 Scheduling Algorithms – Other Types

The two other, most commonly advocated scheduling algorithms are:

1. earliest deadline; and
2. least slack time.

To implement either of them the scheduler needs to know the deadline for each task.

The earliest deadline scheduler, as its name implies, simply chooses the task whose deadline is closest to the current time. Be aware that it chooses from the list of tasks that are ready to run, that is those tasks whose release time is earlier than the current time.

A least slack time scheduler needs to know, in addition to the deadline for each task, the amount of processor time that the task needs in order to complete its execution. Using these values the scheduler calculates which task has the least free or slack time before its deadline.

Little advantage over the priority-based scheduler is obtained from using these algorithms for systems which comprise mainly periodic tasks. However, they perform significantly better for systems with mainly aperiodic tasks as they do not involve the use of any prior assumptions on the rates of occurrence of the events.

The major weakness of the algorithms is that they operate only on the current state of the system and do not look ahead; hence under transient overload conditions decisions may be made in a non-optimum way.

10.7 PRE-RUN-TIME SCHEDULING

For control applications and in some other forms of hard real-time systems there is frequently a requirement for a task to run at exactly T second intervals. This is a constraint that applies to the control algorithms described in Chapter 4. The on-line schedulers described above seek a schedule that ensures that a periodic task runs once per time interval. For example, the priority-based scheduler can ensure that one task, that given the highest priority, will run at exactly T second intervals but no other. For small, hard real-time systems a method which is frequently used to provide precise scheduling for the critical tasks is to build into the scheduler a precalculated schedule.

One method of doing this is to construct a table with the number of columns equal to the number of tasks to be scheduled and the number of rows equal to the lowest common multiple of the task cycle intervals. (The size of the table can be reduced if the cycle intervals are expressed in terms of their greatest common factor.) Each time the scheduler runs it first checks if it is time to read the table, and if it is it reads the appropriate row. It selects those tasks for which the entry in that row is a '1' and runs them. The scheduler uses a counter to keep track of which row it is to read and the counter is reset when all the rows have been read and the sequence restarts. Depending on the ratio between the basic clock tick and the GCF of the cycle times for the task, the scheduler may check the table each time it is entered or only at some multiple of the basic tick.

For example, consider a system with four tasks A, B, C and D with cycle times of 4, 5, 10 and 15 ms respectively. The GCF is 1 and the LCM is 20; hence we need a table with 20 rows as shown in Table 10.5.

As will be seen from the table the approach does not guarantee exact cycle times at every task invocation since in row 16 we find that two tasks A and B are scheduled

Table 10.5   Task scheduling table

	A	B	C	D
0	1	0	0	0
1	0	1	0	0
2	0	0	1	0
3	0	0	0	1
4	1	0	0	0
5	0	0	0	0
6	0	1	0	0
7	0	0	0	0
8	1	0	0	0
9	0	0	0	0
10	0	0	0	0
11	0	1	0	0
12	1	0	1	0
13	0	0	0	0
14	0	0	0	0
15	0	0	0	0
16	1	1	0	0
17	0	0	0	0
18	0	0	0	0
19	0	0	0	0

to run at the same time and this occurrence will occur on every fifth invocation of task A and fourth invocation of task B. By using smaller scheduling time intervals and a much larger table it is possible to reduce the frequency at which tasks coincide but not to eliminate it entirely.

10.8 SCHEDULING – INCLUDING TASK SYNCHRONISATION

So far we have assumed that all the tasks are independent, that is there are no mutual exclusion or precedence relationships between the tasks. Of course in general this is not the case. Introducing these constraints increases the complexity of the problem greatly and methods for determining predictable schedules for a hard time constraint, real-time system with task synchronisation are the subject of much research. There are serious doubts about the use of on-line scheduling for hard time constraint systems and most work is being done on pre-run-time scheduling. The main techniques that are being used are:

1. Model-based techniques, for example Petri net models.
2. Temporal logic and extended states machines.

3. Algorithmic techniques (Xu and Parnas, 1990; Shepard, 1991).

One interesting approach that effectively avoids the scheduling problem is to divide each task into small segments for which the execution times are roughly equal (within an order of magnitude). Critical sections of code must not be split, all intertask communication must be by messages and all synchronisations between tasks must have time-out conditions attached. The segments are prioritised, usually in groups, and the scheduler simply executes the segments in order. That is, it looks at the highest-priority group and executes any segments that are waiting; if there are none it then looks at the next priority level and so on. No event-based tasks are permitted; they are turned into periodic tasks which are used to poll the source of the event. Because there is no pre-emption, each segment, once it starts, runs to completion. The behaviour of the system is predictable (on a worst case basis since the execution time of some tasks will depend on the value of inputs to those tasks).

10.9 SUMMARY

A major weakness of all the system development methodologies that we have examined is the lack of analysis tools. Without such tools the system designer has no means of evaluating design decisions. Resource allocation − which includes partitioning into tasks and scheduling decisions − is obviously a vital area for hard real-time systems. It is no use deciding on a particular task allocation if the resulting task set cannot be scheduled in a way that meets the time constraints.

In this chapter we have briefly examined some of the approaches to modelling and analysing systems for the purposes of evaluating design decisions. Some techniques − processor utilisation, for example − are simple and easily carried out but the information they offer is limited; other methods are more complex. Methods for carrying out the analysis of schedulability and evaluating the safeness of systems are being developed rapidly as are tools to support the methods. Information on techniques and tools, including simulation tools, can be found in Berryman and Sommerville (1991), Harel *et al.* (1990), Liu and Shyamasundar (1990), McCabe *et al.* (1985) and Pressman (1992). Such tools will gradually be added to CASE environments.

The problems become more difficult when distributed systems are used. Information on such systems can be found in Burns and Wellings (1990) and Levi and Agrawala (1990).

Some important points to remember are:

● Petri net models can be used to find out if a system can enter an unsafe state.
● Predictable performance is more important than efficiency in hard systems.
● Processor utilisation calculations give a simple check that can demonstrate immediately if a system cannot be scheduled. The check does not prove that it can be scheduled except for a particular limited set of conditions.

EXERCISES

10.1   Draw a Petri net diagram to represent the state transition diagram shown in Figure 9.6.

10.2   Plot a graph showing how the maximum processor usage for schedulability changes with the number of tasks (n) assuming the rate monotonic scheduling algorithm.

10.3   A system contains four tasks, A, B, C and D. A, C and D are cyclic tasks with periodic times of 0.2, 5 and 0.5 seconds respectively; and B is an aperiodic task with a response time of 0.3 seconds. The execution times for A, B, C and D are 0.08, 0.03, 0.9 and 0.03 seconds respectively. Find the free processor time. Can the tasks be scheduled? What would be the effect of (a) doubling the processor speed; and (b) halving the processor speed?

11

Dependability, Fault Detection and Fault Tolerance

11.1 INTRODUCTION

Methodologies for analysing the reliability of complex systems and techniques for making such systems tolerant of faults, thus increasing the reliability, are well established. In hardware the emphasis is on improved components and construction techniques and the use of redundancy to maintain critical systems functions in the event of single failures (in some systems multiple failures). In software the emphasis has been on improved software construction techniques – 'software engineering' – to reduce latent errors; there has also been work on techniques to introduce redundancy into software systems.

The majority of work on fault tolerance has concentrated on what Anderson and Lee (1981) have termed 'anticipated faults', that is faults which the designer can anticipate and hence 'design in' tolerance. A much more difficult and insidious problem is that of faults in the *design* of the system. These are by definition 'unanticipated faults' (and unanticipatable). Design faults can occur in both hardware and software but are more common in software and much of the effort of software engineering has been directed towards reducing design faults, that is unanticipated faults.

Fault tolerance is just one aspect of constructing *dependable* computer systems. Dependability is defined by Laprie as 'that property of a computing system which allows *reliance* to be justifiably placed on the service it delivers' (Laprie, 1989a). He summarises the attributes of dependability as shown in Figure 11.1 An *error* is the consequence of a *fault* in the system and a *failure* is the effect of an error on the service provided by the system. In this chapter we concentrate on the detection of errors and on fault avoidance and fault tolerance. For background work on measures (reliability and availability) see Shooman (1983), Ham (1984), Liebowitz and Carson (1985) and Laprie (1989b).

11.2 USE OF REDUNDANCY

A well-established technique for increasing the reliability of hardware systems is to

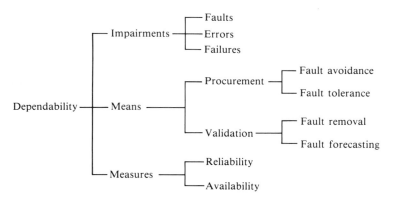

Figure 11.1   Characteristics of dependability.

duplicate or triplicate hardware units. Hardware redundancy can be introduced in two forms: *static* (or masking) redundancy, and *dynamic* redundancy. In static redundancy the duplicate components take over the operation in the event of failure and hence mask the failure of the system from the environment. To detect that a failure has occurred requires some secondary indication of failure. For example, a computer control system with several processors operating in parallel and carrying out (or capable of carrying out) the complete functions of the system with the final action being determined by a majority voting system involves static redundancy. In dynamic redundancy an error detection unit is used to detect an error; once detected the error must be corrected by redundancy elsewhere in the system. An example of dynamic redundancy is a distributed computer system in which each unit carries out a specific subset of the overall functions, but which has some means of detecting when a unit fails and transferring the actions of the failed unit to one or more of the operating units.

The most common example of the static redundancy approach is triple modular redundancy (TMR) in which three identical units are run in parallel and a majority voting system is used to check the outputs. The system is designed to tolerate failure of a single module. It produces output only when two modules agree (2-out-of-3 system).

Although the concepts of multiple channels and majority voting are simple there are some serious practical problems. An obvious problem is that a parallel channel system does not provide any protection against common mode or systematic failures. Another problem is in the decision or voting unit. If the signals entering the unit are two-valued logic signals (Boolean) then a clear unambiguous result is obtained. If any two are true the output will be true; if any two are false then the output will be false. Unambiguous results can normally be obtained also for integer values – they are usually required to be equal. However, if the signals are analog modulating signals or discrete real numbers then the decision unit is more difficult to implement. Three common techniques are:

Average or mean value: this is simple to implement and can be satisfactory if the signals do not deviate from each other greatly or if a large number of channels are used. For a TMR system a fault which causes one of the signals to go to a maximum or minimum value can introduce a significant error in the output.

Average with automatic self-purging: the average is calculated but if one of the channels deviates by more than a specified amount from the average that channel is disconnected. This technique can generate a large transient when the faulty channel is disconnected.

Median selection: instead of taking the average, the median value of three or more channels is selected. This technique avoids large transients.

In the above it should be noted that only the average with automatic self-purge technique provides detection of failure of a channel.

In any system employing parallel channels there will be divergence, that is the individual channels will not give identical values and also the differences in values may change with time without implying failure of any particular channel. To avoid divergence resulting in disconnection of a functioning channel some form of channel equalisation is necessary (Ham, 1984).

There is a fundamental difference between the use of redundancy in hardware and in software. In hardware it is assumed that each unit functions correctly according to its specification and that failure is due to some physical cause – wear, a faulty component, unusual physical stress – that is not common to both units. In software the major causes of failure are design faults – ranging from misinterpretation of the specification to simple coding errors. Simply replacing a software module which has failed with an exact copy will result in an immediate failure of the replacement copy. The approach required for software redundancy is more complicated requiring the production of independent software modules from the same specification.

The production of independent software modules is not a simple task. It is, for example, not adequate to code independent modules from the same design, or to design blind from the same specification. Experiments have shown that most of the errors in the final system stem from ambiguous or misunderstood specifications. Ambiguity in the specification and misunderstanding can be reduced if the independent groups report back to a co-ordinator.

A danger in using redundancy is that agreement between modules can induce a false sense of confidence in the system, or in parts of the system in which redundancy is used, that can result in the failure to diagnose common mode or systematic errors.

11.3 FAULT TOLERANCE IN MIXED HARDWARE–SOFTWARE SYSTEMS

As indicated above one of the basic techniques for introducing fault tolerance is to add redundancy to the system. However, care is required as adding redundancy adds complexity to a system and complexity can increase the likelihood of faults occurring. Thus great care has to be taken in adding redundancy to a system.

Dealing with the reliability and fault tolerance of mixed hardware/software systems introduces difficulties in determining the actual reliability structure of the system. The software is distributed over several hardware elements, for example a software module driving an input device is distributed over the CPU, memory, interface bus and the input device. Although at the software design stage the input driver is (or should be) clearly defined as a module this structure may not be apparent when the software is run. Anderson and Lee (1981) propose a model in which the hardware is viewed as maintaining an interface on which the software is executed. They term this interface the *interpretative interface* and they represent the model as shown in Figure 11.2, where C is a computing system, S is the memory containing the program (code and data) and H is the rest of the hardware. The

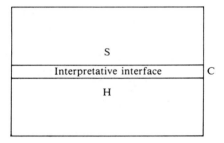

Figure 11.2   Interpretative interface model.

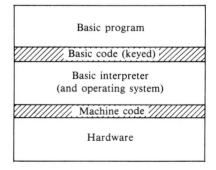

Figure 11.3   Multi-level interpreter system.

interpretative interface represents a language that provides objects and operations to manipulate those objects. At the simplest level this is the machine language of the hardware.

The model is extended in two ways: one is a *multi-level interpreter* system which can be represented as shown in Figure 11.3; the other is the concept of an *extended interpreter* represented as shown in Figure 11.4.

This latter model assumes that the system is required to support an interpreter L_1; however, what is available is some interpreter (language) L_0. If L_0 contains some of the facilities required by L_1 then it is not necessary to provide a complete interpreter; an interpreter which uses some of L_0 but extends it to provide extra facilities can be used. The program can then be considered as being in two parts: part E which is written in L_0 and part P which is written in L_1. The interpreter extension model clearly can be used to represent operating systems. It can be extended to represent *multi-level extended interpreters*.

11.3.1 Mechanisms and Measures

In using this modelling approach two further concepts are helpful. They are used to distinguish between facilities provided by the interpreter and those which are programmed in the interpreted system:

- *mechanism*: provides a specific facility at the interpreter interface and is implemented as part of the interpreter;
- *measure*: performs some specific task and is implemented by means of a set of instructions in the system that is interpreted.

Consider the input of data into a program written in BASIC and suppose that the program has to accept integer values. If it is written using the standard input with a BASIC integer variable given as part of the input statement, then if the user inputs alphabetic characters the error will be detected by a *mechanism* contained within the BASIC interpreter which will force an entry to the error handling system. The programmer will thus have relied on an interpreter mechanism. Alternatively if the input is read into a string variable and string to numeric conversion is performed by a program, then input checking is performed using a *measure*, not a *mechanism*. Mechanisms are likely to be more general purpose and widely applicable than measures and Anderson and Lee place emphasis on generating fault tolerance through the use of mechanisms.

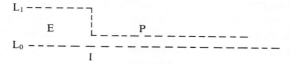

Figure 11.4   Extended interpreter system.

11.3.2 Exceptions

A second basic idea is to separate the normal behaviour of a system from abnormal behaviour. It is assumed that a fault condition will generate an error which will be an exception. A way of reducing complexity is to separate the response to an event into a normal response and an exception response. System designers and implementors should be able to deal with the two responses separately.

For example, a programmer does not normally have to check for 'divide by zero'; the computer hardware is designed to detect this fault and generate an exception response. The normal exception response in this case is a software trap which transfers control to an error handling routine. In many systems the error handler has a fixed, predetermined response – to halt execution of the program. For real-time systems the fixed response is often not wanted and if it cannot be disabled the designer is forced to insert additional coding to trap errors prior to them initiating the built-in mechanism. What is required is a system which allows the designer explicitly to invoke an exception mechanism in response to some predetermined event. This is generally referred to as *raising* an exception. Facilities for enabling and disabling particular handlers are also needed.

A number of high-level languages provide exception handling mechanisms; some of the mechanisms provided by BASIC interpreters designed for real-time applications are both simple and effective. They permit the designer to execute specific instructions in the event of predicted exceptions. A more complex mechanism is provided in the Ada language: when an exception is raised the system backtracks up the calling structure tree until it finds a handler for the specific exception; if no handler has been provided by the implementor then the exception is passed to the built-in run-time handler.

In distributed systems exception handling mechanisms can become complex if handlers are distributed across different processors. The effect of such a distribution is to close-couple the system and hence detract from the improved robustness that can be obtained from a loosely coupled distributed system.

11.4 FAULT DETECTION MEASURES

11.4.1 Replication Checks

The use of replication checks involves duplicating or replicating the activity of the system that is to be checked. A typical example is the use of dual processors each of which runs the system software. The actions of the two systems are compared at specific times or in response to specific events. If it is assumed that the system design is correct and that failures will be independent then no single fault will go undetected. With replication some additional operation is necessary to determine which of the two units is faulty.

The procedure can be extended to include triple redundancy (TMR) and multiple redundancy (N-modular redundancy, NMR). Higher orders of redundancy do not increase the error detection capability but they enable the faulty unit to be identified and the system to continue to run.

Replication is expensive and can increase the complexity of the system. Both can be reduced by limiting replication to selected critical functions. Replication using identical software units cannot detect common mode faults due, say, to errors in the design of the system. To check for design faults the duplicate system would have to be a completely independent design. Adopting such a technique is costly. An alternative is to use a model of the system, for example a Petri-net-based model, to check the high-level control flow of the operational system. Replication can be used in a simple way to detect transient errors by repeated operation of the same system. This technique is frequently used in communication systems and in software for accessing disk drives where several attempts will be made to read or write to a disk before reporting an error and halting the action.

A major problem with replication checks is in deciding on the nature of the comparison. For Boolean and integer values comparison is simple – exact matching is required. For real values some error range must be specified: what should it be? Are cosmetic variations, for example capitalisation and spacing, permitted in strings? For distributed systems where, for example, the replication checks may be performed on different processors, should the data be transmitted in binary form or as ASCII characters? What about complex data structures: do they have to match as a whole or element by element? This is important in triple replication as a structure containing two elements $\{x,y\}$ would generate an error using matching as a whole if the three results returned were $\{a,b\}$, $\{a,d\}$ and $\{e,b\}$ but would return a consensus result $\{a,b\}$ on an element-by-element comparison.

11.4.2 Expected Value Methods

The majority of methods used to detect software errors make use of some form of expected value. The correct progression of a program through a sequence of modules can be checked by using 'baton' passing. Each module passes to its successor module a unique numerical value. If the value received by the successor module does not match the expected value then the sequence of modules is incorrect.

Another technique, sometimes referred to as an 'assertion test', is to build into the program logical tests. Range checking and input/output checks offered in many modern compilers automatically insert coding for assertion testing. Assertion tests can also be incorporated into the application program explicitly by the designer. For example, if on the basis of the software specification, an array or table should at some particular point contain only positive values, then a test to check that this is the case can by inserted in the program code.

Because of the overhead which this type of checking incurs it is often restricted to the testing period. Many compilers provide switches to allow the range checking and other forms of checking to be selected at compile time. Similarly other forms of checking code can be put into conditional compilation segments to allow it to be omitted in the delivered software system. Careful use of assertion testing on entry to and exit from modules can help to stop a fault generated in one module giving rise to faults in other modules.

Adoption of a 'good' module division at the design stage enables 'defensive programming' to be used to check data passing between modules. It is strongly recommended that data passing between modules be carefully checked. If the design has been well done, the properties of data at module boundaries should have been carefully described and documented and hence checks should be capable of giving a clear indication of any faults. For distributed systems expected value techniques should be used as a method for checking information passing between processors. By using such checks the propagation of errors arising from faults can be prevented. For real-time control and data collection applications use of knowledge of the environment enables expected value methods to be extended to include limit and trend checking as well as techniques based on prediction and analysis of the incoming data from the plant.

11.4.3 Watch-dog Timers

Timing checks, that is checks that some function has been carried out within a specified time, can be used to detect errors in a system. It should be noted that timing checks do not indicate that a system is functioning correctly; they can only indicate that it has failed in some way.

Watch-dog timers are commonly used to monitor the detailed behaviour of a real-time system. A watch-dog task 'watchdog_timer' is run at fixed time intervals as a high-priority task. It decrements counters (timers) held in a pool and checks that the counters have not reached zero. If a counter has reached zero the 'watchdog_timer' task signals to the executive that a time out has occurred. Tasks being monitored by the watch-dog timers periodically reset the counters; thus a time out occurs when they fail to reset the counter within a given time interval. Watch-dog timers can also be used to monitor peripherals. By monitoring critical tasks they can also show up a temporary overload on the system, or clearly show that a system has insufficient processing capacity.

The watch-dog timer can also be applied to software communication channels. Problems of synchronisation can often be simplified if a task can restrict its commitment to waiting for information for a predetermined time. In many real-time applications there are circumstances where it is better to continue with out-of-date information (for example, the previous value) or to estimate a data value rather than simply to wait indefinitely. For example, in a feedback control system it is normally better to continue, in the event of a failure of a single instrument, with an estimated value than to halt the controller action. The failure must of course be reported and the instrument repaired or replaced within a reasonable time.

11.4.4 Reversal Checks

For systems in which there is a one-to-one relationship between input and output a useful error check is to compute what the input should be for the actual output and compare the computed input value with the actual input value. For example, disk drivers often read back the data segment which has just been written and compare the data read back with that which was sent. Mathematical computations often lend themselves to reversal checks; an obvious example is to check the computation of a square root by squaring the answer. However, care is needed in such checks to take into account the finite accuracy of the computation.

A variation on this technique can be used in systems in which there is a fixed relationship between the output and input. For example, for matrix inversion the product of the input matrix and the output matrix should be the unit matrix.

11.4.5 Parity and Error Coding Checks

Parity and error coding checks are well-known techniques for detecting and correcting specific types of memory and data transmission errors. The most widely known is the single-bit parity check applied to memory storage and to asynchronous data transmission. This will detect the loss of an odd number of bits in a storage or transmission unit. A simple parity is an effective and efficient solution to detecting errors when they occur as single-bit errors in a unit; it is not an effective method if typical failures result in multiple-bit errors.

More complicated codes such as the Hamming, cyclic redundancy, and M-out-of-N codes are used to detect multiple-bit errors and some of these codes allow reconstruction, that is correction of the error. The use of these codes involves adding a greater amount of redundancy than the use of a simple parity code.

The above codes are generally used to detect errors due to hardware component failure. Codes can also be applied to data to detect software errors. For example, the use of a checksum computed on a block of data and held with it can be used to detect both hardware and software errors. A combination of parity and checksum added to a data unit and a block of data respectively provides an efficient method of error detection for large and complex sets of data.

11.4.6 Structural Checks

Checksum codes are typically applied to data structures and provide a check on the consistency and integrity of the data contained in the structure. Errors can also occur in the data structure itself, for example pointers in linked lists can be corrupted. Thus checks on the structural integrity are also required.

A very simple check in a linked list is to maintain in the list header a pointer to the last item in the list as well as one to the first item. If the pointer to the last item does not correspond to the last item identified by following through the chain of pointers from the first item then an error has occurred. A simple check of this form, however, provides little information on the error and the possible damage. Also there is little information to help with recovery. A frequently used alternative scheme is the doubly linked list in which each element contains both a forward and a backward pointer.

11.4.7 Diagnostic Checks

Diagnostic checks are used to test the behaviour of components used to construct the systems, not the system itself. The checks use a set of inputs for which the correct outputs are known. They are usually programs used to test for faults in the hardware of the system and typically are expensive in terms of the time and resources used. As a consequence they are rarely used as primary error detection measures but are restricted to attempts to locate more precisely faults detected by other means. A difficulty widely experienced is that it is not easy to design diagnostic checks that impose on a component conditions as stringent as those imposed by the actual system.

11.5 FAULT DETECTION MECHANISMS

Some mechanisms for fault detection based on interface exceptions have already been mentioned. These include illegal instruction, arithmetic overflow and underflow, protection violation and non-existent memory. Few systems offer mechanisms beyond these. Although, for example, a compiler for a strongly typed language will detect and flag as errors attempts to perform arithmetic operations on variables declared of type character, the underlying hardware will not distinguish between words holding integers, characters, reals, etc. Hence a run-time fault that results in an attempt to add a word containing a character to a word containing a real would not result in an exception being signalled. One mechanism to overcome this problem, the use of tagged storage, has been offered on a few computer systems. With tagged storage a few extra bits in each word are allocated to be used to identify the type of object being held in the word.

Other useful mechanisms for fault detection are ones which detect attempts to exceed array bounds or to exceed ranges specified for integers and these are provided in a few computers.

11.6 DAMAGE CONTAINMENT AND ASSESSMENT

Damage results from faults or errors propagating throughout the system. A major problem in assessing the extent of the damage arises through the presence of parallel operations. This is true for both standard sequential programs running on a single processor and for multi-tasking and multi-processor distributed systems. The larger the number of tasks and processors the greater the problem.

One approach to the problem is to use techniques similar to those used for preventing clashes between competing concurrent processes. Sections of code are treated as critical sections, or certain objects are permitted to belong only to one task at any particular time. Access control mechanisms – monitors, guards, locks – are used to protect critical sections or objects.

Obvious containment measures are to limit access to a file while it is being updated by a task. If checks on data consistency and structural checks are carried out prior to releasing the file then the effects of a faulty update can be prevented from spreading.

An important technique for assessing damage is to mark bad data in some way. This technique can also be used for containment if incoming data to a task is checked to see if it is marked as bad. In distributed systems it is important to test thoroughly all incoming data messages since the detection of errors or bad data at the input can prevent damage spreading between processors.

11.7 PROVISION OF FAULT TOLERANCE

11.7.1 Redundancy

A method for introducing redundant modules into a system is the use of *recovery blocks*. Associated with a check point is a recovery block which contains one or more alternative code modules. If an error is detected the primary alternative is used. At the end of the recovery block there is an acceptance test: if, as a result of running the alternative code module, the acceptance test is passed the recovery block is exited. If the acceptance test fails then the next alternative module is tried. If all code modules are exhausted before an acceptable result is achieved then an error is reported.

The general structure is:

```
establish recovery point
primary module
acceptance test
alternate module
acceptance test
```

and this is normally expressed using the syntax:

```
ensure <acceptance test>
by        <primary module>
else by<alternate module 1>
else by<alternate module 2>
.
.
.
else by<alternate module n>
else error
```

A characteristic of real-time systems that can be used to simplify the problem of fault recovery is that tasks are repeated at frequent intervals with new data sets supplied at the system boundary. For example, for a feedback control loop a single set of bad data values can be ignored if it can be assumed that the next set of readings will be correct. If the control loop is stable and well designed the effect will be minor. The effect of a single bad set can be reduced further by replacing the bad set with a predicted value based on some form of extrapolation. The use of predicted values can allow for a series of bad values until some other corrective action can be taken, for example switching to another instrument.

A simple recovery block could be:

```
ensure <data good>
by      <normal value module>
else by<predicted value module>
else error
```

A more complex error recovery block would need to take into account in the acceptance test the time for which the data was bad, that is for how long it is acceptable to use predicted data. And also in practice care would be required to avoid generating a spurious error if the next good data value had diverged from the predicted value.

11.7.2 Deadline Mechanisms

The provision of fault tolerance for real-time computer systems has to take into account that the time to perform some operation enters into the specification of the system and failure to complete in some specified time constitutes an error. Therefore there may not be time to carry out some of the recovery procedures outlined above unless special techniques are used. One such technique is the *deadline mechanism*.

This is an attempt to deal with the requirement that service must be completed within a specified time. The problem is illustrated in the following code segment for a fault-tolerant navigation system:

```
every 1 second
within 10 milliseconds
calculate by
     (* primary module *)
     read sensors
     calculate new position
else by
     (* alternate module *)
     approximate new position from
     old position
```

The every statement is used to specify the repeat time of the particular task. The within statement specifies the maximum amount of time that can be permitted to elapse between starting the task and getting the results back from it. Two modules are specified, a primary module which reads the sensors and calculates the new position and a single alternate module which estimates the position from the previous position value. It is assumed that the alternate module is error free and requires less time to run than the primary module.

In order to meet the overall time deadline the system implementor has to determine accurately the execution time for the alternate module in order to determine how long can be allowed for the primary module to produce a good result. For the 10 ms deadline if the alternate module is estimated to take 3 ms then the primary module must return a result within 7 ms.

11.7.3 Bad Data Marking

The use of modular design techniques facilitates the use of defensive programming to detect bad or faulty data. The major problem arises not in detecting bad data but in deciding what action to take. An interesting solution to this problem is that adopted by the designers of the CUTLASS system. Associated with each data variable is a tag bit which indicates whether or not the data is bad. When a test on the data is carried out the tag bit is set or reset according to the result of the test. Bad data is allowed to propagate through the system but carries with it an indication that it is bad. The language support system contains known rules for evaluating expressions containing bad data. For example, if two logic signals are combined using the OR function then both have to be bad for the result to be tagged as bad, whereas for the AND combination if either signal is bad then the result is tagged as bad. With this approach the program module which determines that a data value is bad does not need to know which modules it should inform since the information is automatically conveyed with the data.

CUTLASS was designed to support distributed control and the use of tagged data avoids some of the problems of synchronising recovery in distributed systems.

The tagging of data allows easy implementation of recovery blocks and gradual degradation of the system as a particular task can be programmed to select alternate function modules according to the status of the input data set.

For example, if the control signal calculated for a particular actuator is tagged as bad the output module can be programmed to leave the actuator set at its last value, move the actuator to a predicted value, or move the actuator to a predicted value unless the bad data time exceeds a preset amount at which time the actuator position is frozen. In the recovery block notation this can be expressed as:

```
ensure            <data good, timeout=false>
by
    ensure        <data good>
        by        <primary module - output data>
        else by   <predict output>
    else error
else by           <do not move actuator>
else error
```

The module in the inner block used to predict the output needs to have an acceptance check in it to avoid large movements of the value if the bad data indication comes at a time when a large-value movement is occurring. Such an acceptance check could force an error which would cause the outer block to institute the <do not move actuator> module.

11.7.4 Recovery Measures – Check Points

A standard technique which has been used for a long time in data processing systems is to insert check points in the program. At a check point a copy of data which would be required to restart the program at that point is written to backing store or to some protected area of memory. If a fault is detected at a test point or the next check point in the program then the program is 'rolled back' to the previous check point and restarted.

The strategy can be readily adapted for use in real-time systems. In a control system, for example, back-up copies of values such as set points, controller parameters and possibly a 'history' of selected plant variables are held either on backing store or in battery backed-up memory. In the event of failure they can be reloaded.

Recovery mechanisms involving the storage of data values can provide protection against hardware faults or software faults which do not recur. However, in the majority of cases if a software module fails once it will continue to fail. The only way in which the system can continue is if a replacement module can be activated. This technique implies redundancy.

11.8 SUMMARY

We have described some of the problems of designing and producing reliable software and how these problems are different from those associated with hardware reliability. The importance of being able to produce reliable and fault-tolerant software will continue to increase as the use of devices incorporating computers grows. Improved production methods will increase software reliability; however, it will remain difficult to make software-based systems tolerant of unanticipated errors. Safe failure of systems incorporating software will continue to depend in the last resort on hardware safety provisions.

For many real-time engineering applications loosely coupled distributed computer systems can be used. The advantage of such systems is that they can be made robust through the use of redundant processors and communication systems. The loose coupling permits the containment of faults to specific parts of the system thus reducing the extent of the performance failure and easing the problem of reinstatement. Most applications of this type are well defined and hence it is possible to specify closely actions to be taken in the event of expected failure modes and most systems can continue to operate in a degraded mode. For example, it is normal for tasks to be allocated at construction time to a specific node (processor) in the system and the action to be taken in the event of failure of that node is normally predetermined.

As with any distributed system there is the problem of system consistency and steps must be taken to ensure that at the end of the recovery procedure the system is returned to a consistent state. For many real-time engineering applications much of the data decreases in value as it ages and hence much of the system will automatically return to a consistent state as new plant data is obtained. The problem areas concern environmental data that is input in incremental rather than absolute form; in such cases consistency can only be regained by restarting the device or system from some known absolute datum.

Bibliography

ABBOTT, C., 'Intervention schedule for real-time programming', *IEEE Trans. Software Engineering*, SE-10(3): 268–74 (1984).

AHSON, S.I., 'A microprocessor-based multi-loop process controller', *IEEE Trans. Industrial Engineering*, IE-30(1): 34–9 (1983).

ALEXANDER, H., JONES, V., *Software Design and Prototyping Using 'Me too'*, Prentice Hall, Englewood Cliffs, NJ (1990).

ALFORD, M.W., 'A requirements engineering methodology', *IEEE Trans. Software Engineering*, SE-3: 180–93 (1977).

ALLWORTH, S.T., ZOBEL, R.N., *Introduction to Real-time Software Design* (2nd edition), Macmillan, London (1987).

ANDERSON, T. (editor), *Resilient Computing Systems*, Wiley, New York (1985).

ANDERSON, T., LEE, P.A., *Fault Tolerance, Principles and Practice*, Prentice Hall, Englewood Cliffs, NJ (1981).

ANDERSON, T., RANDELL, B. (editors), *Computer Systems Reliability*, Cambridge University Press, Cambridge (1979).

ANDREWS, G.R., 'Synchronising resources', *ACM Trans. Programming Languages and Systems*, 3(4): 405–31 (1981).

ANDREWS, G.R., 'The distributed programming language SR – mechanisms, design and implementation', *Software Practice and Experience*, 12(8): 719–54 (1982).

ANDREWS, M., *Programming Microprocessor Interfaces for Control and Instrumentation*, Prentice Hall, Englewood Cliffs, NJ (1982).

ANON, 'Computing control – a commercial reality', *Control Engineering*, 9(5): 40 (1959).

ARZEN, K.E., 'An architecture for expert system based feedback control', *Artificial Intelligence in Real-Time Control, IFAC*, pp. 15–20 (1988).

ASTROM, K.J., WITTENMARK, B., *Computer Controlled Systems: Theory and design*, Prentice Hall, Englewood Cliffs, NJ (1984).

ASTROM, K.J., ANTON, J.J., ARZEN, K.E., 'Expert control', *Automatica*, 22(3): 277–86 (1986).

AURICOSTE, J.G., 'Applications of digital computers to process control', in Coales, J. (editor) *Automation in the Chemical, Oil and Metallurgical Industries*, Butterworth, London (1963).

AUSLANDER, D.M., SAGUES, P., *Microprocessors for Measurement and Control*, Osborne/McGraw-Hill, New York (1981).

AUSLANDER, D.M., TAKAHASHI, Y., TOMIZUKA, M., 'The next generation of single loop controllers: hardware and algorithms for the discrete/decimal process controller', *ASME J. of Dynamic Systems, Measurement and Control*: 280–2 (1975).

BAKER, T.P., SCALLON, G.M., 'An architecture for real-time software systems', *IEEE Software Magazine*, 3(3): 50–8 (1986).

BARNES, J.G.P., *RTL/2 Design and Philosophy*, Heyden, London (1976).

BARNES, J.G.P., *Programming in Ada*, Addison-Wesley, Wokingham (1982).

BARNEY, G.C., *Intelligent Instrumentation*, Prentice Hall, London (1985). 2nd edition 1988.

BELL, D., MORREY, I., PUGH, J., *Software Engineering*, Prentice Hall, Englewood Cliffs, NJ (1987). 2nd edition 1992.

BEN-ARI, M., *Principles of Concurrent Programming*, Prentice Hall, Englewood Cliffs, NJ (1982).

BENNETT, S., *Real-time Computer Control: An introduction*, Prentice Hall, Englewood Cliffs, NJ (1988).

BENNETT, S., 'Expert systems in real-time control', in Lamba, S.S., Singh, Y.P. (editors) *Distributed Computer Control Systems*, Tata McGraw-Hill, New Delhi (1992).

BENNETT, S., LINKENS, D.A. (editors), *Computer Control of Industrial Processes*, Peter Peregrinus, Stevenage (1982).

BENNETT, S., LINKENS, D.A. (editors), *Real-time Computer Control*, Peter Peregrinus, Stevenage (1984).

BENNETT, S., VIRK, G.S. (editors), *Computer Control of Real-Time Processes*, Peter Peregrinus, Stevenage (1990).

BERRYMAN, S.J., SOMMERVILLE, I., 'Modelling real-time constraints', *Third International Conference on Software Engineering for Real Time Systems*, IEE, London, pp. 164–7 (1991).

BIBBERO, R.J., *Microprocessors in Instruments and Control*, Wiley, New York (1977).

BOOCH, G., *Software Engineering with Ada*, Benjamin Cummings, Menlo Park, CA (1983).

BRINCH HANSEN, P., 'Structured multi-programming', *Communications of the ACM*, 15(7): 574–7 (1972).

BRINCH HANSEN, P., *Operating System Principles*, Prentice Hall, Englewood Cliffs, NJ (1973).

BRINCH HANSEN, P., 'The programming language concurrent Pascal', *IEEE Trans. Software Engineering*, SE-1(2): 199–206 (1975).

BRODIE, L., *Starting Forth*, Prentice Hall, Englewood Cliffs, NJ (1986) 2nd edition.

BROOKS, F., *The Mythical Man-month*, Addison-Wesley, New York (1975).

BROWN, G.S., CAMPBELL, D.P., 'Instrument engineering: its growth and promise in process-control problems', *Mechanical Engineering*, 72(2): 124 (1950).

BUDGEN, D., 'Combining MASCOT with Modula-2 to aid the engineering of real-time systems', *Software – Practice and Experience*, 15: 767–93 (1985).

BUDGEN, D., *Software Development with Modula-2*, Addison-Wesley, Wokingham (1989).

BULL, G., LEWIS, A., 'Real-time BASIC', *Software – Practice and Experience*, 13: 1075 (1983).

BURKITT, J.K., 'Reliability performance of an on-line digital computer when controlling a plant without the use of conventional controllers', *Automatic Control in the Chemical Process and Allied Industries, Society of Chemical Industry*, pp. 125–40 (1965).

BURNS, A., *Programming in Occam 2*, Addison-Wesley, Wokingham (1988).

BURNS, A., WELLINGS, A., *Real-Time Systems and their Programming Languages*, Addison-Wesley, Wokingham (1990).

CAMERON, J.R., 'An overview of JSD', *IEEE Trans. Software Engineering*, SE-12(2): 222–40 (1986).

CAMERON, J.R., 'The modelling phase of JSD', *Software and Information Technology*, 30(6): 373–83 (1988).

CAMERON, J.R., *JSP & JSD: The Jackson Approach to Software Development* (2nd edition), IEEE Computer Society Press, New York (1989).

CASSELL, D.A., *Microcomputers and Modern Control Engineering*, Prentice Hall, Englewood Cliffs, NJ (1983).

CHARD, R.A., *Software Concepts in Process Control*, NCC Publications, Manchester (1983).

CHETTO, H., SILLY, M., BOUCHENTOUF, T., 'Dynamic scheduling of real-time tasks under precedence constraints', *J. of Real-Time Systems*, 2: 181–94 (1990).

CIVERA, P., DEL CORSO, D., GREGORETTI, F., 'Microcomputer systems in real-time applications', in Tzafestas, S.G. (editor) *Microprocessors in Signal Processing, Measurement and Control*, Reidel, Dordrecht (1983).

COHEN, G.H., COON, G.A., 'Theoretical consideration of retarded control', *Trans. ASME*, 75(5): 827–34 (1953).

CONSTANTINE, L.L., YOURDON, E., *Structured Design*, Prentice Hall, Englewood Cliffs, NJ (1979).

COOLING, J.E., *Software Design for Real-Time Systems*, Chapman & Hall, London (1991).

CORE – *Controlled Requirements Expression*, Systems Designers plc, Fleet, Hampshire, document no. 1986/0786/500/PR/0158 (1986).

CROLL, P., NIXON, P., 'Developing safety-critical software within a CASE environment', *IEE Colloquium No. 1991/087, Computer aided software engineering tools for real-time control*, April (1991).

CULLYER, W.J., 'Implementing safety-critical systems: the VIPER microprocessor', *VLSI Specification, Verification and Synthesis*, Kluwer Academic Press, Brentford (1988).

CULLYER, W.J., PYGOTT, C.H., 'Application of formal methods to the VIPER microprocessor', *Proc. IEE*, 134 (Part E): 133–41 (1987).

CULSHAW, B., 'Smart structures – a concept or reality?', *Proc. I.Mech.E., Part I J. of Systems and Control Engineering*, 206(I1): 1–8 (1992).

DAHLIN, E.B., 'Designing and tuning digital controllers', *Instruments and Control Systems*, 41(June): 77–83, 87–91 (1968).

DEMARCO, T., *Structured Analysis and System Specification*, Prentice Hall, Englewood Cliffs, NJ (1978).

DESHPANDE, P.B., ASH, R.H., *Elements of Computer Process Control*, Prentice Hall, Englewood Cliffs, NJ (1983).

DETTMER, R., 'The VIPER microprocessor', *Electronics and Power*, October (1986).

DIJKSTRA, E.W., 'Cooperating sequential processes', *Programming Languages*, Academic Press, London (1968).

DOTAN, Y., BEN-ARIEH, D., 'Modeling flexible manufacturing systems: the concurrent logic programming approach', *IEEE Trans. Robotics and Automation*, 7(1): 135–48 (1991).

DOWNS, E., CLARE, P., COE, I., *Structured Systems Analysis and Design Method*, Prentice Hall, Englewood Cliffs, NJ (1988).

EDWARDS, J.B., 'Process control by computer', in Bennett, S., Linkens, D.A. (editors) *Computer Control of Industrial Processes*, Peter Peregrinus, Stevenage (1982).

EFSTATHIOU, J., *Expert Systems in Process Control*, Longman, Harlow (1989).

FAULK, S.R., PARNAS, D.L., 'On the uses of synchronisation in hard-real-time systems', *Proceedings of the Real-time Systems Symposium, IEEE, Arlington, VA, 6–8 Dec.*, pp. 101–9 (1983).

FEUER, A., GEHANI, N.H., *Comparing and Assessing Programming Languages, Ada, C. and Pascal*, Prentice Hall, Englewood Cliffs, NJ (1984).

FFYNLO-CRAINE, J., MARTIN, G.R., *Microcomputers in Engineering and Science*, Addison-Wesley, Reading, MA (1985).

FOSTER, S.C., SOLOWAY, I., *Real-time Programming: Neglected topics*, Addison-Wesley, Reading, MA (1981).

FRANKLIN, G.F., POWELL, J.D., *Digital Control of Dynamic Systems*, Addison-Wesley, Reading, MA (1980). Edition with M.L. Workman, 1990.

FREEDMAN, A.L., LEES, R.A., *Real-time Computer Systems*, Edward Arnold, London (1977).

GANE, C., SARSON, T., *Structured Systems Analysis*, Prentice Hall, Englewood Cliffs, NJ (1979).

GAWTHROP, P.J., 'Automatic tuning of commercial PID controllers', Chapter 3 in Bennett, S., Virk, G.S. (editors) *Computer Control of Real-Time Processes*, Peter Peregrinus, Stevenage (1990); also in *IEEE Control Systems Magazine*, January (1990).

GERTLER, J., SEDLAK, J., 'Software for process control – a survey', in Glass, R.L. (editor) *Real-time Software*, Prentice Hall, Englewood Cliffs, NJ (1983).

GLASS, R.L., *Real-time Software*, Prentice Hall, Englewood Cliffs, NJ (1983).

GODFREY, K., 'Digital control systems', in Holdsworth, B., Martin, G.R. (editors) *Digital Systems Reference Book*, Butterworth Heinemann, Oxford (1991).

GOFF, K.W., 'Dynamics in direct digital control', *J. of Instrument Society of America*, 1: 45–9, 44–54 (1966).

GOLDSMITH, S., *A Practical Guide to Real Time System Development*, Prentice Hall, Englewood Cliffs, NJ (1993).

GOLTEN, J., VERWER, A., *Control System Design and Simulation*, McGraw-Hill, London (1991).

GOMAA, H., 'A software design method for real-time systems', *Communications of the ACM*, 27: 938–49 (1984).

GOMAA, H., 'Software development of real-time systems', *Commununications of the ACM*, 29: 657–68 (1986).

GORSLINE, G.W., *Computer Organisation* (2nd edition), Prentice Hall, Englewood Cliffs, NJ (1986).

GOSCINSKI, A., *Distributed Operating Systems*, Addison-Wesley, Reading, MA (1991).

GUTH, R., *Computer Systems for Process Control*, Plenum Press, New York (1986).

HALANG, W.A., SACHA, K.M., *Real-time Systems*, World Scientific, Singapore (1992).

HAM, P.A.L., 'Reliability in computer control of turbine generator plant', in Bennett, S., Linkens, D.A. (editors) *Real-time Computer Control*, Peter Peregrinus, Stevenage (1984).

HAREL, D., *et al.*, 'STATEMATE: a working environment for the development of complex reactive systems', *IEEE Trans. Software Engineering*, SE-16(3): 403–14 (1990).

HATLEY, D.J., PIRBHAI, I.A., *Strategies for Real-time System Specification*, Dorset House, New York (1988).

HATONO, I., YAMAGATA, K., TAMURA, H., 'Modeling and on-line scheduling of flexible manufacturing systems using stochastic Petri nets', *IEEE Trans. Software Engineering*, SE-17(2): 126–32 (1991).

HAUPTMANN, S., REINIG, G., 'Portable Modula-2 based real-time operating system', *Microprocessors & Microsystems*, 14(3): April (1990).

HEATH, W.S., *Real-time Software Techniques*, Van Nostrand, New York (1991).

HENIGER, K.L., 'Specifying software requirements for complex systems: new techniques and their application', *IEEE Trans. Software Engineering*, SE-6: 3–13 (1980).

HENRY, R.M., 'Generating sequences', in Linkens, D.A., Virk, G.S. (editors) *Computer Control*, Institute of Measurement and Control for SERC, London (1987).

HINE, D., BURBRIDGE, L., 'A microcomputer algorithm for open-loop step-motor control', *Transactions of the Institute of Measurement and Control*, 1: 233–9 (1979).

HOARE, C.A.R., 'Monitors: an operating system structuring concept', *Communications of the ACM*, 17(10): 549–57 (1974).

HOARE, C.A.R., 'Communicating sequential processes', *Communications of the ACM*, 21(8): 666–77 (1978).

HOLLAND, R.C., *Microcomputers for Process Control*, Pergamon, Oxford (1983).

HOLLOWAY, L.E., KROGH, B.H., 'Synthesis of feedback control logic for a class of controlled Petri-nets', *IEEE Trans. Automatic Control*, AC-35(5): 514–23 (1990).

HOLT, R.C., GRAHAM, G.S., LOZOWSKA, E.D., SCOTT, M.A., *Structured Concurrent Programming with Operating Systems Applications*, Addison-Wesley, Reading, MA (1978).

HOOD, *HOOD Reference Manual*, Issue 3.0 Sept. 1989, Document Reference WME/89-173/JB, European Space Agency (1989).

HOPCROFT, J.E., ULLMAN, J.D., *Introduction to Automata Theory, Languages, and Computation*, Addison-Wesley, Reading, MA (1979).

HOPPER, A., TEMPLE, S., WILLIAMSON, R., *Local Area Network Design*, Addison-Wesley, Reading, MA (1986).

HOUPIS, C.H., LAMONT, G.B., *Digital Control Systems: Theory, hardware, software*, McGraw-Hill, New York (1985)

HUGHES, J.G., *Database Technology: A software engineering approach*, Prentice Hall, Englewood Cliffs, NJ (1988).

IEEE, *IEEE Trial-use Standard Specifications for Microprocessor Operating Systems Interfaces*, Wiley, New York (1985).

ISERMAN, R., *Digital Control Systems*, Springer, Berlin (1981).

ISERMAN, R., 'Process fault detection based on modelling and estimation methods – a survey', *Automatica*, 20(4): 387–404 (1984).

JACKSON, K., SIMPSON, H.R., 'MASCOT – a modular approach to software construction operation and test', *RRE Tech. Note*, No. 778 (1975).

JACKSON, M., *System Development*, Prentice Hall, Englewood Cliffs, NJ (1983).

JANICKI, R., LAUER, P.E., *Specification and Analysis of Concurrent Systems: the COSY approach*, EACTS Monographs on Theoretical Computer Science, vol. 26, Springer, Berlin (1992).

JANSON, P.A., *Operating Systems: Structures and mechanisms*, Academic Press, London (1985).

JOHNSON, C.D., *Microprocessor-based Process Control*, Prentice Hall, Englewood Cliffs, NJ (1984).

JONES, A.H., 'Real-time expert controllers for industrial plants', in Linkens, D.A., Virk, G.S. (editors) *Computer Control*, Institute of Measurement and Control for SERC, London (1987).

JONES, A.H., PORTER, B., 'Expert tuners for PID controllers', *Proceedings of IASTED, International Conference on Computer-Aided Design and Applications, Paris* (1985).

JOVIC, F., *Process Control Systems: Principles of design and operation*, Kogan Page, London (1986).

KAISLER, S.H., *The Design of Operating Systems for Small Computers*, Wiley, New York (1983).

KALANI, G., *Microprocessor Based Distributed Control Systems*, Prentice Hall, Englewood Cliffs, NJ (1989).

KANDLUR, D.D., KISKIS, D.L, SHIN, K.G., 'HARTOS: A distributed real-time operating system', *ACM Operating Systems Review*, 23(July): 72–89 (1989).

KATZ, P., *Digital Control Using Microprocessors*, Prentice Hall, Englewood Cliffs, NJ (1981).

KISSELL, T.E., *Understanding and Using Programmable Controllers*, Prentice Hall, Englewood Cliffs, NJ (1986).

KOIVO, H.N., PELTOMAA, A., 'Micro-computer real-time multi-tasking operating systems in control applications', *Computers in Industry*, 5: 31–9 (1984).

KOWAL, J.A., *Analyzing Systems*, Prentice Hall, Englewood Cliffs, NJ (1988).

KRAMER, J., MAGEE, J., SLOMAN, M., 'Conic: an integrated approach to distributed computer control'. *Proc. IEE Part E*, 130: 1–10 (1983).

KRAUS, T.W., MYRON, T.J., 'Self-tuning PID controller uses pattern recognition approach', *Control Engineering*, June: 106–11 (1984).

KRIJGSMAN, A.J., VERBRUGGEN, H.B., BRUIJN, P.M., 'Knowledge-based real-time control', in *Artificial Intelligence in Real-Time Control, Proceedings of the IFAC Workshop, Swansea*, pp. 7–13 (1988).

KUO, B.C., *Digital Control Systems*, Holt Saunders, New York (1980).

LAMB, D., *Software Engineering*, Prentice Hall, Englewood Cliffs, NJ (1988).

LAPRIE, J.C., 'Dependability: a unifying concept for reliable computing and fault tolerance', in Anderson, T. (editor) *Dependability of Resilient Computers*, BSP Professional Books, Oxford, pp. 1–28 (1989a).

LAPRIE, J.C., 'Dependability evaluation: hardware and software', in Anderson, T. (editor) *Dependability of Resilient Computers*, BSP Professional Books, Oxford, pp. 44–67 (1989b).

LARDENNOIS, R., 'Using Modula-2 for safety critical control in an urban transportation system', *Microprocessors & Microsystems*, 14(3): 177–80 (1990).

LAWRENCE, P.W., MAUCH, K., *Real-time Microcomputer Design*, McGraw-Hill, New York (1987).

LEIGH, J.R., *Applied Digital Control* (2nd edition), Prentice Hall, Englewood Cliffs, NJ (1992).

LEVI, S.-T., AGRAWALA, A.K., *Real-time System Design*, McGraw-Hill, New York (1990).

LIEBOWITZ, B.H., CARSON, J.H., *Multiple Processor Systems for Real-time Applications*, Prentice Hall, Englewood Cliffs, NJ (1985).

LIGHTFOOT, D., *Formal Specification Using Z*, Macmillan, London (1991).

LINKENS, D.A., VIRK, G.S. (editors), *Computer Control*, Institute of Measurement and Control for SERC, London (1987).

LISKOV, B., SCHEIFLER, R., 'Guardians and actions: linguistic support for robust, distributed programs', *ACM Transactions on Programming Languages and Systems*, 5(3): 381–404 (1983).

LISTER, A.M., *Fundamentals of Operating Systems* (2nd edition), Macmillan, London (1979).

LIU, C.L., LAYLAND, J.W., 'Scheduling algorithms for multiprogramming in a hard real-time environment', *JACM*, 20(1): 46–61 (1973).

LIU, L.Y., SHYAMASUNDAR, R.K., 'Static analysis of real time distributed systems', *IEEE Trans. Software Engineering*, SE-16(3): 373–88 (1990).

LONGBOTTOM, R., *Computer System Reliability*, Wiley, New York (1980).

LOWE, E.I., HIDDEN, A.E., *Computer Control in Process Industries*, Peter Peregrinus, Stevenage (1971).

LUYBEN, W.L., *Process Modelling, Simulation and Control for Chemical Engineers*, McGraw-Hill, New York (1974).

MCCABE, T.J., *et al.*, 'Structured real-time analysis and design', *COMPSAC-85*, IEEE, New York, pp. 40–51 (1985).

MCCLURE, C., *CASE is Software Automation*, Prentice Hall, Englewood Cliffs, NJ (1989).

MACKENZIE, D., 'Negotiating arithmetic, constructing proof: the sociology of mathematics and information technology', *Social Studies of Science*, 23: 37–65 (1993).

MANO, M.M., *Digital Logic and Computer Design*, Prentice Hall, Englewood Cliffs, NJ (1979).

MASCOT, *The Official Handbook of MASCOT*, version 3.1, Computing Division, RSRE Malvern (1987).

MASON, A.J., 'Practical implementation of large real time expert systems for process and plant management', *ISA Advance Control Conference, Birmingham, England*, pp. 11.1–11.15 (1989).

MELLICHAMP, D. (editor), *Real-time Computing with Applications to Data Acquisition and Control*, Van Nostrand Reinhold, New York (1983).

MICHEL, G., *Programmable Logic Controllers*, Wiley, New York (1990).

MOORHOUSE, T.J., 'MDSE Concepts', Ferranti Computer Systems, Alvey Project Document MDSE/GEN/TN/F3.4 July (1986).

MULLERY, G.P., 'CORE – a method for controlled requirements expression', *Fourth International Conference on Software Engineering*, Washington, DC, IEEE Computer Society Press, Los Angeles, CA (1979).

MYERS, G.J., *Composite Structured Design*, Van Nostrand Reinhold, New York (1978).

MYERS, G.L., *Software Reliability*, Wiley, New York (1976).

NELSON, V.P., CARROLL, W.D., *Tutorial: Fault-tolerant computing*, IEEE Computer Society Press, Los Angeles, CA (1987).

NIELSEN, K., *Ada in Distributed Real-time Systems*, Intertext Books, McGraw-Hill, New York (1990).

OLSON, G., PIANI, G., *Computer Systems for Automation and Control*, Prentice Hall, Englewood Cliffs, NJ (1992).

OSTROFF, J.S., *Temporal Logic for Real-time Systems*, Research Studies Press, Taunton (1989).

OSTROFF, J.S., WONHAM, W.M., 'A framework for real-time discrete event control', *IEEE Trans. Automatic Control*, AC-35(4): 386–97 (1990).

PARNAS, D.L., 'A technique for software module specification with examples', *Communications of the ACM*, 15: 330–6 (1972a).

PARNAS, D.L., 'On the criteria to be used in decomposing systems into modules', *Communications of the ACM*, 15: 1053–8 (1972b).

PARNAS, D.L., CLEMENTS, P.C., WEISS, D.M., 'The modular structure of complex systems', *IEEE Trans. Software Engineering*, SE-11(3): 259–66 (1985).

PERDU, D.M., LEVIS, A.H., 'A Petri net model for evaluation of expert systems in organizations', *Automatica*, 27(2): 225–38 (1991).

PETERSON, J.L., *Petri Net Theory and Modelling of Systems*, Prentice Hall, Englewood Cliffs, NJ (1981).

POMBERGER, G., *Software Engineering and Modula-2*, Prentice Hall, Englewood Cliffs, NJ (1986).

PORTER, B., JONES, A.H., MCKEOWN, C.B., 'Real-time expert tuners for PI controllers', *Proc. IEE, Control Theory and Applications*, 134(D): 260–3 (1987).

POWER, H.M., SIMPSON, R.J., *Introduction to Dynamics and Control*, McGraw-Hill, London (1978).

PRESSMAN, R.S., *Software Engineering* (3rd edition), McGraw-Hill, New York (1992).

PROCK, J., 'A new technique for fault detection using Petri nets', *Automatica*, 27(2): 239–46 (1991).

PYLE, I.C., 'Methods for the design of control software', in *Software for Computer Control. Proceedings Second IFAC/IFIP Symposium on Software for Computer Control, Prague 1979*, Pergamon, Oxford (1979).

PYLE, I.C., *Developing Safety Systems: A guide using Ada*, Prentice Hall, Englewood Cliffs, NJ (1991).

RAI, S., AGRAWAL, D.P. (editors), *Advances in Distributed System Reliability*, IEEE Computer Press, Los Angeles, CA (1990).

RATCLIFF, B., *Software Engineering*, Blackwell, Oxford (1987).

REISIG, W., *Petri Nets: An introduction*, EATCS Monographs on Theoretical Computer Science, vol. 4, Springer, Berlin (1982).

RICARDO, C.M., *Database Systems: Principles, Design, and Implementation*, Macmillan, New York (1990).

ROBINSON, P.J., *Hierarchical Object Oriented Design*, Prentice Hall, New York, London (1993).

RODD, M.G., 'RTMMS – An OSI-based real-time messaging system', *J. of Real-Time Systems*, 2: 213–34 (1990).

RODD, M.G., FARZIN, D., *Communication Systems for Industrial Automation*, Prentice Hall, New York, London (1989).

SANDOZ, D.J., 'A survey of computer control', in Bennett, S., Linkens, D.A. (editors), *Computer Control of Industrial Processes*, Peter Peregrinus, Stevenage (1984).

SARGEANT, M., SHOEMAKER, R.L., *Interfacing Microcomputers to the Real World*, Addison-Wesley, Reading, MA (1981).

SAVAS, E.S., *Computer Control of Industrial Processes*, McGraw-Hill, New York (1965).

SEARS, K.H., MIDDLEDITCH, A.E., 'Software concurrency in real-time control systems: a software nucleus', *Software – Practice and Experience*, 15: 739–59 (1985).

SHARP, J.A., *An Introduction to Distributed and Parellel Processing*, Blackwell Scientific, Oxford (1987).

SHEPARD, T., 'Scheduling real-time systems', *IEEE Trans. Software Engineering*, SE-17(7): 669 (1991).

SHOOMAN, M.L., *Software Engineering: Design, reliability, and management*, McGraw-Hill, New York (1983).

SHRIVASTAVA, S.K. (editor), *Reliable Computer Systems: Collected papers of the Newcastle Reliability Project*, Springer, Berlin (1985).

SIMMONDS, W.H., 'Representation of real knowledge for real-time use', *Artificial Intelligence in Real-Time Control, IFAC*, pp. 63–7 (1988).

SIMPSON, H.R., JACKSON, K., 'Process synchronization in MASCOT', *Computer Journal*, 22: 332–45 (1979).

SLOMAN, M., KRAMER, J., *Distributed Systems and Computer Networks*, Prentice Hall, London (1987).

SMITH, C.L., *Digital Computer Process Control*, Intertext Educational, Scranton, PA (1972).

SMITH, L.S., 'Practical aspects of implementing PID controllers', Chapter 4 in Bennett, S., Virk, G.S. (editors) *Computer Control of Real-Time Processes*, Peter Peregrinus, Stevenage (1990).

SOMMERVILLE, I., *Software Engineering*, Addison-Wesley, London (1982).

STANKOVIC, J.A., RAMAMRITHAM, K., *Hard Real-Time Systems*, IEEE Computer Society Press, Los Angeles, CA (1988).

STEPHANOPOULOS, G., *Chemical Process Control*, Prentice Hall, Englewood Cliffs, NJ (1985).

STEVENS, W., *Software Design: Concepts and methods*, Prentice Hall, Englewood Cliffs, NJ (1991).

STEVENS, W., MYERS, G., CONSTANTINE, L., 'Structured design', *IBM Systems Journal*, 13(2): 115–39 (1974).

STIRE, T.G. (editor), *Process Control Computer Systems: Guide for managers*, Ann Arbor Science, Ann Arbor, MI (1983).

STONE, H.S., *Microcomputing Interfacing*, Addison-Wesley, Reading, MA (1983).

TAKAHASHI, Y., RABINS, M.J., AUSLANDER, D.M., *Control and Dynamic Systems*, Addison-Wesley, Reading, MA (1970).

TAKUNDA, H., MERCER, C.W., 'ARTS: A distributed real-time kernel', *ACM Operating Systems Review*, 23: 29–39 (1989).

TANENBAUM, A., *Modern Operating Systems*, Prentice Hall, Englewood Cliffs, NJ (1992).

TAYLOR, N., 'Dover's smart bridge', *Proc. I.Mech.E., Part I J. of Systems and Control Engineering*, 206(I1): 9–18 (1992).

THOMAS, H.W., SANDOZ, D.J., THOMSON, M., 'New desaturation strategy for digital PID controllers', *Proc. IEE, Part D, Control Theory and Applications*, 130(4): 188–92 (1983).

THOMSON, A.C., 'Real-time artificial intelligence for process monitoring and control', *Artificial Intelligence in Real-Time Control, IFAC*, pp. 89–94 (1988).

TOCCI, R.J., *Digital Systems: Principles and applications*, Prentice Hall, Englewood Cliffs, NJ (1980).

TUCKER, A.B., *Programming Languages*, McGraw-Hill, New York (1985).

TZAFESTAS, S.G. (editor), *Microprocessors in Signal Processing, Measurement and Control*, Reidel, Dordrecht (1983).

VALVANIS, K.S., 'On hierarchical analysis and simulation of flexible manufacturing systems with extended Petri-nets', *IEEE Trans. Systems, Man and Cybernetics*, SMC-20(1): 94–110 (1990).

VIRK, G.S., 'Parallel processing for computer control', Chapter 7 in Bennett, S., Virk, G.S. (editors) *Computer Control of Real-Time Processes*, Peter Peregrinus, Stevenage (1990).

VITINS, M., SIGNER, K., 'Performance modelling of control systems', in Guth, R. (editor) *Computer Systems for Process Control*, Plenum Press, New York (1986).

WARD, P.T., MELLOR, S.J., *Structured Development for Real-time Systems*, Yourdon Press, New York (1986).

WARWICK, K., THAM, M.T., *Failsafe Control Systems*, Chapman & Hall, London (1990).

WIENER, R., SINOVEC, R., *Software Engineering with Modula-2 and Ada*, Wiley, New York (1984).

WILKIE, J.D.F., 'Batch process control', Chapter 16 in Bennett, S., Virk, G.S. (editors) *Computer Control of Real-Time Processes*, Peter Peregrinus, Stevenage (1990).

WILLIAMS, T.J., 'Two decades of change', *Control Engineering*, 24(9): 71–6 (1977).

WILLIAMSON, D., *Digital Control and Implementation: Finite wordlength considerations*, Prentice Hall, Englewood Cliffs, NJ (1991).

WIRTH, N., 'Towards a discipline of real-time programming', *Communications of the ACM*, 22: 577–83 (1977).

WIRTH, N., *Programming in Modula-2*, Springer, Berlin (1982). Second edition 1983.

WITTING, P.A., 'Digital controllers for process control applications', in Warwick, K., Rees, D. (editors) *Industrial Digital Control Systems*, Peter Peregrinus, Stevenage (1988).

WOODWARD, P.M., WETHERALL, P.R., GORMAN, B., *The Official Definition of CORAL 66*, HMSO, London (1970).

WOOLVET, G.A., *Transducers in Digital Systems*, Peter Peregrinus, Stevenage (1977).

XU, J., PARNAS, D.L., 'Scheduling processes with release times, deadlines, precedence and exclusion relations', *IEEE Trans. Software Engineering*, SE-16(3): 360–9 (1990).

YOUNG, S.J., *Real-time Languages: Design and development*, Ellis Horwood, Chichester (1982).

YOUNG, S.J., *An Introduction to Ada*, Ellis Horwood, Chichester (1983).

YOURDON, E.N. (editor), *Classics in Software Engineering*, Yourdon Press, New York (1979).

YOURDON, E., CONSTANTINE, L.L., *Structured Design – Fundamentals of a Discipline of Computer Program and Systems Design*, Prentice Hall, Englewood Cliffs, NJ (1979).

ZAVE, P., 'An operational approach to specification for embedded systems', *IEEE Trans. Software Engineering*, SE-8: 250–69 (1982).

ZAVE, P., 'An overview of the PAISLey project', *AT & T Lab. Tech. Report* (1984).

ZAVE, P., 'The anatomy of a process control system', *AT & T Lab. Tech. Report* (1984).

ZAVE, P., 'The operational versus conventional approach to software development', *Commun. ACM*, 27: 104–18 (1984).

ZAVE, P., 'Case study: the PAISLey approach applied to its own software tools', *Comput. Lang.*, 11(1): 15–28 (1986).

ZAVE, P., 'A compositional approach to multiparadigm programming', *IEEE Software*, 6(5): 15–25 (1989).

ZAVE, P., 'An insider's evaluation of PAISLey', *IEEE Trans. Software Engineering*, SE-17(3): 212–25 (1991).

ZAVE, P., SCHELL, W., 'Salient features of an executable specification language and its environment', *IEEE Trans. Software Engineering*, SE-12(2): 312–25 (1986).

ZIEGLER, J.G., NICHOLS, N.B., 'Optimum settings for automatic controllers', *Trans. ASME*, 64(8): 759–68 (1942).

ZIEGLER, J.G., NICHOLS, N.B., 'Process lags in automatic control circuits', *Trans. ASME*, 65: 433–44 (1943).

Index